Large (C> = 24) Polycyclic Aromatic Hydrocarbons

CHEMICAL ANALYSIS

A SERIES OF MONOGRAPHS ON ANALYTICAL CHEMISTRY
AND ITS APPLICATIONS

Editor
J. D. WINEFORDNER

VOLUME 158

WILEY-INTERSCIENCE

A JOHN WILEY & SONS, INC., PUBLICATION

New York • Chichester • Weinheim • Brisbane • Singapore • Toronto

Thanks for
your assistance.
Bill

Large (C> = 24) Polycyclic Aromatic Hydrocarbons: Chemistry and Analysis

JOHN C. FETZER

Chevron Research and Technology Company
Richmond, California

WILEY-INTERSCIENCE

A JOHN WILEY & SONS, INC., PUBLICATION

New York • Chichester • Weinheim • Brisbane • Singapore • Toronto

This book is printed on acid-free paper. ♾

Published simultaneously in Canada.

For ordering and customer service, call 1-800-CALL-WILEY

Library of Congress Cataloging-in-Publication Data:

Fetzer, John.
Large (C>=24) polycyclic aromatic hydrocarbons: chemistry and analysis / John Fetzer.
p. cm. — (Chemical analysis ; v. 158)
Includes index.
ISBN 0-471-36354-5 (acid-free paper)
1. Polycyclic aromatic hydrocarbons. I. Title. II. Series.
QD341.H9 F35 2001
547′.61—dc21 00-026054

Printed in the United States of America.

10 9 8 7 6 5 4 3 2 1

CONTENTS

PREFACE AND ACKNOWLEDGMENTS

Over the many years of my research on large polycyclic aromatic hydrocarbons, I have been fortunate to collaborate with many fine scientists. Several have also become good friends. I would like to thank these and the members of their research groups. The following listing follows no particular order of importance or chronology: Wilt Biggs, Kiyokatsu Jinno, Joe Pesek, Steve Wise, Lane Sander, Philippe Garrigues, Helene Budzinski, Werner Schmidt, Josef Michl, Jacek Waluk, Bernie Simoneit, Mihaly Beck, Anders Colmsjo, Tuan Vo-Dinh, Bill Acree, Ken Street, Art LaFleur, Judy Wornat, Bill Herndon, Klaus Muellen, Ken Laali, Lou Allamandola, Doug Hudgins, John Kershaw, and Milton Lee. If I have inadvertently left out any of my colleagues, I ask them to please excuse this oversight.

I would like to mention the advice of Dr. L. B. (Buck) Rogers, my graduate research advisor, who told me to get into interesting, new fields and not to stick with whatever I had been doing. I almost laugh at the evolution of this advice when I look at the many areas to which it has taken me. After the first few years, my research efforts into the large PAHs had spread from their initial roots in the chromatographic separations of the large PAHs to many other fields, including far-flung ones such as synthesis. Dr. Rogers was almost appalled that I had branched out so far that I was actually doing a lot of synthetic organic chemistry.

In particular, I would like to extend special thanks to Max Zander, who, besides working with me on several interesting studies, has given me two of the greatest compliments that someone working in this field could receive. These comments, coming from one of the world's acknowledged experts, have inspired me. First, Max very kindly compared my work to that of Eric Clar. On a later date, he compared it to his own. I could only have hoped to use these statements as inspiration, and I hope there is some reality in what appears to me to be a great over-assessment of my impact on the field of large PAH research.

Most of all, I thank my wife Lori for putting up with all the discussions on the content of this book, for giving me the time that I spent on assembling the information given here and in writing it, and for her thoughtfulness and love.

John C. Fezter
Richmond, CA

CHEMICAL ANALYSIS

A SERIES OF MONOGRAPHS ON ANALYTICAL CHEMISTRY AND ITS APPLICATIONS

J. D. Winefordner, *Series Editor*

Vol. 1. **The Analytical Chemistry of Industrial Poisons, Hazards, and Solvents**. *Second Edition*. By the late Morris B. Jacobs

Vol. 2. **Chromatographic Adsorption Analysis**. By Harold H. Strain (*out of print*)

Vol. 3. **Photometric Determination of Traces of Metals**. *Fourth Edition*
Part I: General Aspects. By E. B. Sandell and Hiroshi Onishi
Part IIA: Individual Metals, Aluminum to Lithium. By Hiroshi Onishi
Part IIB: Individual Metals, Magnesium to Zirconium. By Hiroshi Onishi

Vol. 4. **Organic Reagents Used in Gravimetric and Volumetric Analysis**. By John F. Flagg (*out of print*)

Vol. 5. **Aquametry: A Treatise on Methods for the Determination of Water**. *Second Edition (in three parts)*. By John Mitchell, Jr. and Donald Milton Smith

Vol. 6. **Analysis of Insecticides and Acaricides**. By Francis A. Gunther and Roger C. Blinn (*out of print*)

Vol. 7. **Chemical Analysis of Industrial Solvents**. By the late Morris B. Jacobs and Leopold Schetlan

Vol. 8. **Colorimetric Determination of Nonmetals**. *Second Edition*. Edited by the late David F. Boltz and James A. Howell

Vol. 9. **Analytical Chemistry of Titanium Metals and Compounds**. By Maurice Codell

Vol. 10. **The Chemical Analysis of Air Pollutants**. By the late Morris B. Jacobs

Vol. 11. **X-Ray Spectrochemical Analysis**. *Second Edition*. By L. S. Birks

Vol. 12. **Systematic Analysis of Surface-Active Agents**. *Second Edition*. By Milton J. Rosen and Henry A. Goldsmith

Vol. 13. **Alternating Current Polarography and Tensammetry**. By B. Breyer and H. H. Bauer

Vol. 14. **Flame Photometry**. By R. Herrmann and J. Alkemade

Vol. 15. **The Titration of Organic Compounds** (*in two parts*). By M. R. F. Ashworth

Vol. 16. **Complexation in Analytical Chemistry: A Guide for the Critical Selection of Analytical Methods Based on Complexation Reactions**. By the late Anders Ringbom

Vol. 17. **Electron Probe Microanalysis**. *Second Edition*. By L. S. Birks
Vol. 18. **Organic Complexing Reagents: Structure, Behavior, and Application to Inorganic Analysis**. By D. D. Perrin
Vol. 19. **Thermal Analysis.** *Third Edition*. By Wesley Wm. Wendlandt
Vol. 20. **Amperometric Titrations**. By John T. Stock
Vol. 21. **Reflectance Spectroscopy**. By Wesley Wm. Wendlandt and Harry G. Hecht
Vol. 22. **The Analytical Toxicology of Industrial Inorganic Poisons**. By the late Morris B. Jacobs
Vol. 23. **The Formation and Properties of Precipitates**. By Alan G. Walton
Vol. 24. **Kinetics in Analytical Chemistry**. By Harry B. Mark, Jr. and Garry A. Rechnitz
Vol. 25. **Atomic Absorption Spectroscopy**. *Second Edition*. By Morris Slavin
Vol. 26. **Characterization of Organometallic Compounds** (*in two parts*). Edited by Minoru Tsutsui
Vol. 27. **Rock and Mineral Analysis**. *Second Edition*. By Wesley M. Johnson and John A. Maxwell
Vol. 28. **The Analytical Chemistry of Nitrogen and Its Compounds** (*in two parts*). Edited by C. A. Streuli and Philip R. Averell
Vol. 29. **The Analytical Chemistry of Sulfur and Its Compounds** (*in three parts*). By J. H. Karchmer
Vol. 30. **Ultramicro Elemental Analysis**. By Günther Tölg
Vol. 31. **Photometric Organic Analysis** (*in two parts*). By Eugene Sawicki
Vol. 32. **Determination of Organic Compounds: Methods and Procedures**. By Frederick T. Weiss
Vol. 33. **Masking and Demasking of Chemical Reactions**. By D. D. Perrin
Vol. 34. **Neutron Activation Analysis**. By D. De Soete, R. Gijbels, and J. Hoste
Vol. 35. **Laser Raman Spectroscopy**. By Marvin C. Tobin
Vol. 36. **Emission Spectrochemical Analysis**. By Morris Slavin
Vol. 37. **Analytical Chemistry of Phosphorus Compounds**. Edited by M. Halmann
Vol. 38. **Luminescence Spectrometry in Analytical Chemistry**. By J. D. Winefordner, S. G. Schulman, and T. C. O'Haver
Vol. 39. **Activation Analysis with Neutron Generators**. By Sam S. Nargolwalla and Edwin P. Przybylowicz
Vol. 40. **Determination of Gaseous Elements in Metals**. Edited by Lynn L. Lewis, Laben M. Melnick, and Ben D. Holt
Vol. 41. **Analysis of Silicones**. Edited by A. Lee Smith
Vol. 42. **Foundations of Ultracentrifugal Analysis**. By H. Fujita
Vol. 43. **Chemical Infrared Fourier Transform Spectroscopy**. By Peter R. Griffiths
Vol. 44. **Microscale Manipulations in Chemistry**. By T. S. Ma and V. Horak
Vol. 45. **Thermometric Titrations**. By J. Barthel
Vol. 46. **Trace Analysis: Spectroscopic Methods for Elements**. Edited by J. D. Winefordner
Vol. 47. **Contamination Control in Trace Element Analysis**. By Morris Zief and James W. Mitchell
Vol. 48. **Analytical Applications of NMR**. By D. E. Leyden and R. H. Cox

Vol. 49. **Measurement of Dissolved Oxygen**. By Michael L. Hitchman
Vol. 50. **Analytical Laser Spectroscopy**. Edited by Nicolo Omenetto
Vol. 51. **Trace Element Analysis of Geological Materials**. By Roger D. Reeves and Robert R. Brooks
Vol. 52. **Chemical Analysis by Microwave Rotational Spectroscopy**. By Ravi Varma and Lawrence W. Hrubesh
Vol. 53. **Information Theory as Applied to Chemical Analysis**. By Karl Eckschlager and Vladimir Stepanek
Vol. 54. **Applied Infrared Spectroscopy: Fundamentals, Techniques, and Analytical Problem-solving**. By A. Lee Smith
Vol. 55. **Archaeological Chemistry**. By Zvi Goffer
Vol. 56. **Immobilized Enzymes in Analytical and Clinical Chemistry**. By P. W. Carr and L. D. Bowers
Vol. 57. **Photoacoustics and Photoacoustic Spectroscopy**. By Allan Rosencwaig
Vol. 58. **Analysis of Pesticide Residues**. Edited by H. Anson Moye
Vol. 59. **Affinity Chromatography**. By William H. Scouten
Vol. 60. **Quality Control in Analytical Chemistry**. *Second Edition*. By G. Kateman and L. Buydens
Vol. 61. **Direct Characterization of Fineparticles**. By Brian H. Kaye
Vol. 62. **Flow Injection Analysis**. By J. Ruzicka and E. H. Hansen
Vol. 63. **Applied Electron Spectroscopy for Chemical Analysis**. Edited by Hassan Windawi and Floyd Ho
Vol. 64. **Analytical Aspects of Environmental Chemistry**. Edited by David F. S. Natusch and Philip K. Hopke
Vol. 65. **The Interpretation of Analytical Chemical Data by the Use of Cluster Analysis**. By D. Luc Massart and Leonard Kaufman
Vol. 66. **Solid Phase Biochemistry: Analytical and Synthetic Aspects**. Edited by William H. Scouten
Vol. 67. **An Introduction to Photoelectron Spectroscopy**. By Pradip K. Ghosh
Vol. 68. **Room Temperature Phosphorimetry for Chemical Analysis**. By Tuan Vo-Dinh
Vol. 69. **Potentiometry and Potentiometric Titrations**. By E. P. Serjeant
Vol. 70. **Design and Application of Process Analyzer Systems**. By Paul E. Mix
Vol. 71. **Analysis of Organic and Biological Surfaces.** Edited by Patrick Echlin
Vol. 72. **Small Bore Liquid Chromatography Columns: Their Properties and Uses**. Edited by Raymond P. W. Scott
Vol. 73. **Modern Methods of Particle Size Analysis**. Edited by Howard G. Barth
Vol. 74. **Auger Electron Spectroscopy**. By Michael Thompson, M. D. Baker, Alec Christie, and J. F. Tyson
Vol. 75. **Spot Test Analysis: Clinical, Environmental, Forensic and Geochemical Applications**. By Ervin Jungreis
Vol. 76. **Receptor Modeling in Environmental Chemistry**. By Philip K. Hopke
Vol. 77. **Molecular Luminescence Spectroscopy: Methods and Applications** (*in three parts*). Edited by Stephen G. Schulman
Vol. 78. **Inorganic Chromatographic Analysis**. Edited by John C. MacDonald

Vol. 79. **Analytical Solution Calorimetry**. Edited by J. K. Grime
Vol. 80. **Selected Methods of Trace Metal Analysis: Biological and Environmental Samples**. By Jon C. VanLoon
Vol. 81. **The Analysis of Extraterrestrial Materials**. By Isidore Adler
Vol. 82. **Chemometrics.** By Muhammad A. Sharaf, Deborah L. Illman, and Bruce R. Kowalski
Vol. 83. **Fourier Transform Infrared Spectrometry**. By Peter R. Griffiths and James A. de Haseth
Vol. 84. **Trace Analysis: Spectroscopic Methods for Molecules**. Edited by Gary Christian and James B. Callis
Vol. 85. **Ultratrace Analysis of Pharmaceuticals and Other Compounds of Interest**. Edited by S. Ahuja
Vol. 86. **Secondary Ion Mass Spectrometry: Basic Concepts, Instrumental Aspects, Applications and Trends.** By A. Benninghoven, F. G. Rüdenauer, and H. W. Werner
Vol. 87. **Analytical Applications of Lasers**. Edited by Edward H. Piepmeier
Vol. 88. **Applied Geochemical Analysis**. By C. O. Ingamells and F. F. Pitard
Vol. 89. **Detectors for Liquid Chromatography**. Edited by Edward S. Yeung
Vol. 90. **Inductively Coupled Plasma Emission Spectroscopy: Part 1: Methodology, Instrumentation, and Performance; Part II: Applications and Fundamentals**. Edited by J. M. Boumans
Vol. 91. **Applications of New Mass Spectrometry Techniques in Pesticide Chemistry**. Edited by Joseph Rosen
Vol. 92. **X-Ray Absorption: Principles, Applications, Techniques of EXAFS, SEXAFS, and XANES**. Edited by D. C. Konnigsberger
Vol. 93. **Quantitative Structure-Chromatographic Retention Relationships**. By Roman Kaliszan
Vol. 94. **Laser Remote Chemical Analysis**. Edited by Raymond M. Measures
Vol. 95. **Inorganic Mass Spectrometry**. Edited by F. Adams, R. Gijbels, and R. Van Grieken
Vol. 96. **Kinetic Aspects of Analytical Chemistry**. By Horacio A. Mottola
Vol. 97. **Two-Dimensional NMR Spectroscopy**. By Jan Schraml and Jon M. Bellama
Vol. 98. **High Performance Liquid Chromatography**. Edited by Phyllis R. Brown and Richard A. Hartwick
Vol. 99. **X-Ray Fluorescence Spectrometry**. By Ron Jenkins
Vol. 100. **Analytical Aspects of Drug Testing**. Edited by Dale G. Deustch
Vol. 101. **Chemical Analysis of Polycyclic Aromatic Compounds**. Edited by Tuan Vo-Dinh
Vol. 102. **Quadrupole Storage Mass Spectrometry**. By Raymond E. March and Richard J. Hughes
Vol. 103. **Determination of Molecular Weight**. Edited by Anthony R. Cooper
Vol. 104. **Selectivity and Detectability Optimization in HPLC**. By Satinder Ahuja
Vol. 105. **Laser Microanalysis**. By Lieselotte Moenke-Blankenburg
Vol. 106. **Clinical Chemistry**. Edited by E. Howard Taylor

Vol. 107. **Multielement Detection Systems for Spectrochemical Analysis**. By Kenneth W. Busch and Marianna A. Busch

Vol. 108. **Planar Chromatography in the Life Sciences**. Edited by Joseph C. Touchstone

Vol. 109. **Fluorometric Analysis in Biomedical Chemistry: Trends and Techniques Including HPLC Applications**. By Norio Ichinose, George Schwedt, Frank Michael Schnepel, and Kyoko Adochi

Vol. 110. **An Introduction to Laboratory Automation**. By Victor Cerdá and Guillermo Ramis

Vol. 111. **Gas Chromatography: Biochemical, Biomedical, and Clinical Applications**. Edited by Ray E. Clement

Vol. 112. **The Analytical Chemistry of Silicones**. Edited by A. Lee Smith

Vol. 113. **Modern Methods of Polymer Characterization**. Edited by Howard G. Barth and Jimmy W. Mays

Vol. 114. **Analytical Raman Spectroscopy**. Edited by Jeanette Graselli and Bernard J. Bulkin

Vol. 115. **Trace and Ultratrace Analysis by HPLC**. By Satinder Ahuja

Vol. 116. **Radiochemistry and Nuclear Methods of Analysis**. By William D. Ehmann and Diane E. Vance

Vol. 117. **Applications of Fluorescence in Immunoassays**. By Ilkka Hemmila

Vol. 118. **Principles and Practice of Spectroscopic Calibration**. By Howard Mark

Vol. 119. **Activation Spectrometry in Chemical Analysis**. By S. J. Parry

Vol. 120. **Remote Sensing by Fourier Transform Spectrometry**. By Reinhard Beer

Vol. 121. **Detectors for Capillary Chromatography**. Edited by Herbert H. Hill and Dennis McMinn

Vol. 122. **Photochemical Vapor Deposition**. By J. G. Eden

Vol. 123. **Statistical Methods in Analytical Chemistry**. By Peter C. Meier and Richard Zünd

Vol. 124. **Laser Ionization Mass Analysis**. Edited by Akos Vertes, Renaat Gijbels, and Fred Adams

Vol. 125. **Physics and Chemistry of Solid State Sensor Devices**. By Andreas Mandelis and Constantinos Christofides

Vol. 126. **Electroanalytical Stripping Methods**. By Khjena Z. Brainina and E. Neyman

Vol. 127. **Air Monitoring by Spectroscopic Techniques**. Edited by Markus W. Sigrist

Vol. 128. **Information Theory in Analytical Chemistry**. By Karel Eckschlager and Klaus Danzer

Vol. 129. **Flame Chemiluminescence Analysis by Molecular Emission Cavity Detection**. Edited by David Stiles, Anthony Calokerinos, and Alan Townshend

Vol. 130. **Hydride Generation Atomic Absorption Spectrometry**. Edited by Jiri Dedina and Dimiter L. Tsalev

Vol. 131. **Selective Detectors: Environmental, Industrial, and Biomedical Applications**. Edited by Robert E. Sievers

Vol. 132. **High-Speed Countercurrent Chromatography**. Edited by Yoichiro Ito and Walter D. Conway

Vol. 133. **Particle-Induced X-Ray Emission Spectrometry**. By Sven A. E. Johansson, John L. Campbell, and Klas G. Malmqvist

Vol. 134. **Photothermal Spectroscopy Methods for Chemical Analysis**. By Stephen E. Bialkowski

Vol. 135. **Element Speciation in Bioinorganic Chemistry**. Edited by Sergio Caroli

Vol. 136. **Laser-Enhanced Ionization Spectrometry**. Edited by John C. Travis and Gregory C. Turk

Vol. 137. **Fluorescence Imaging Spectroscopy and Microscopy**. Edited by Xue Feng Wang and Brian Herman

Vol. 138. **Introduction to X-Ray Powder Diffractometry**. By Ron Jenkins and Robert L. Snyder

Vol. 139. **Modern Techniques in Electroanalysis**. Edited by Petr Vanýsek

Vol. 140. **Total-Reflection X-Ray Fluorescence Analysis**. By Reinhold Klockenkamper

Vol. 141. **Spot Test Analysis: Clinical, Environmental, Forensic, and Geochemical Applications**. *Second Edition*. By Ervin Jungreis

Vol. 142. **The Impact of Stereochemistry on Drug Development and Use**. Edited by Hassan Y. Aboul-Enein and Irving W. Wainer

Vol. 143. **Macrocyclic Compounds in Analytical Chemistry**. Edited by Yury A. Zolotov

Vol. 144. **Surface-Launched Acoustic Wave Sensors: Chemical Sensing and Thin-Film Characterization**. By Michael Thompson and David Stone

Vol. 145. **Modern Isotope Ratio Mass Spectrometry**. Edited by T. J. Platzner

Vol. 146. **High Performance Capillary Electrophoresis: Theory, Techniques, and Applications**. Edited by Morteza G. Khaledi

Vol. 147. **Solid Phase Extraction: Principles and Practice**. By E. M. Thurman

Vol. 148. **Commercial Biosensors: Applications to Clinical, Bioprocess and Environmental Samples**. Edited by Graham Ramsay

Vol. 149. **A Practical Guide to Graphite Furnace Atomic Absorption Spectrometry**. By David J. Butcher and Joseph Sneddon

Vol. 150. **Principles of Chemical and Biological Sensors**. Edited by Dermot Diamond

Vol. 151. **Pesticide Residue in Foods: Methods, Technologies, and Regulations**. By W. George Fong, H. Anson Moye, James N. Seiber, and John P. Toth

Vol. 152. **X-Ray Fluorescence Spectrometry**. *Second Edition*. By Ron Jenkins

Vol. 153. **Statistical Methods in Analytical Chemistry**. *Second Edition*. By Peter C. Meier and Richard E. Zünd

Vol. 154. **Modern Analytical Methodologies in Fat- and Water-Soluble Vitamins**. Edited by Won O. Song, Gary R. Beecher, and Ronald R. Eitenmiller

Vol. 155. **Modern Analytical Methods in Art and Archaeology.** Edited by Enrico Ciliberto and Guiseppe Spoto

Vol. 156. **Shpol'skii Spectroscopy and Other Site Selection Methods: Applications in Environmental Analysis, Bioanalytical Chemistry and Chemical Physics**. Edited by C. Gooijer, F. Ariese and J. W. Hofstraat

Vol. 157. **Raman Spectroscopy for Chemical Analysis.** By Richard L. McCreery

Vol. 158. **Large (C>=24) Polycyclic Aromatic Hydrocarbons: Chemistry and Analysis**. By John C. Fetzer

CHAPTER

1

INTRODUCTION

1.1 INTRODUCTION AND GENERAL DISCUSSION

Interest in the large polycyclic aromatic hydrocarbons (LPAHs) has grown over the past two decades. This has been reflected by numerous papers in synthetic organic chemistry, chemical analysis, occurrence, theoretical studies and modeling, and several other research areas.

In this book, this class of chemical compounds will be defined as those with aromatic cores (the portion of the molecule containing the fused aromatic rings, as opposed to any alkyl substituents) of 24 carbons. This corresponds to molecular masses of 300 daltons or more. This limit is somewhat arbitrary when the properties of the PAHs are examined. As described in this and later chapters, there is a continuum in the behavior of the PAHs. The very small PAHs, such as phenanthrene, anthracene, or fluorene (*compounds 1–3*), do behave very differently than the larger PAHs, such as pyranthrene or ovalene (*compounds 4 and 5*). The trends in between, however, do not show a sudden change in properties but rather show gradual changes from carbon number to ever-increasing carbon number.

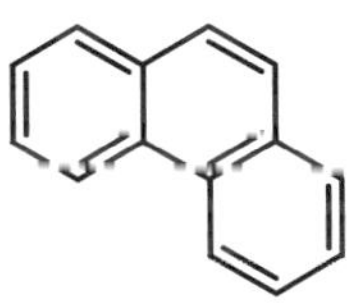

1

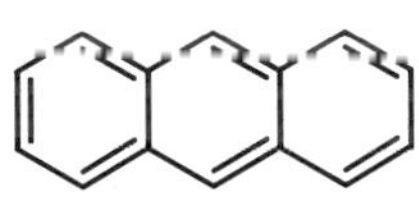

2

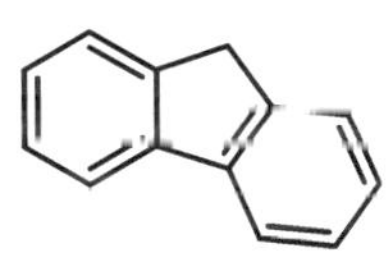

3

The definition of what will be called an LPAH, and therefore the compound classes covered in this book, is also limited to those ring structures containing only five- and six-member rings, what Herndon calls the one-boundary-edge or magic-boundary PAHs (1). The ones with other numbers of carbon atoms in the ring can be synthesized in the laboratory but are very, very rarely found elsewhere. Some novel structures containing six-member rings are not also

Large (C>=24) Polycyclic Aromatic Hydrocarbons: Chemistry and Analysis, by John C. Fetzer.
ISBN 0-471-36354-5

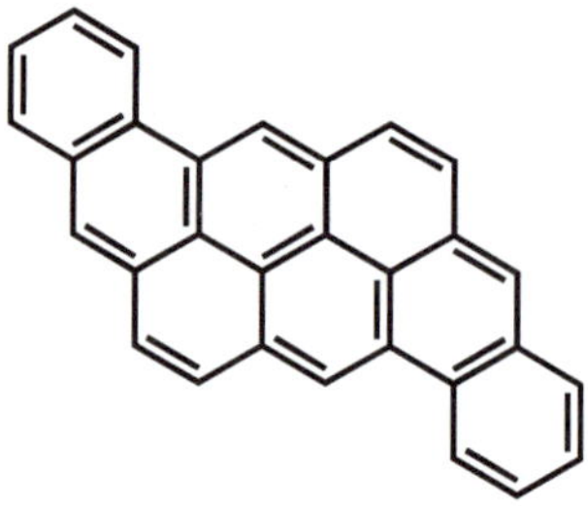

4

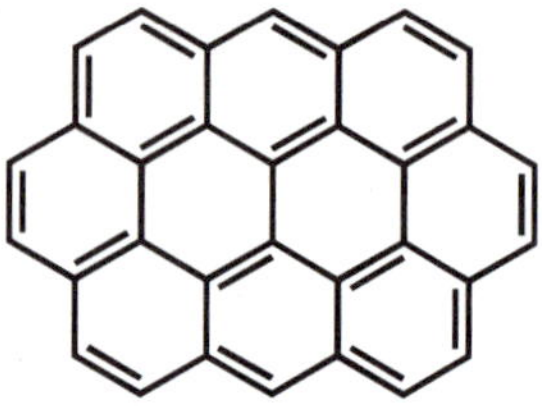

5

covered. These include the circulenes that have an internal "hole" of seven or more bonds (although coronene, with six internal bonds, is defined as an LPAH); kekulene; the wide variety of polyphenyls, polynaphthyls, cyclophanes, annulenes, and other compounds with small aromatic parts joined by single bonds; and the numerous other structures with more than one boundary edge. An example of a circulene is shown as *compound 6*.

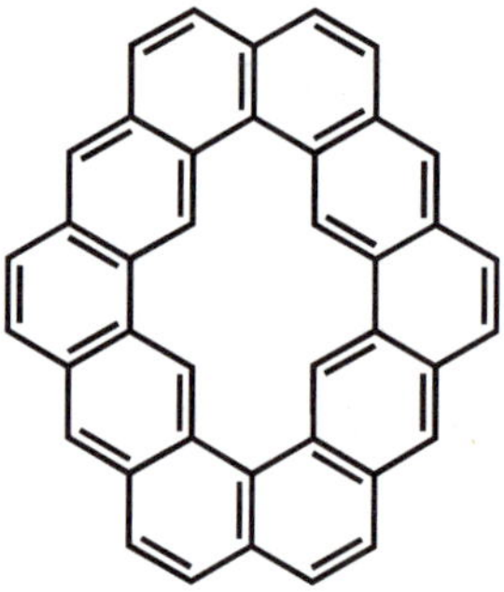

6

This choice of 24 carbons as the dividing line between the smaller and larger PAHs is not unreasonable. This is because, until recently, the more commonly available reference compounds [and analytical procedures, such as those mandated by the United States Environmental Protection Agency, (USEPA)] were dibenz[*a,c*]anthracene, dibenz[*a,h*]anthracene, dibenz[*a,j*]anthracene, picene, indeno[1,2,3-*cd*]pyrene, and benzo[*ghi*]perylene (*compounds 7–12*). These have 22 carbons and molecular masses of 276 or 278 daltons. Although coronene (*compound 13*), at 300 daltons and a formula of $C_{24}H_{12}$, was commercially available, the other LPAHs of this carbon number generally were not easy to acquire. Those of more than 24 carbons were seldom available from commercial sources.

Throughout this volume the acronym LPAHs will be used as the plural noun for this subgroup of the polycyclic aromatic hydrocarbons (PAHs). Also, in line

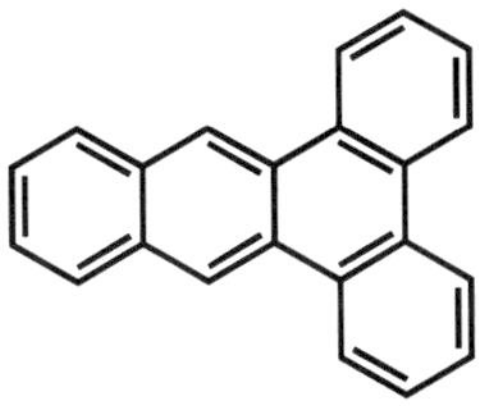

7

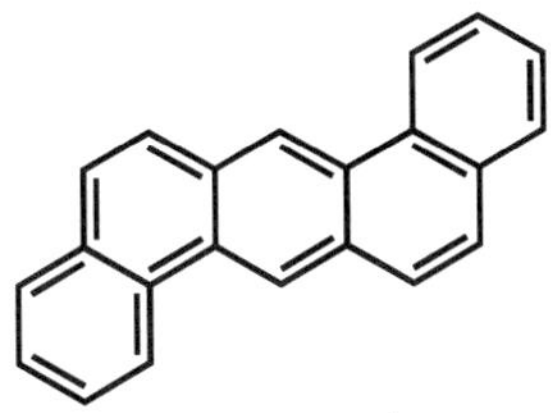

8

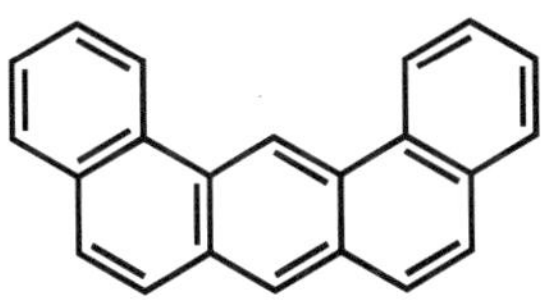

9

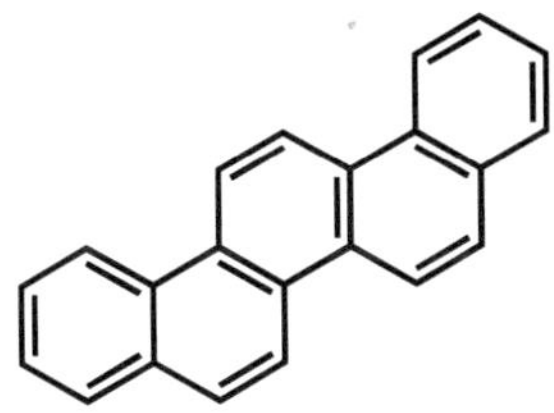

10

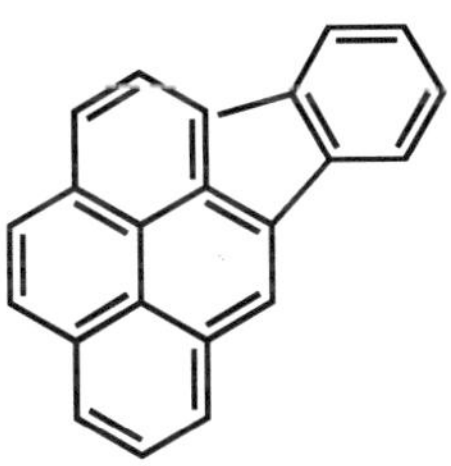

11

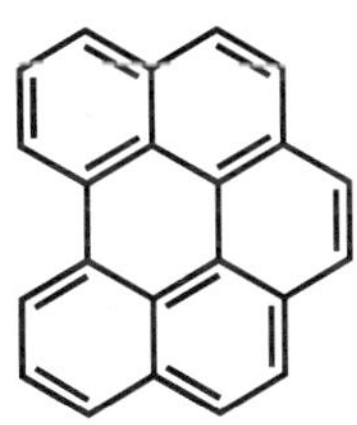

12

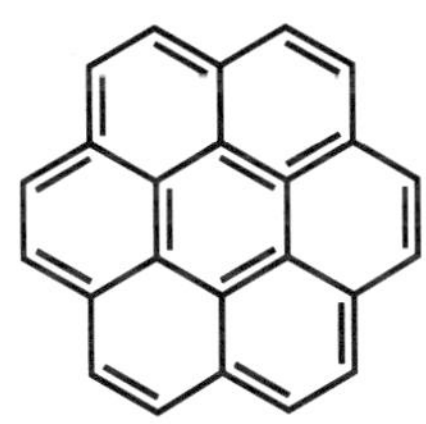

13

with the practice of the American Chemical Society (2), LPAH will either refer to an individual compound in this class or be used as an adjective (such as in LPAH structure or LPAH spectrum). The term LPAHs is used as the plural, collective noun.

As one example of the growth in interest in LPAHs, the contents of two conferences on PAH research can be compared. There has been a series of symposia on research on the polycyclic aromatic compounds, PACs. (3, 4). The focus of these meetings has predominantly been on their chemistry and biology. The symposium volume for one of the early meetings, held in 1978, had 56 papers. Only one of these dealt with LPAHs, and that was only as a part of work on the Shpol'skii spectroscopy of PAHs. Only 3 of the 33 PAHs studied in that work were LPAHs. In contrast, at a recent meeting held in 1997, there were 6 oral presentations on LPAHs out of a total of 77. Additionally, this conference had poster sessions and there were 9 LPAH-related poster presentations out of 68 total posters. Thus there were 15 LPAH-related presentations out of a total of 145.

About one-half of the 15 LPAH-related presentations dealt solely with LPAH research. It is also interesting to note the overall growth in the research in PAHs.

Although there had been a few studies involving analysis of LPAHs before 1980, the large majority of work has been done since then. Many factors have limited the study of LPAHs. A few of these have been partially addressed recently, leading to an upsurge in studies of LPAHs. The most dominant reason for this was a prior unavailability of reference compounds (well-defined, usually synthesized samples of a pure individual LPAH). In the past twenty years, many new LPAHs have been synthesized, isolated, and characterized and thus have become available for use by analytical chemists. A few others have also been isolated from nonsynthetic sources and characterized. Without pure materials, many secondary reasons then arise that kept researchers from studying LPAHs. These include the paucity of information on biological activities, occurrence, and the proper means for analysis.

The various factors limiting research into LPAHs will be discussed below, but first a review of the general properties of the smaller PAHs is necessary. This is in order to make it easier to understand the specific factors governing the chemistry and analysis of the LPAHs.

1.2 THE SMALLER POLYCYCLIC AROMATIC HYDROCARBONS

Any discussion of the LPAHs must first begin with an extensive review of the smaller PAHs. This is because the principles that generally govern the LPAHs are the same as those of the smaller PAHs. These smaller PAHs have been studied much more, and so the understanding of their behavior is much, much greater.

There have been many books written on the chemistry of the PAHs (5–17); the most notable was the two-volume set written by Eric Clar (5). Although the focus of these volumes was the synthesis of PAHs, they also contain ultraviolet (UV) absorbance spectra of hundreds of PAHs, information about their reactivities and fluorescence colors, and chapters on structure, carcinogenicity, and aromaticity. Some of the subsequent books on the PAHs have provided general coverage, but the bulk of the books have been focused on specific areas, such as theoretical calculations, carcinogenicity, organic synthesis, spectroscopy, and analysis.

The chemistry and analysis of the PAHs have received increasing amounts of attention over the past few years. This research has mainly been focused on those PAHs of less than two dozen ring carbons. Many of these have been found to be mutagenic or carcinogenic. Sixteen of these PAHs have been designated as priority pollutants by the USEPA, with the regulatory requirements that they be analyzed for in air and water emissions, in contaminated soil and other materials, as well as in specific manufactured products. This EPA listing has focused a phenomenal amount of attention on these specific 16 PAHs.

Many other smaller PAHs have been shown to be highly active mutagens or carcinogens. No LPAHs are currently on the USEPA or any other regulatory listings, and only a very few have been fully tested and their biological activities determined.

1.3 STRUCTURE AND PROPERTIES

The simplest multi-ring aromatic hydrocarbon is the 2-ring naphthalene, with 10 ring carbons. The number of isomeric structures for PAHs greatly increases with increasing carbon number. Also, the types of PAHs become more varied. There are two broad classes of PAHs, those consisting of only six-member rings and those containing at least one five-member ring. These are called alternant and nonalternant PAHs. The name alternant arises because all the π bonds can be arranged in an alternating, continuous fashion when the π electron Kekule-type structures are drawn. This cannot be done for the complete structure of a nonalternant PAH.

To illustrate these isomeric structural differences, when one takes the progression starting with benzene at one ring, only one structure can result by addition of one ring, the two-ring naphthalene. Adding another ring to this structure results in the two isomeric 14-carbon PAHs, phenanthrene and anthracene, if the ring is added by fusion to a single face.

Addition of another ring can proceed in a similar fashion to produce structures such as tetracene, chrysene, tetraphene (benz[*a*]anthracene), and triphenylene (*compounds 14–17*). These types of PAHs with structures in which the rings are joined only by one shared face for each ring connection are known as the ortho-fused PAHs.

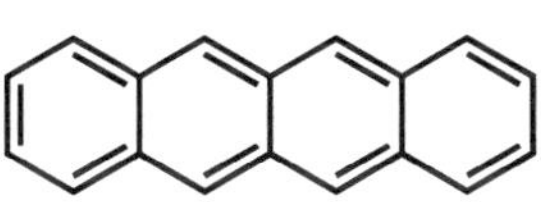

14

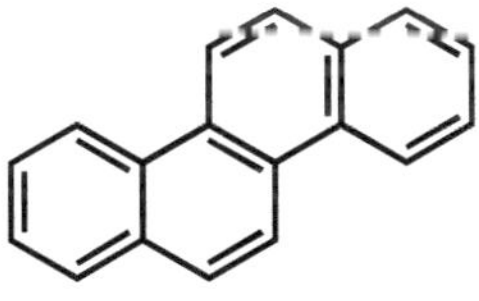

15

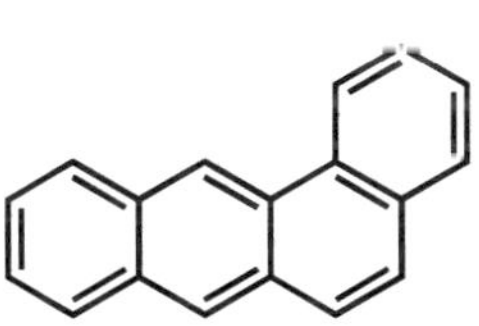

16

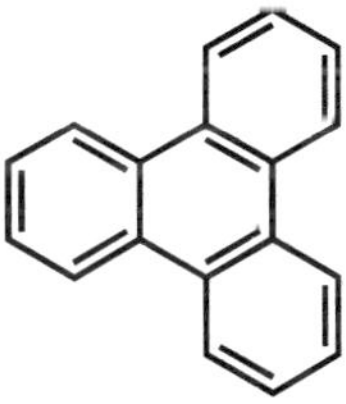

17

The number of isomeric LPAHs for five ortho-fused rings (22 carbons) is 15 and for six rings (26 carbons) it is 37. Above this number, the calculations are cumbersome and complicating issues, such as chiral isomer pairs, make enumeration difficult.

As the number of fused rings increases, one important structural factor emerges. Addition of a ring to phenanthrene at its "c" face gives benzo[*c*]phenanthrene (*compound 18*). The peripheral hydrogen atoms of this molecule would overlap if the molecule were planar. This "steric strain" of not wanting to have such spatially adjacent atoms is relieved by the carbon skeleton of the molecule assuming a nonplanar configuration. Addition of another ring to give dibenzo[*c*,*g*]phenanthrene (*compound 19*) increases this. Bjorseth (6)

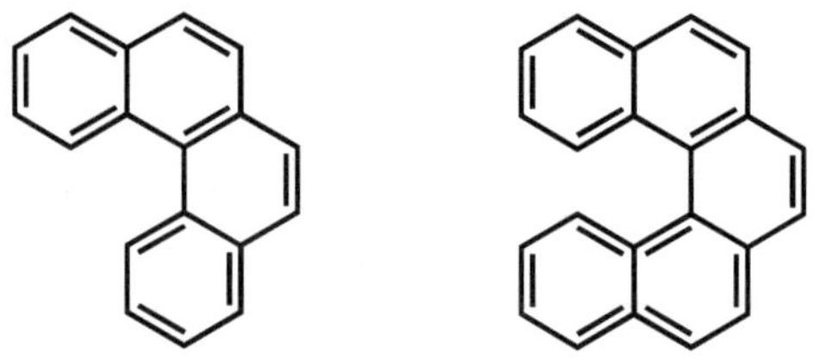

18 **19**

coined the names of these three concave arrangements of rings that cause steric strain. He calls them bay, cove, and fjord regions for these structures of three, four, and five faces. These three structures are shown in Figure 1.

When the additional ring is added by filling in the bay region of the phenanthrene structure, another class of PAHs is created. A bay region is the three-sided internally concave structure highlighted. This ring addition gives the

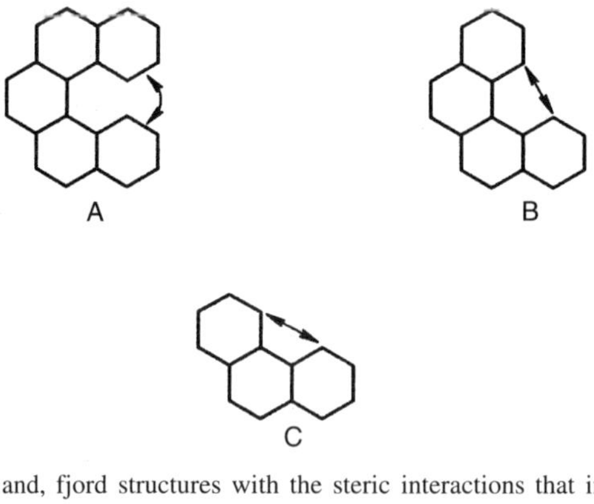

Figure 1. Bay, cove, and, fjord structures with the steric interactions that induce nonplanarity in LPAHs.

PAH pyrene (*compound 20*), the first member of the peri-fused PAHs. The peri-fused PAHs have rings that are connected to each other by more than one face; also, they have carbon atoms that are not on the periphery of the molecule. Other common peri-fused PAHs are perylene (*compound 21*) and benzo[*ghi*]perylene. The peri-fused PAHs have some carbon atoms that are not on the periphery of the structure. Some of these highly condensed structures, which are also composed of only six-member rings, are the most thermodynamically stable class of PAHs. They are, therefore, the most commonly found PAHs.

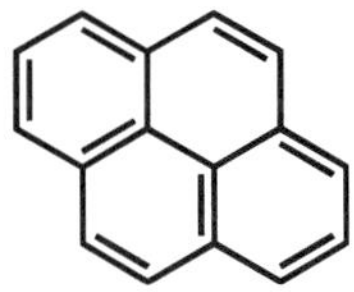

20

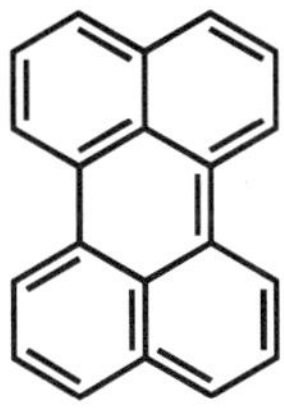

21

By definition the peri-fused PAHs also include those in which any portion of the PAH is a peri-fused structure, despite the addition to this of numerous ortho-fused structures. Therefore, the structure shown, benzo[*e*]pyrene (*compound 22*), is peri-fused because it has a portion that is the peri-fused PAH pyrene.

The other major class of PAHs is those containing at least one five-member ring. Common examples are fluorene, fluoranthene (*compound 23*), and indeno[1,2,3-*cd*]pyrene. PAHs with five-member rings are much less thermodynamically favored than those with six-member rings because the π electrons are not readily delocalized across a five-member ring. The resulting resonance stability, therefore, is lower.

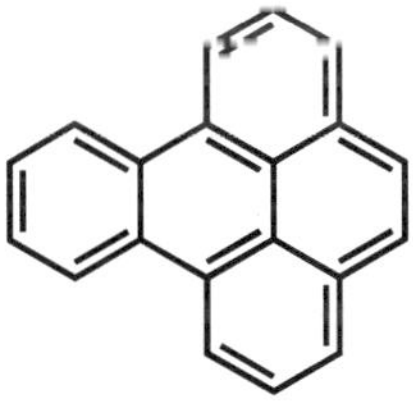

22

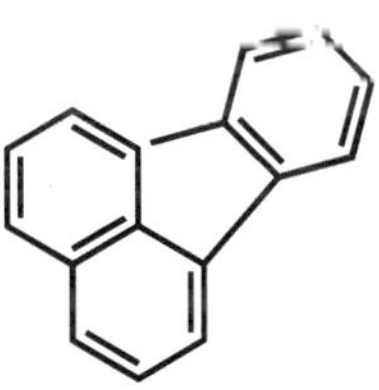

23

1.4 PAH NOMENCLATURE

The International Union for Pure and Applied Chemistry (IUPAC) has formulated a set of systematic rules for the naming of PAHs (18). These were first formalized in the early 1960s, so any work before that time relies on unofficial, often inconsistent, naming rules. For example, benzo[1,2]pyrene was frequently used for what is now called benzo[*e*]pyrene and benzo[3,4]pyrene referred to benzo[*a*]pyrene (*compound 24*). When the systematic IUPAC rules were created, the "1,2" positions of pyrene were designated as the "a" face.

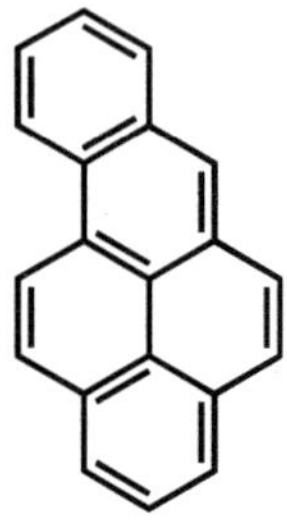

24

In using the IUPAC rules, the process of naming a PAH starts with finding which of a set of defined core structures are contained in the PAH. These core structures include naphthalene, phenanthrene, chrysene, pyrene, and certain other structures. These are numbered in ascending order and are shown in Figure 2. The naming of a PAH is done by examining the PAH structure and finding the highest-number core structure that can be completely contained as part of it.

There had been some inconsistencies in IUPAC nomenclature that were only recently modified (by the 1998 recommendations of the organic chemistry nomenclature committee). The four-ring linear structure is now called tetracene rather than the previously accepted naphthacene. The naming cores had included the bent ortho-fused structures phenanthrene, pentaphene (*compound 25*), hexaphene (*compound 26*), heptaphene (*compound 27*), octaphene (*compound 28*), etc. There was no four-ring core structure. The core structure tetraphene was added.

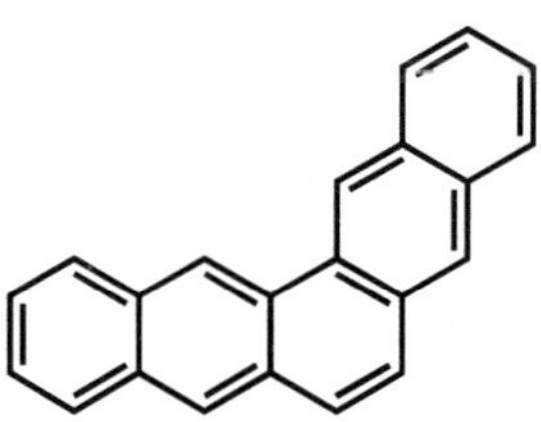

25

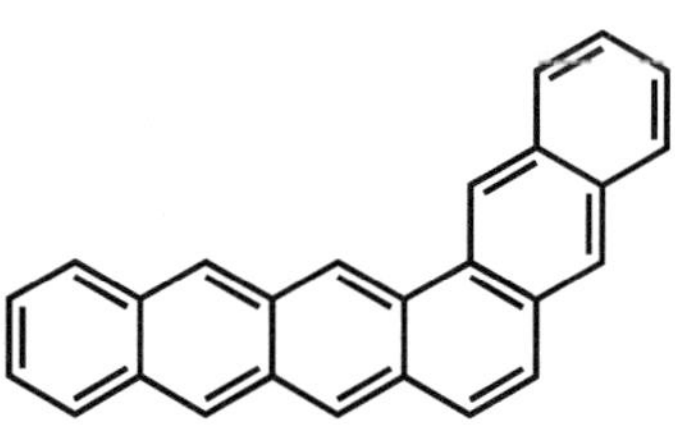

26

The next step in naming a PAH is establishing the orientation of the core structure, which is defined by placing the structure so that its longest row of rings as the *x*-axis. The *y*-axis is then defined so that there is the highest possible number of rings in the upper right-hand quadrant. If there is more than one possible configuration, the preferred one should have the fewest possible rings

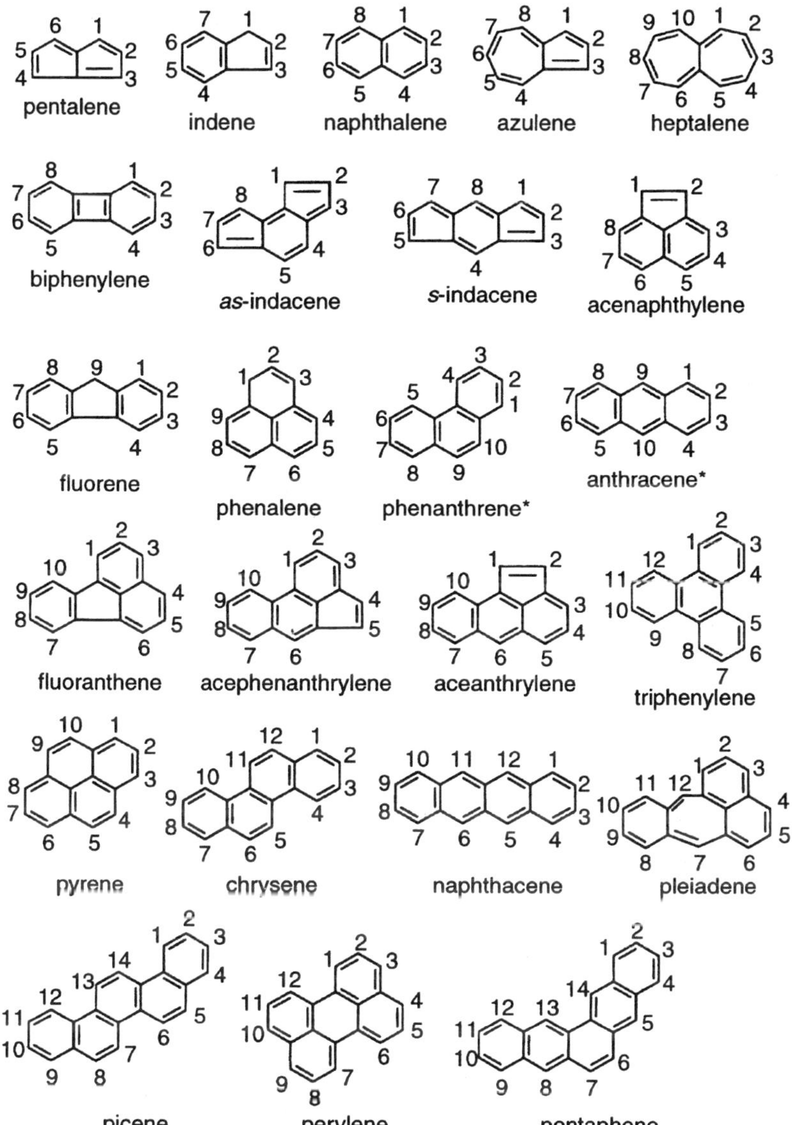

Figure 2. IUPAC core naming structures with the defined numbering system for each. For faces during fusion, the lettering starts analogously with the "a" face being the face between 1 and 2.

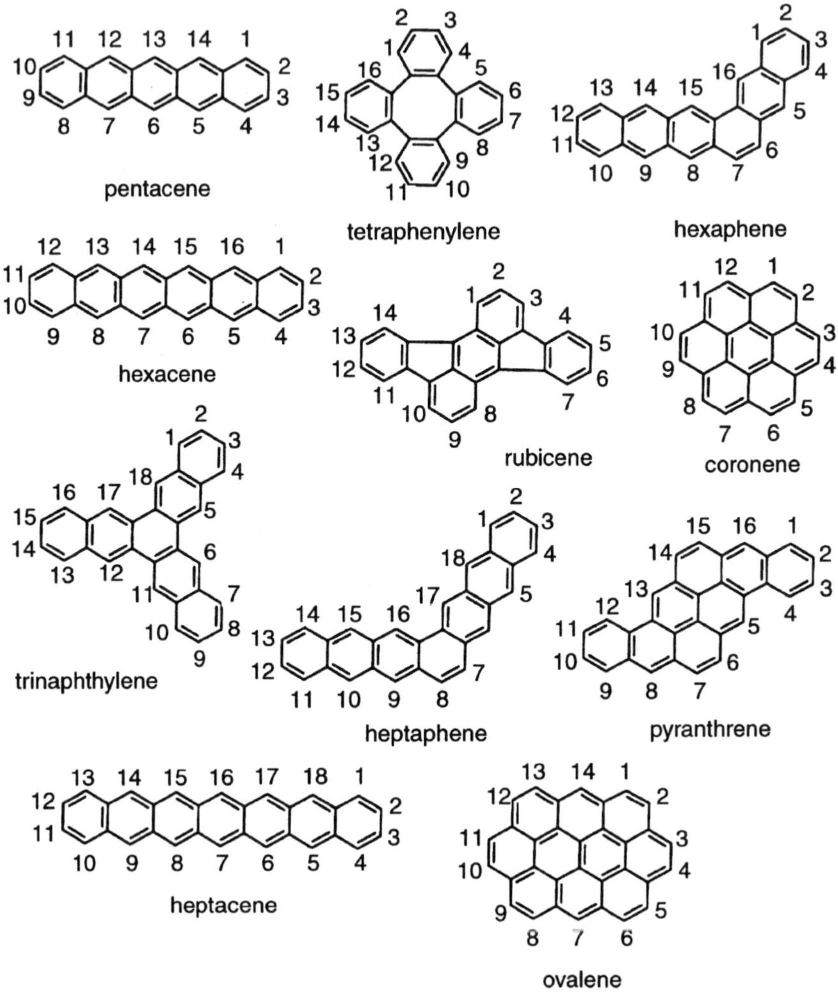

Figure 2. (*Continued*)

in the lower left-hand quadrant. Thus, for the bent phene series, hexaphene is oriented so that the four linear-ring segment is the *x*-axis. The row of three rings is pointed towards the upper right-hand quadrant (as shown for *compound 26*).

The fusing group endings are generally shortened by dropping the final syllable or "e" on the name and replacing with an "o", so phenanthrene fused is phenanthro and pyrene fused is pyreno. The IUPAC nomenclature for fused acenaphthylene and perylene has changed. The fusing forms are acenaphtho and perylo, rather than the earlier accepted acenaphthyleno and peryleno forms.

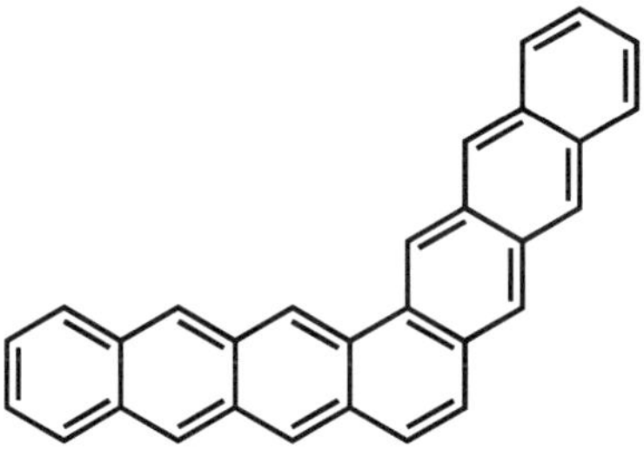

27

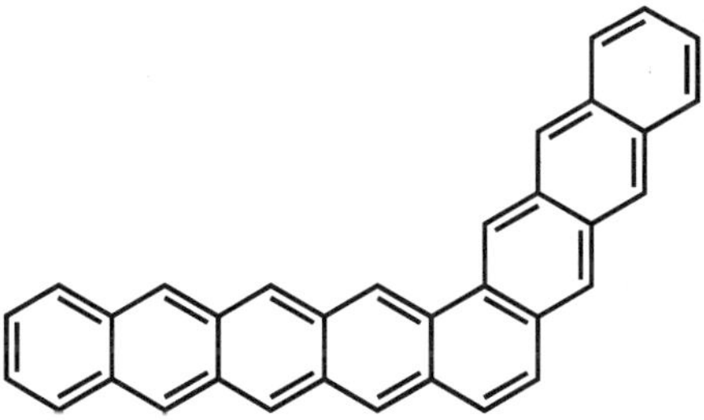

28

The faces on the periphery of the core structure are sequentially lettered, with the "a" face being the one at the upper right-hand side (see examples). The letters proceed clockwise around the periphery. If there are more than 26 faces, the sequence continues with "a1", "b1", etc., until all the faces have been designated. The letters of the contiguous faces of one fused group are listed in series without separating commas, for example in benzo[*ghi*]perylene. For multiple fusion, the noncontiguous faces would be separated by commas, as in dibenz[*a,h*]anthracene. Note that for core structures that begin with a vowel, the ending vowel of the fused group is dropped.

The faces of the structures that are fused onto the core structure are numbered based on the numbering system of the structure being fused, for example, when a naphthalene structure is fused to coronene through three faces. The three fusing faces on the naphthalene could be numbered any number of ways, 1,2,3, or 2,3,4, or 8,1,2 ,or even 2,1,8. The IUPAC rule states that the lowest set of numbers that are possible is used. The correct choice depends on the orientation of the naphthalene moiety, relative to the direction of the lettering on the core structure. In the case of the three-face fused naphthalene on coronene, the numbering would be 8,1,2 because the direction of the lettering where it is fused increases from the "a" face to the "b" face on the core structure. The alternative configurations giving 3,4,5 or 5,4,3 are not chosen because the number set has higher values. The lettering of coronene would be "abc". The full IUPAC name would be naphtho[8,1,2-*abc*]coronene, giving the structure shown as *compound 29*.

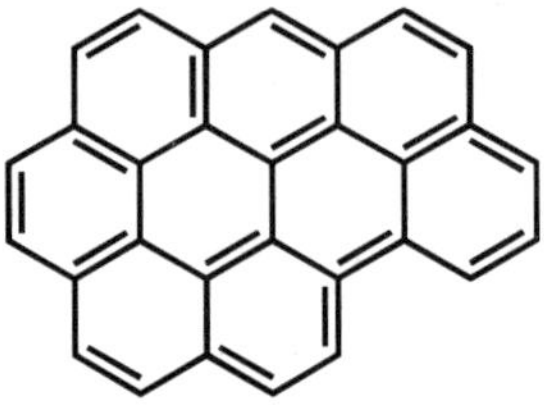

29

Each group of numbers and letters (describing the fusion of structures of a particular type) are set off by square braces ([]) and not parentheses. The numbers are separated from the letters by hyphens. When more than one set is contained in the square braces (showing more than 1 fusion of that type of group), they are separated from each other by colons.

When there are multiple fusing groups, two main rules govern the naming. First, if the same core structure can be found with different arrangements of fusing groups, the preferred combination of fusing groups is the one made up of the simpler structures (those toward the beginning of the structures list). This is in contrast to the preference in core structures that aims towards the structure highest on the list. Thus, two benzo fusions are preferred over an alterative with a naphtho; or a benzo and naphtho is preferred over a phenanthro, anthra, or phenaleno.

The second rule for fusing groups is that the lettering of the core structure will be such that the higher fused group has the alphabetical preference. Therefore, for the structure shown (*compound 30*), the naphtho fusion is "bcd" and the benzo fusion is "pqr".

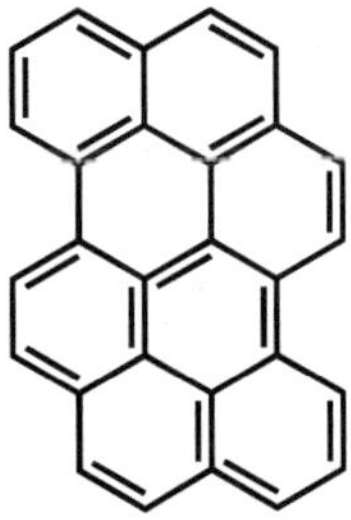

30

For multiple fusing groups on a core structure, the order of naming is alphabetical. Thus an anthra is listed before a benzo, which is before a coroneno, which is before a naphtho, and so on. The prefixes denoting the number of each type of fusing group (di-, tri-, tetra-, etc.) are ignored in the alphabetization; only the lettering of the fusing groups are used. If one fusing group must be fused to another, then this becomes the alphabetical prefix for the whole group. Thus, for the structures shown (*compounds 31 and 32*), the sixteen-ring dinaphtho[2,1,8,7-*defg*:2′,1′,8′,7′-*mnop*]pentapheno[3,4,5,6,7,8,9,10-$uvwxyza_1b_1c_1d_1$]heptaphene and seventeen-ring benzo[13,14] pentapheno[3,4,5,6,7,8,9,10-$uvwxyza_1b_1c_1d_1$]-dinaphtho[2,1,8,7-*defg*:2′,1′,8′,7′-*mnop*]heptaphene, the benzo group fused to the pentapheno group in the second structure is listed as a benzo-fused group (which can be thought of as consisting of the six total rings).

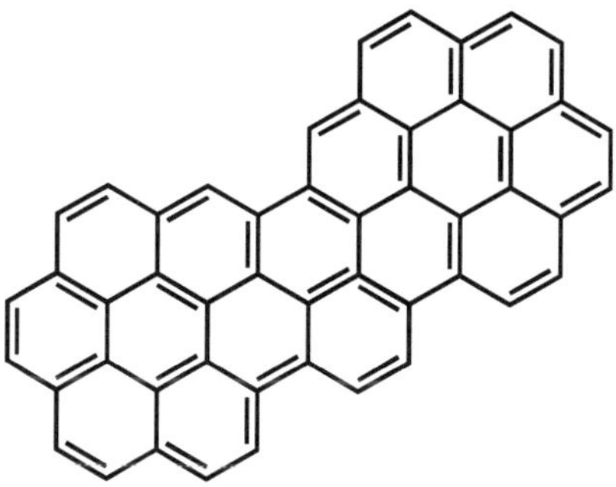

31

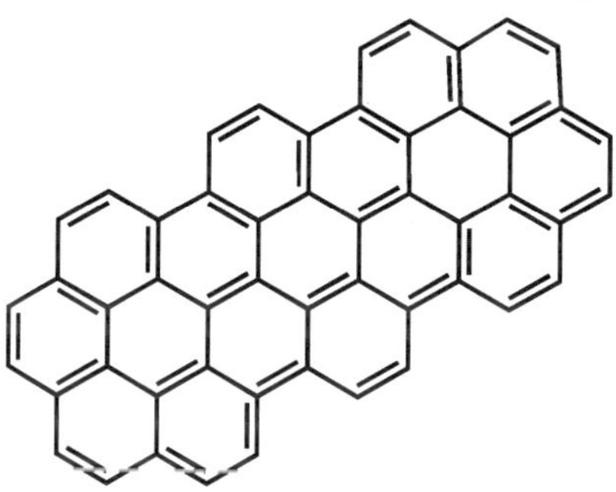

32

There are few core-naming structures of greater than five rings. This severely limits the naming of the LPAHs. Because both the fused structure(s) and the core structure are chosen from this list, with the core structure being the one completely found in the compound to be named, a combination of only small structures leads to very long and complicated names. A LPAH structure can have 20 or more rings, for example, but the ovalene, pyranthrene, coronene, or other large core structures may not be part of it. Then the IUPAC name may be based on a very small core structure such as perylene or picene with lots of fused benzo , naphtho , or other small ring structures.

A few other examples of LPAH nomenclature and structures are also shown to give the reader a sense of the naming conventions. Note that because the selection of core structures is so limited, even very similar isomers that differ only in the positioning of one ring, usually end up with drastically different names. Sander and Wise (19) have published a compilation of 660 PAH structures, of which over 500 are LPAHs of up to 38 carbon atoms. With each structure, they also list the Chemical Abstract Service registry number, formula, molecular weight, IUPAC name (and some alternant or trivial names that have been used in the literature for some of the PAHs), and calculated molecular dimensions. The positional numbering is also shown on each structural diagram, enabling the reader to name derivatives of the PAH.

Several trivial names that had been acceptable have now been dropped. These include acenaphthene, cholanthrene, aceanthrene, and acephenanthrene. Before the recent naming changes there was one anomaly in the naming of LPAHs. Clar and others reduced the large ketone dyestuff violanthrone and its isomer isoviolanthrone. They obtained the two fully aromatic nine-ring LPAHs shown. They naturally called these violanthrene and isoviolanthrene (*compounds 33 and 34*). Previously, the IUPAC nomenclature committee, however, chose to use

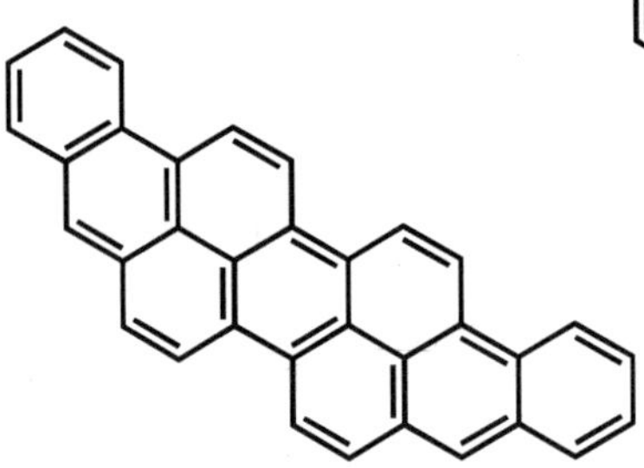

33

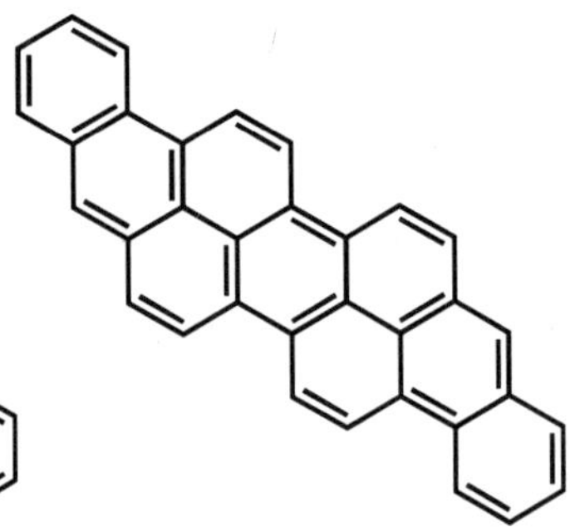

34

these two names as acceptable "trivial" names when referring to the analogous structures without the carbonyl groups, i.e., with methylene bridges in some positions (*compounds 35 and 36*). These reduced forms could be used to name the similar LPAHs, because hydrogen abstraction of a naming group is not allowed. Therefore, the two LPAHs are officially named benzo[*rst*]anthra[9,1,2-*cde*]pentaphene and benzo[*rst*]phenanthro[10,1,2-*cde*]pentaphene. The recent nomenclature changes have also eliminated these two trivial names, and the inconsistency no longer exists.

The numerous recent changes in IUPAC nomenclature could drastically affect the overall naming of LPAHs. The addition of the core name tetraphene makes the new IUPAC nomenclature inconsistent with either Beilstein's Compendium or the Chemical Abstracts Service usage. Common PAHs will now be named differently. Tetraphene is the structure commonly called benz[*a*]anthracene, and

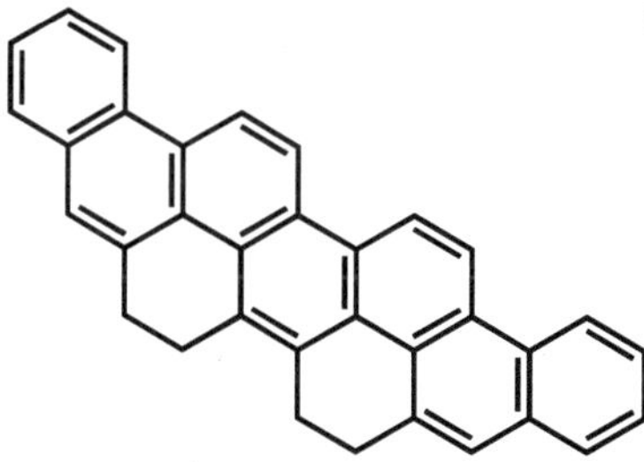

35

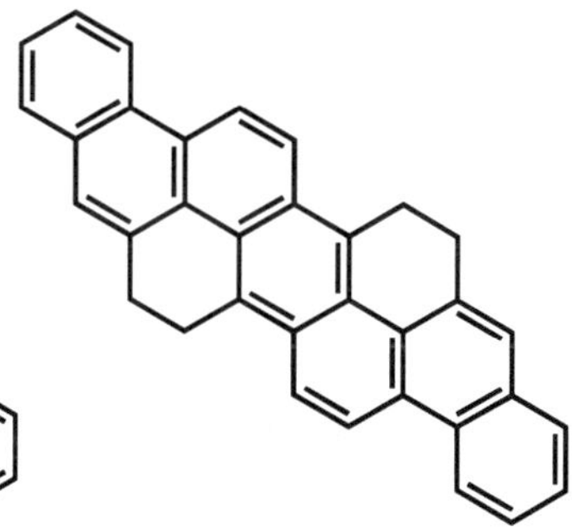

36

benzo[*pqr*]tetraphene is the IUPAC-preferred name for benzo[*a*]pyrene. The effects of this change have not yet become evident, but there is the potential for much confusion as future references using IUPAC names will not match the names widely used for the same compounds. This will only have a limited effect on a few structures that will have tetraphene as a core structure, but the effect of its being a new fusing group with a preference over some of the older groups may be great. The systematic rules of IUPAC stress consistency in nomenclature; however, past usage often was not consistent but usable.

1.5 CARCINOGENICITY, MUTAGENICITY, AND OTHER BIOLOGICAL PROPERTIES OF PAHS

PAH biological activity has been shown to be very structurally dependent. PAHs, or their degradation products or metabolites, must have certain three-dimensional (3D) structures to fit into DNA and either chemically alter it or bind to specific sites that effect the DNA's activity. Many PAH structural features have been found to meet these spatial criteria. These include the presence of the so-called "bay regions" and areas of more π electron localization or susceptibility toward oxidation or reduction. The various members of a set of isomers can thus range from extremely highly biologically active to no activity at all.

The first chemical to be directly tied as a cancer-causing compound was the PAH benzo[*a*]pyrene. Throughout the decades afterwards, both many other PAHs and the wide range of PAH-containing materials have been tested and shown to have deleterious biological activities. The list of active PAHs is quite long and includes many structural types and even many of the methyl derivatives. Several reviews and books have been written on PAH carcinogenicity and other biological activities.

The area of LPAH biological activities, as with all the other areas of LPAH research, has scarcely been studied, and the vast bulk of the work has only been done in the past two decades. The same reason, a lack of available standard compounds, caused this. Additionally, researchers believed that increasing size would both reduce solubility (and thus lower availability for metabolic transformation in the body) and increase stability (and thus decrease any reactivity toward epoxide, hydroxyl, keto, or other intermediate formations that leads to the ultimate carcinogen that interacts with DNA. Recent research has shown that these trends may be true in general, but specific instances in which they are not true are not uncommon.

Some benzologues (benzologues are the PAH structures that result when a series of ortho-fused rings are added to the beginning structure) of benzo[*a*]pyrene have shown deleterious biological activities. Most notable is dibenzo[*a,l*]pyrene, properly named dibenzo[*def,p*]chrysene, *compound 37* (20). This LPAH is one of the most carcinogenic substances so far studied. This compound possesses the structure, electron densities, and reactivity to be both an initiator and a promoter in carcinogenesis, resulting in its extremely high potency.

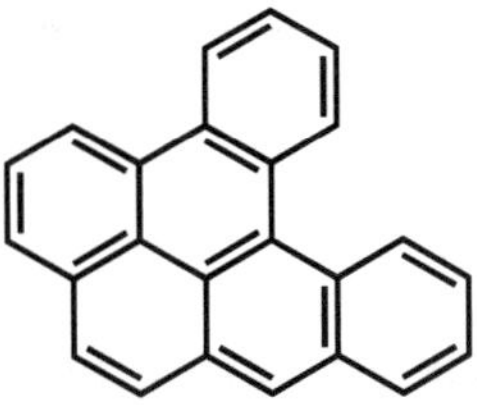

37

Lacassagne and co-workers (21, 22) showed that dibenzo[*cd,lm*]perylene (*compound 38*) had a weak response for tumorigenic activity. This LPAH is interesting in this context because the currently popular model for predicting PAH mutagenicity relies on a structural assessment of the bay regions in the PAH. This seven-ring LPAH does not contain the correct geometry under this model for mutagenic activity, and it is predicted to be inactive. In the same study, the benzo[*a*]pyrene benzologue benzo[*a*]naphtho[8,1,2-*cde*]tetracene (*compound 39*) was also found to cause tumors during mouse skin painting

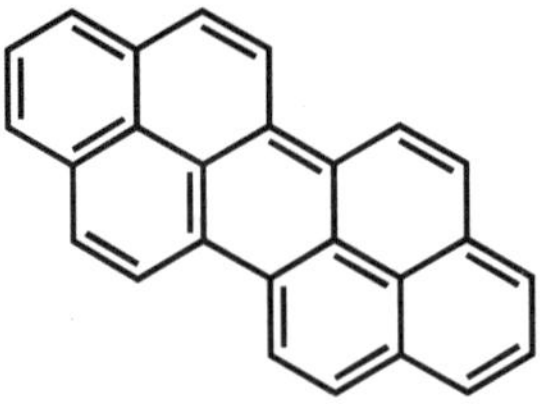

38

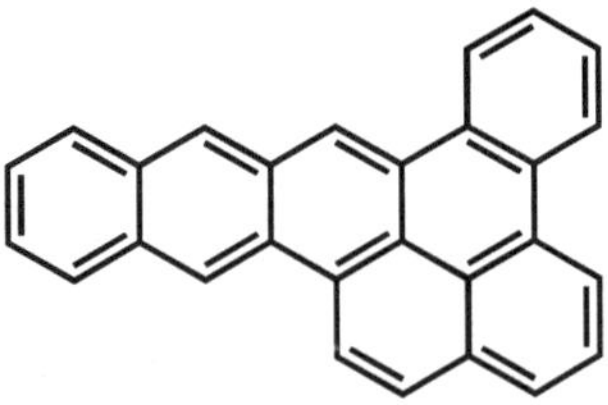

39

experiments. Benzo[*a*]coronene, dibenzo[*a,l*]pentacene, diindeno[1,2,3-*cd*:1′,2′,3′-*lm*]perylene (periflanthene), and six benzologues of fluorene were found not to be active (the first 3 structures are *compounds 40–42*). Surprisingly, the benzo-[*a*]pyrene benzologue benzo[*rst*]pentaphene also did not show activity.

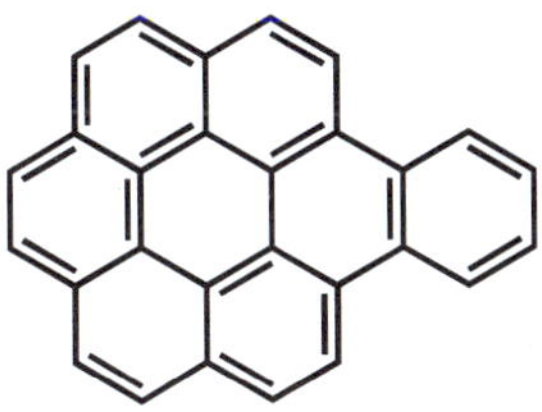

40

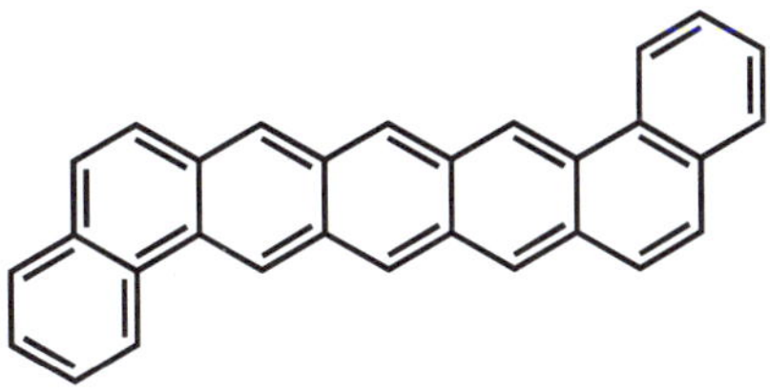

41

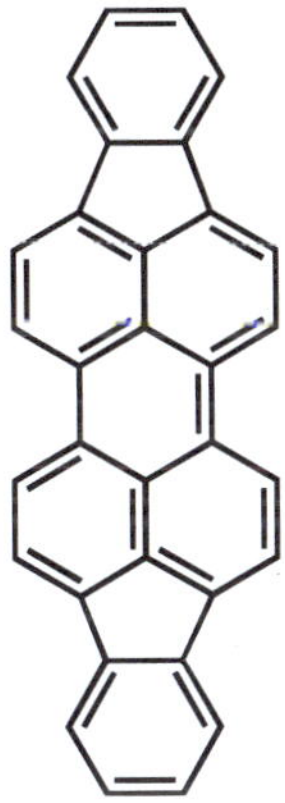

42

1.6 OCCURRENCE OF PAHS

The PAHs are widely occurring. They are produced in a wide variety of ways, generally as the products of the incomplete combustion of carbon-containing material (often fossil fuels or organic matter). The LPAHs have been less studied, and their occurrence will be reviewed in a later chapter.

Combustion at lower energies, or in an oxygen-poor atmosphere, cannot completely convert the carbon into carbon dioxide. The PAHs are preferred products under these conditions because of kinetic pathways and thermodynamics.

Coal tar and coal tar pitches are predominantly PAHs or the hetero-atom containing PACs. The types of PAHs are ortho- and peri-fused, alternant and nonalternant, and a wide range of individual isomers for each carbon number and degree of condensation.

In petroleum (crude oil), aromatic molecules can be found that result from removal of functional groups and aromatization of steroidal (and other similar) structures. These original molecules were produced biologically. The subsequent transformations occurred over many millions of years. Addition of long-chain alkyl groups onto these aromatic molecules also commonly occurs. Substituted naphthalene, phenanthrene, chrysene, and picene can be found in many crude oils (23–25).

Certain oils contain substituted perylenes. This specific PAH is thought to result from biodegradation from either of two pigments, an octahydroxybisanthrone that has been found in fossilized sea lilies, and erythroapin. The first is suspected of yielding perylenes with three saturated rings. The latter can occur in a variety of materials (insects, fungi, and marine organisms). Soma et al. (26) report on the concentration of perylene and the methylperylene isomers in sediments from Lake Baikal in central Asia. They ascribe these PAHs to the degradation of algal pigments. They also analyzed for chlorophylls A and B and saw rough correlation. They propose that the perylene is an indicator of anoxic sediment conditions. The origin of perylene in geological materials is still unresolved, and the possible sources have been recently reviewed. Jacob, Alawi, and Grimmer (27) examined the perylene distributions in a Jordanian shale oil. They cite numerous references on the possible origins of perylene in fossil fuels, noting its prevalence in marine sediment-derived materials and the postulation that the quinones xanthoaphin, erythoaphin, and rhodoaphin are the source material.

There are a few reports of larger PAHs in crude oil but only at ultra-trace levels. These were concentrated by distillation methods, and the heaviest fractions were analyzed for LPAHs. These will be discussed in later chapters.

Petroleum processing converts crude oil into products such as gasoline, diesel and jet fuel, mineral oil, wax, and asphalt. Except for the last, all of these are processed so that their sulfur, oxygen, nitrogen, and metals contents are very low compared to the material in the crude oils that has the same distillation temperature. This cleanup is done predominantly by catalytic treatment under high temperatures. As part of this processing, either the naturally occurring PAH-type molecules are destroyed or the carbon-skeletons are rearranged. This rearrangement yields trace levels of PAHs. The different types of processing use different catalysts and temperatures and also often hydrogen pressures (because for most of the products the desired composition is a mixture of saturated hydrocarbons, an aim is to destroy or remove the alkenes and aromatics). The specific PAHs that are produced in the processes are determined by these processing variables. Almost all processed petroleum products contain trace levels of these PAHs. Used and re-refined (regenerated) lubricating oils also contain PAHs.

When the hydrocarbon fuels, including coal and fuel oil, are burned more PAHs are made. Soots contain a wide variety of PAHs. Similarly, PAHs are produced by the charcoal and wood used to cook or smoke foods. A large number of studies have been done on foods prepared in this fashion (8). Coal tar and coal tar pitch are notorious materials that are almost totally PAHs or their hetero-atom-containing analogues.

The PAHs not only end up on the solid materials resulting from combustion but also are volatilized or emitted while absorbed on fine, airborne particulates.

Certain of the chapters of this work, such as the chromatography and fluorescence chapters, will contain many specific references to LPAH occurrences, and a final chapter will be devoted to a summary.

REFERENCES

1. W. C. Herndon, in *Polynuclear Aromatic Hydrocarbons*, L. Ebert, Ed., Advances in Chemistry 217, American Chemical Society, Washington, DC, 1988, p. 1–12.
2. *The ACS Style Guide: A Manual for Authors and Editors*, J. S. Dodd, Ed., American Chemical Society, Washington, DC, 1986, p. 45.
3. *Polynuclear Aromatic Hydrocarbons: Third International Symposium on Chemistry and BiologyCarcinogenesis and Mutagenesis*, P. W. Jones and P. Leber, Eds., Ann Arbor Science, 1979.
4. Full Program and Abstracts, 17th International Symposium on Polycyclic Aromatic Compounds, Nov. 4–6, 1997, Charlotte, NC.
5. E. Clar, *Polycyclic Hydrocarbons*, John Wiley and Sons, New York, 1964, Vols. 1 and 2.
6. A. Bjorseth, Ed., Marcel Dekker, New York, 1983. *Handbook of Polycyclic Aromatic Hydrocarbons*, Vols. 1 and 2.
7. R. A. Hites and W. J. Simonsick, *Calculated Molecular Properties of Polycyclic Aromatic Hydrocarbons*, Elsevier, Amsterdam, 1987.
8. J. R. Dias, *Handbook of Polycyclic Aromatic Hydrocarbons*, Parts A and B, Elsevier, Amsterdam, 1987 and 1988.
9. *Advances in the Theory of Benzenoid Hydrocarbons II*, Gutman, I. Ed., Springer-Verlag, Berlin, 1992.
10. R. G. Harvey, *Polycyclic Aromatic Hydrocarbons: Chemistry and Carcinogenicity*, Cambridge University Press, Cambridge, UK, 1991
11. R. G. Harvey, *Polycyclic Aromatic Hydrocarbons*, John Wiley and Sons, New York, 1996.
12. K. D. Bartle, M. L. Lee, and M. Novotny, *The Analytical Chemistry of Polycyclic Aromatic Compounds*, Academic, New York, 1981.
13. *Spectral Atlas of Polycyclic Aromatic Compounds*, W. Karcher, R. J. Fordham, J. J. Dubois, P. D. J. M. Glaude, and J. A. M. Ligthart, Eds., Reidel, Dordrecht, The Netherlands, 1983, Vol. 1.

14. *Spectral Atlas of Polycyclic Aromatic Compounds*, W. Karcher, Ed., Kluwer, Dordrecht, The Netherlands, 1988, Vol. 2.
15. *Spectral Atlas of Polycyclic Aromatic Compounds*, W. Karcher, J. Devillers, and P. Garrigues, Eds., Kluwer, Dordrecht, The Netherlands, 1991, Vol. 3.
16. L. Nakhimovsky, M. Lamotte, and J. Joussot-Dubien, *Handbook of Low Temperature Electronic Spectra of Polycyclic Aromatic Hydrocarbons*, Elsevier, Amsterdam, 1989.
17. *Chemical Analysis of Polycyclic Aromatic Compounds*, T. Vo-Dinh, Ed., John Wiley and Sons, New York, 1989.
18. *1998 Recommendations of the IUPAC Committee on Organic Nomenclature: Fused Ring Systems*, G. P. Moss.
19. L. C. Sander and S. A. Wise, "Polycyclic Aromatic Hydrocarbon Structure Index", NIST Special Publication 922, National Institute of Standards and Technology, U. S. Department of Commerce, Washington, DC, 1997.
20. E. Cavalieri, S. Higginbotham, and E. Rogan, *Polycyclic Arom. Cmpds.* **6** 177–183 (1994).
21. A. Lacassagne, H. P. Buu-Hoi, F. Zajdela, and F. Vingiello, *Naturwissenschaften* **55** 43 (1968).
22. A. Lacassagne, H. P. Buu-Hoi, F. Zajdela, G. Saint-Ruf, and C. R. Hebd, *Séance Acad. Sci., Ser. D* **266** 301–304 (1968).
23. M. Blumer, *Sci. Amer.* **234** 34 (1976).
24. Z. Alzenshtat, *Geochim. Cosmochim. Acta* **31** 559 (1973).
25. J. E. Silliman, P. A. Meyers, and B. J. Eadie, *Org. Geochem.* **29** 1737–1744 (1999).
26. Y. Soma, A. Tanaka, M. Soma, and T. Kawai, *Org. Geochem.* **24** 553–561 (1996).
27. J. Jacob, M. Alawi, and G. Grimmer, *Polycyclic Arom. Cmpds.* **2** 13–17 (1991).

CHAPTER

2

THE LARGE PAHs

2.1 GENERAL DISCUSSION

As mentioned in Chapter 1, the large PAHs (or LPAHs) are defined in this volume as those with 24 or more aromatic ring carbons. These compounds have been much less studied than the smaller PAHs. Historically, all of the major reasons for this can be traced back to the paucity of available, authentic standard compounds. Without standards, no work can be done into the biological effects, analysis, occurrence, and physical properties of the LPAHs. Without these types of work, interest in this class of compounds has never grown beyond a few isolated studies.

The number of isomers of the LPAHs increases rapidly with carbon number, as the molecules get larger. The structural regularity of the PAHs due to their valency and the requirements in structure due to π bonding, can be arranged in a sort of periodic table for the PAHs, arranged by the carbon number and the degree of condensation. Table 1 is an updated version of a table that was created in 1989 by the author for a chapter on PAH separations (1). It shows the enumeration for the alternant PAHs, and includes syntheses that have been performed in the past dozen years. Although there are some differences between this and the earlier version of the table, there have been no gross changes in the numbers of known isomers. The change amounts only to about 20 new LPAHs out of the well over 1000 that had yet to be synthesized in 1989. Thus the overwhelming number of LPAHs remain to be synthesized. A more positive view, however, might be taken if the number of characterized or synthesized very large PAHs (VLPAHs) is examined. Arbitrarily choosing this classification of VLPAHs to start at 40 carbons, the number has more than doubled since 1989 and more than quadrupled since Clar's two volumes were published.

For every carbon number, the most condensed formula possible has the fewest number of isomers. In some cases, such as for $C_{24}H_{12}$ and $C_{32}H_{14}$, there is only one possible alternant structure. For these two formulae, these are the LPAHs coronene and ovalene.

Large (C>=24) Polycyclic Aromatic Hydrocarbons: Chemistry and Analysis, by John C. Fetzer.
ISBN 0-471-36354-5

Table 1. Number of Synthesized PAH's (Theoretically Possible Number)

Number of Carbons	Number of Hydrogens 8	10	12	14	16	18	20
10	1 (1)	–	–	–	–	–	–
12	–	–	–	–	–	–	–
14	–	2 (2)	–	–	–	–	–
16	1 (1)	–	–	–	–	–	–
18	–	–	5 (5)	–	–	–	–
20	–	–	3 (3)	–	–	–	–
22	–	–	2 (2)	12 (12)	–	–	–
24	–	–	1 (1)	13 (13)	–	–	–
26	–	–	–	6 (9)	27 (37)	–	–
28	–	–	–	–	4 (8)	31 (62)	–
30	–	–	–	2 (3)	11 (58)	30 (123)	–
32	–	–	–	1 (1)	8 (46)	16 (289)	–
34	–	–	–	–	2 (37)	10 (333)	8 (411)

Why is it better to classify the LPAHs by their carbon number rather than by their ring number? Even though they may all have the same number of rings, the various types of LPAHs can be very different from each other. The seven-ring LPAHs will be compared. Ortho-fused structures, such as heptacene (*compound 43*), heptaphene, and their isomers, have 30 carbon atoms. The seven-ring peri-condensed structures dibenzo[*h,rst*]pentaphene (*compound 44*) and its other isomers have 28 carbon atoms, dibenzo[*cd,lm*]perylene and its isomers have 26, and coronene has 24. It can become very confusing to talk of LPAHs classified by ring number when dealing with complex mixtures or formation mechanisms.

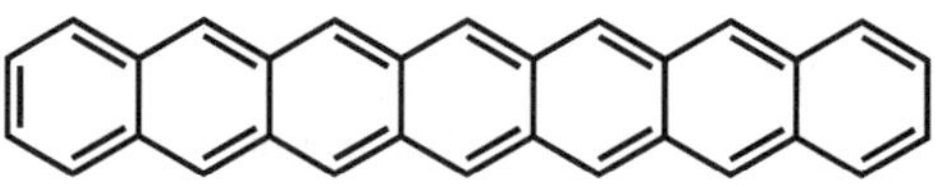

43

A differentiation of the LPAHs by carbon number and the degree of condensation is much simpler. As an example (2), it is much clearer to see the successive steps in the condensation of hexaphenylbenzene (*compound 45*, a seven-ring aromatic molecule by a ring-oriented classification but not an LPAH by the definition used here) to hexabenzo[*bc,ef,hi,kl,pq,st*]coronene (*compound 46*). From a carbon number view, this series of reactions only changes in the degree of condensation of a 42-carbon structure by formation of additional

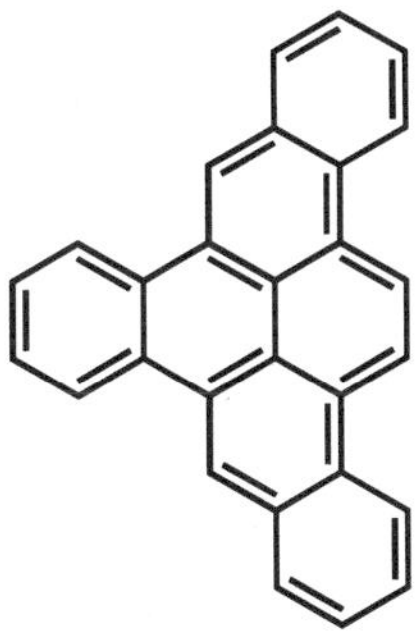

44

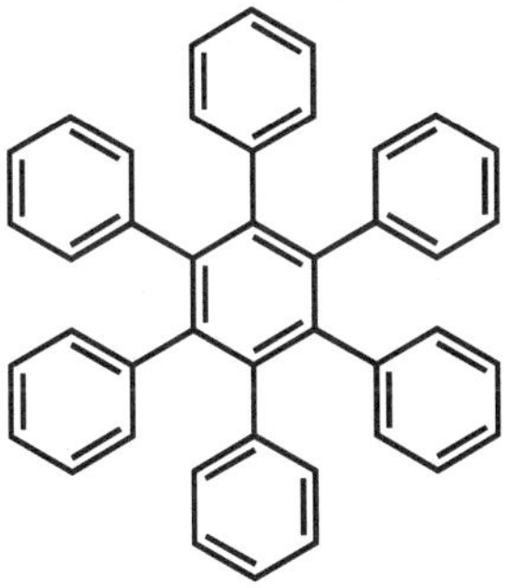

45

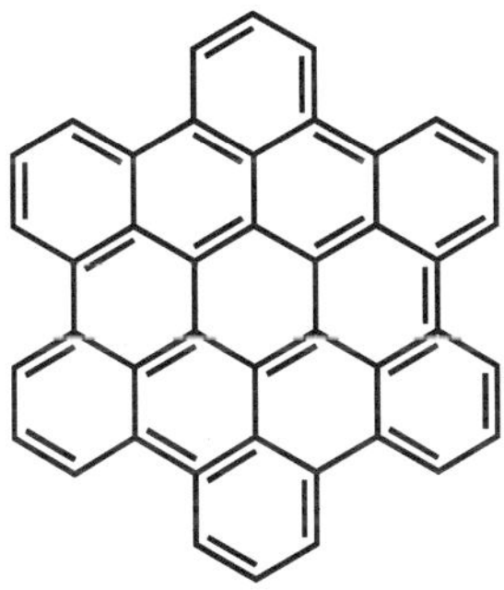

46

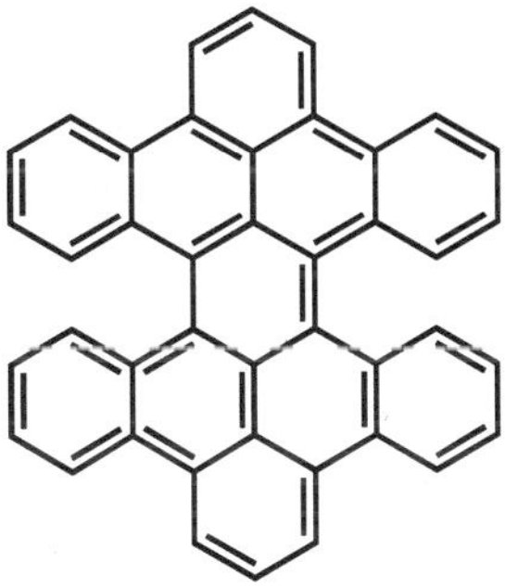

47

carbon-carbon bonds. This is especially true for the condensations resulting from the LPAH hexabenzo[*a*,*cd*,*f*,*j*,*lm*,*o*]perylene (*compound 47*), where it and each successive product of a further bond formation are LPAHs.

An interesting sidelight to ring number discussions has been whether to count the internal rings in structures such as coronene or ovalene. Are these really holes in the structure, as in a doughnut or bagel, or are they rings? Thus, is an LPAH such as coronene six or seven rings? Extending this confusion to the benzologue of coronene yields a straightforward answer to this question. Benzo[*a*]coronene is isomeric to structures such as benzo[*pqr*]naphtho[8,1,2-*bcd*]perylene. This latter compound has no internal rings and is obviously an eight-ring LPAH. Therefore, its isomers must also have eight rings. Coronene then must have seven, and so there are no holes in this or any other LPAH structures.

For specific isomeric structures, the synthetic steps needed to make them increase and get more specific as the carbon number increases. The isolation of a single LPAH from a reaction mixture also gets much more complicated. Concomitantly, the number of known (well characterized) standards drops off much more rapidly than the carbon number increases. These two facts can be highlighted by one example. A recent synthesis (3) produced one new $C_{36}H_{16}$ isomer. Previously, only 2 of the 20 possible structures had been synthesized. Thus the known number of isomers had been increased by 50 % over what had been done before, but this additional isomer was only 5 % of the total possible alternant isomers. In the same series of syntheses, the number of known $C_{36}H_{18}$ isomers was increased by four, more than doubling the number of known LPAHs of this formula.

The importance of standards in research is shown by work done with the 24-carbon LPAHs. All of the 33 possible alternant structures and many of the nonalternant structures are available (albeit in some cases from very diverse and obscure sources) or known (with published UV absorbance or fluorescence spectra for comparison). There have been three very-detailed studies of the 24-carbon LPAHs in different materials (4–6) that have used this large set of standard compounds to identify many of the LPAHs seen. One of these studies also extended the work to partially do the 26-carbon LPAHs, but there were few standards and only some of the components could be identified for this carbon number.

These three studies also highlight the fact that some analytical techniques, such as gas chromatography, work well for lower carbon numbers but cannot do the higher carbon numbers anywhere near as well.

This small number of known standards not only means that analytical chemists cannot identify and quantify the LPAHs in their samples but also that the occurrence of LPAHs in many materials have not been determined. Additionally, methods that are both sensitive and selective for individual LPAHs have also not been developed. There are several examples of newly characterized LPAHs being found in previously analyzed samples. For example, phenanthro-[5,4,3-*abcd*]perylene (*compound 48*) was found in extracts from hydrothermal vent bitumens very soon after this LPAH was first characterized (7). The products of anthracene pyrolysis also contained several peaks that were identified after a collection of new nonalternant LPAHs became available for comparison. Most LPAH analyses suffer from a large number of unidentified peaks.

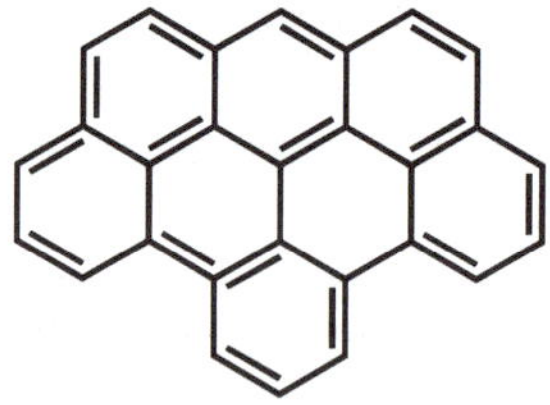

48

The synergy of the lack of available LPAH standards and the dearth of study of LPAHs reaches beyond the boundaries of analytical chemistry. Without pure standards, the epidemiologist also cannot determine whether LPAHs are of keener interest because of potential biological activity, and people interested in the production of PAHs, or soots, fullerenes, or any other carbonaceous material cannot identify specific structures that may help elucidate both production mechanisms and the nature of the materials produced.

2.2 LPAH PROPERTIES

Many PAH properties extend continuously upward with increasing molecular weight. Thus volatility and solubility diminish as the number of carbons goes up. A higher degree of condensation for any particular carbon number also decreases these two properties.

The one major exception to these trends is when steric factors induce nonplanarity in the LPAH carbon skeleton. The steric strain produced by bay, cove, and fjord regions is much more common in these large structures.

As the number of rings increases, the chances increase for steric strain of spatially adjacent rings within a LPAH structure. The strain is commonly reduced by the structure, assuming nonplanar conformations through twisting or bending of portions of the structure. Bridging rings, where two larger portions of the LPAH are joined by one ring, are an especially common area in which nonplanarity is found. Herndon and others estimate that these nonplanar LPAHs greatly outnumber the planar ones.

The most extreme instance of nonplanar PAHs is the family of helicenes. This series of compounds starts with benzo[*c*]phenanthrene, which is only slightly contorted out of plane. The next members of the series, dibenzo[*c*,*g*]phenanthrene and phenanthro[3,4-*c*]phenanthrene (*compounds 49 and 50*), are progressively more nonplanar. Helicenes of up to 58 carbons (14 rings) have been synthesized (8, 9). The larger of these LPAHs possess spiral configurations,

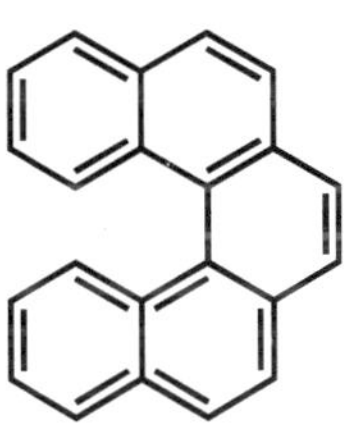

49

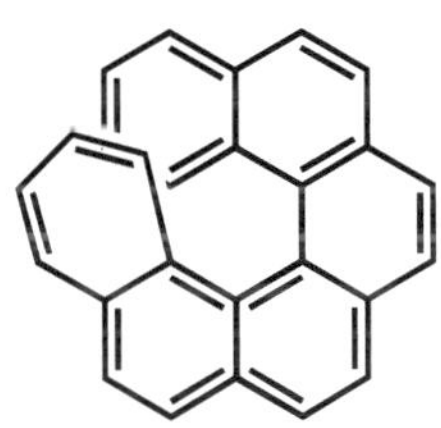

50

giving this class of LPAHs its trivial name. In fact, there are clockwise and counterclockwise conformers. Chromatography, using special shape- and chiral-selective phases, can partially separate the conformers from each other. These two types of helicenes possess phenomenally high optical activities, each in an opposite direction relative to the incident polarized light.

For solubilization, increasing carbon number results in the use of more exotic solvents to obtain solutions. For LPAHs of 10 rings or more, solvents like *o*-dichlorobenzene, 1,2,4-trichlorobenzene, or 1-methylnaphthalene are commonly used. Most LPAHs are exceedingly insoluble in most common solvents. Only aromatic solvents and those with strongly polarizable functionalities, such as chloro, bromo, or nitro groups, are good solvents.

For vaporization of LPAHs, higher temperatures (very often in concert with lower pressures through the use of vacuum pumps) are needed. The sublimation temperatures for LPAHs are very high. As part of a study on the thermal reactivity of PAHs, Lewis and Edstrom (10) measured the melting and boiling (sublimation) points of a large number of PAHs, including over a dozen LPAHs, using differential scanning calorimetry. These included benzo[*a*]coronene, dibenzo[*def,p*]chrysene, pyranthrene, and benzo[1,2,3-*cd*:4,5,6-*c′d′*]diperylene (trivially called quaterrylene; *compound 51*). The largest PAH with a published (11) experimentally-determined sublimation temperature is tribenzo[*a,g,m*]coronene (correctly named tribenzo[*fgh,pqr,za*$_1$*b*$_1$]trinaphthylene; *compound 52*), at 537-539°C. Zander (12) reported that benzo[1,2,3-*bc*:4,5,6-*b′c′*]dicoronene, $C_{48}H_{20}$, (*compound 53*) melts in the range of 600–700°C in a sealed evacuated quartz tube. Its sublimation temperature was reported to be higher than the onset of the quartz tube melting.

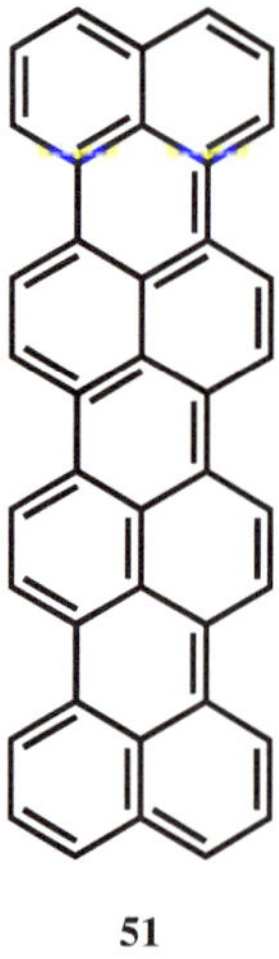

51

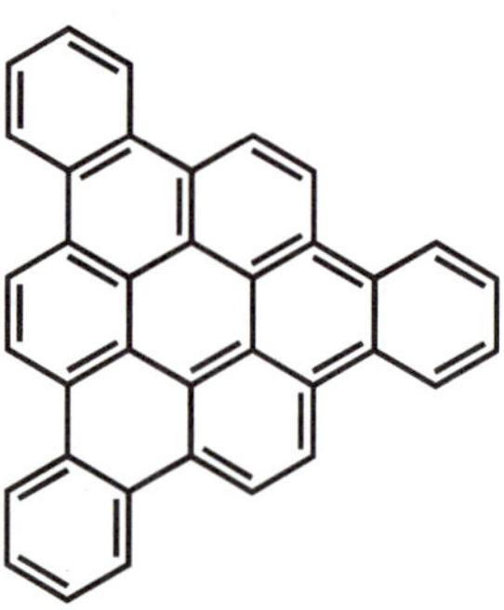

52

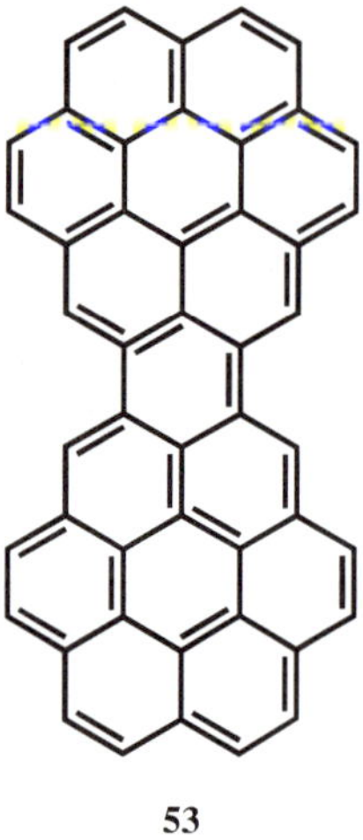

53

Efficient crystal packing influences the sublimation temperature of the LPAHs (13). Crystal packing for the LPAHs is disproportionately higher than for the smaller PAHs because of the stronger molecular attractions between the larger structures. Rapid crystal formation and the presence of impurities from the solvents and LPAHs create dislocations that disrupt the orderly arrangement of the molecules in the crystals. This drastically effects the measured melting and sublimation temperatures. The more nonplanar an LPAH structure is, the less likely it is that there will be an efficient arrangement in the crystal structure (the stacking of planar LPAHs, is in contrast, very efficient).

Another type of steric strain that leads to nonplanar structures is possible in the larger PAHs as well. Curved LPAHs occur if the LPAH structures have internally-fused, incised five-member rings. The simplest example of this is corannulene (*compound 54*), a bowl-shaped highly symmetrical structure. There has been a lot of work recently in this type of LPAH, mainly in the synthetic organic chemistry area relating to the total synthesis of fullerenes. This will be discussed in detail in the following section.

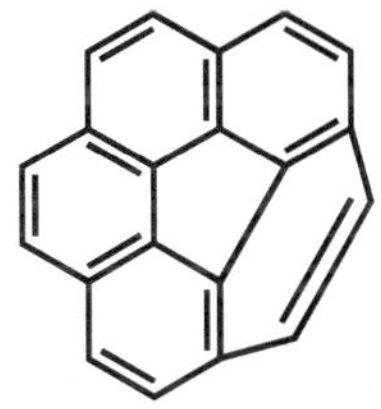

54

Nonplanarity drastically effects LPAH properties, with increased solubility and heightened volatility being two major physical manifestations. Thus the highly nonplanar LPAH hexabenzo[*a*,*cd*,*f*,*j*,*lm*,*o*]perylene is much more soluble in weak solvents, like methanol or *n*-hexane, than the LPAH consisting of its core structure, dibenzo[*cd*,*lm*]perylene. Other effects found in spectral and chromatographic behavior will be discussed separately in later chapters.

Generally, LPAH stability decreases from the most condensed structures. These are structures that are composed either only of interconnected fully-aromatic sextets (this is by definition the case only for carbon numbers divisible by 6) or of the most condensed structure for a particular carbon number, e.g., coronene and ovalene. Less stable are the other peri-fused LPAHs. The nonalternant and ortho-fused LPAHs are generally less stable, but this depends on specific structural features and the carbon number. The larger the LPAH, the more stable for most classes.

The lone exception is those LPAHs with the linear ortho-fused rings. The longer the linear portion of an ortho-fused LPAH is, the less stable it is when compared to analogous structures. Pentacene can be studied if the sample is freshly purified from the pentacen-6,13-dione that forms during storage. Hexacene (*compound 55*) can only be studied in extremely fast experiments before oxidation degrades it (14–16). Even then, the rapid oxidation makes any determination of certain spectral features or molar absorptivities impossible. This is also very likely true for pentacene, because even with its slower oxidation rate the molar absorptivity of its highest wavelength band, that is near 580 nm, is 5900, 9900, or 12,000. There have been fleeting reports of heptacene chemistry or spectroscopy, and octacene still exists only in the theoretical chemists' models.

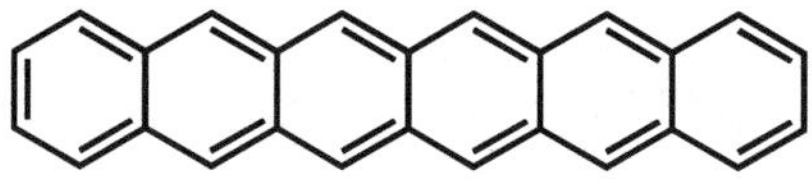

55

Ortho-fused LPAHs that are nonlinear are more stable than the linear isomers. Thus dibenzo[*a*,*l*]pentacene is more stable than heptacene. It is even more stable than pentacene. This extends the trend that phenanthrene is more stable than anthracene and that triphenylene and chrysene are more stable than naphthacene.

The fusion of rings to peri-fused cores also shows this trend. Anthra[2,3-*a*]coronene (*compound 56*) is less stable than naphtho[2,3-*a*]coronene (*compound 57*). The latter is less stable than benzo[*a*]coronene and also the isomeric dibenzo[*a*,*j*]coronene (*compound 58*).

As opposed to the PAHs, little research into the structure or behavior of LPAH radicals or carbocations has been done. Laali, Tanaka, and Fetzer (17)

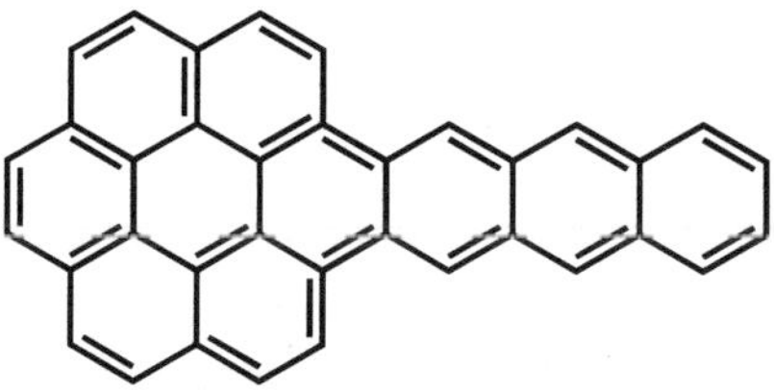

56

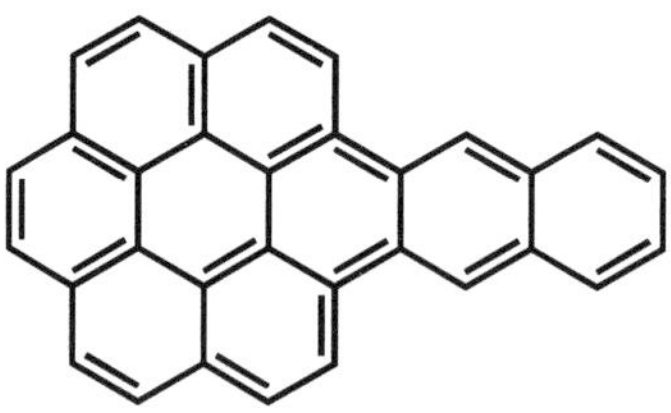

57

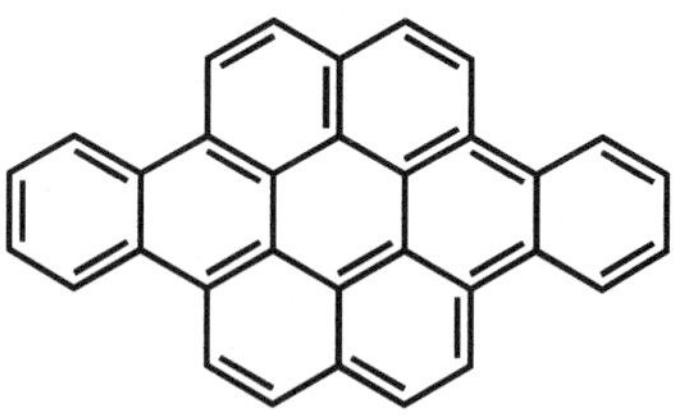

58

studied the dications of dibenzo[*cd,lm*]perylene and compared them to those of perylene and some methyl perylene derivatives. In contrast to perylene, which gives bright red solutions, the LPAH gives dark blue solutions. It also shows paramagnetic behavior because it has 24 π electrons. Perylene, having 18, a number not divisible by 4, shows diamagnetic behavior.

Suenaga, Miyahara, Shimizu, and Inazu (18) synthesized the stable "trinaphthophenalenium" ion (the IUPAC-preferred naming is benzo[*cd*]phenaleno[1,2,3,4-*fghi*]perylene cation; *compound 59*). The base LPAH is not stable, having an extra "odd" electron. Therefore, this cation is stable only when with an appropriate anion in the salt form or in solution. Suenaga et al. report that the trifluoroacetate salt is a dark violet solid. This symmetrical (and thus very resonance stabilized) cation was examined by cyclic voltametry, and the results were found to compare similarly to the phenalenium cation.

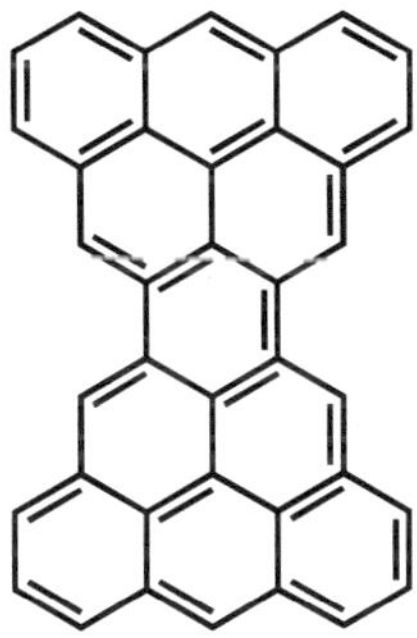

59

As an extension to the earlier work of Clar (19), Dias and Koehler (20) studied the color-forming reaction of LPAHs and concentrated sulfuric acid. The cation or dication of the LPAH is thought to be the colored species. In general, there was only one initial color, with the predominant colors being purple, blue, or green. Other colors, red, orange, yellow, or pink, were seen for a few LPAHs. Many of the samples showed color changes and most ended up as brownish precipitates. Dias and Koehler ascribed these to the sulfonation of the LPAH. They studied nine compounds specifically to determine the onset of color as a function of acid concentration. The percentage of sulfuric acid that gave color correlated to the highest occupied molecular orbital (HOMO) energy. There was some confusion over tones; for example, is pink only a very dilute shade of the ultimate violet seen for some solutions?

Another color reaction of the LPAHs has been reported but involves complexation of the neutral species (21). Iron(III) chloride has been used for many years to separate nitrogen-containing compounds from petroleum samples. Jewett reports that some PAHs, including the LPAH coronene, also form colored complexes with iron(III). It forms a green complex both when solutions are passed through chromatographic adsorption columns loaded with iron(III) chloride and as the free complex in 1,2-dichloroethane. From the colors of other PAHs (for example perylene, which gives brown and violet under the two conditions), the LPAHs should also give different-colored complexes.

2.3 RECENT LPAH SYNTHESES

In the decades since Clar's comprehensive work, only three reviews of LPAH synthesis have appeared. Unfortunately, one is a book written in German (2), another is a more comprehensive book on PAH synthesis that only covers the LPAHs of seven rings as only a small part of the total work on syntheses of all PAHs of up to seven rings (22), and the third is a very brief review article (23). Several new LPAHs have been produced since this last review. Many of these new syntheses will be described, as well as the inclusion of the reactions covered in the other three reviews because the material in these reviews has not been readily available in one source. This review does not attempt to be an all-encompassing description of LPAH syntheses. It is aimed at giving the reader a feel for some of the synthetic approaches that have been used to produce new LPAHs and some of the issues that arise in this field.

In a recent book (22), Harvey thoroughly describes many PAH syntheses, with one chapter on many of the general reactions that can be used to make PAHs. Unfortunately, the subject matter stops at seven rings and the chapters are arranged in terms of ring number only. Thus ortho-fused LPAH structures of 30 or fewer carbons are described. For the peri-fused LPAHs, only the syntheses of $C_{24}H_{12}$, $C_{24}H_{14}$, $C_{26}H_{14}$, and $C_{28}H_{16}$ LPAHs are covered. For details on the synthesis of specific LPAHs of these formulae, the reader is encouraged to refer to this book. The LPAHs of the same carbon numbers that are more condensed ($C_{28}H_{14}$) have eight rings and were left out. Harvey does give references to UV absorbance, fluorescence, NMR, and photoelectron spectra for the PAHs covered.

Friedel-Crafts reactions have been widely used to produce LPAHs (17). This reaction relies on a catalyst, such as aluminum(III) chloride, to abstract a hydrogen atom from the hydrocarbon creating a carbocation. This reactive site then can form a new carbon-carbon bond with an adjacent carbon atom. Using this reaction, Vingiello and Ojakaar (24) synthesized four 7-ring $C_{28}H_{16}$ isomers. For one product, benzo[*mno*]naphtho[1,2-*g*]chrysene, sodium chloride was used with the aluminum(III) chloride to condense a naphthylbenz[*a*]-anthracene. A tin(IV) chloride mixture yielded a different isomer, naphtho-[2,1-*a*]perylene, from the identical starting material (Fig. 3). A different naphthylbenz[*a*]anthracene gave benzo[*mn*]naphtho[1,2,3-*hi*]tetracene and benzo[*a*]naphtho[2,1,8-*defg*]chrysene (*compounds 60 and 61*) in a sodium chloride-

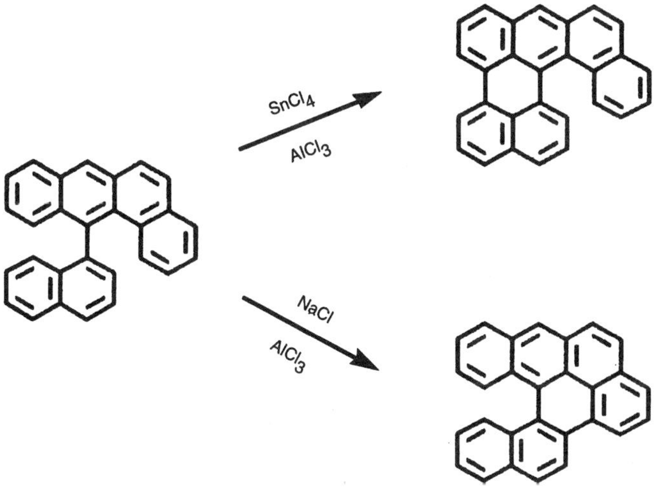

Figure 3. Two of the synthetic routes used by Ojakaar and Vingiello that involve Friedel-Crafts condensation.

aluminum(III) chloride melt. A fifth LPAH, phenaleno[1,2,3,3*a*,4-*fghi*]perylene (*compound 62*), was obtained by using the sodium chloride-aluminum chloride melt on the naphtho-[2,1-*a*]perylene or by using the tin(IV) chloride-aluminum(III) chloride melt on the benzo[*mno*]naphtho[1,2-*g*]chrysene (using the other reaction on the individual products of the first pair of syntheses). Unfortunately, the authors did not reproduce the UV absorbance spectra of these five LPAHs but only a table of UV absorbance wavelengths of the bands found.

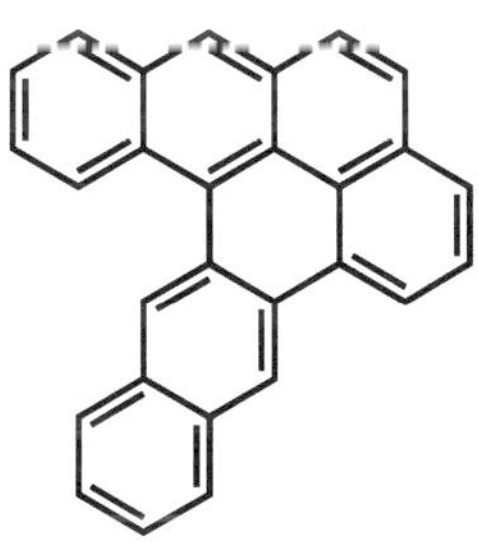

60

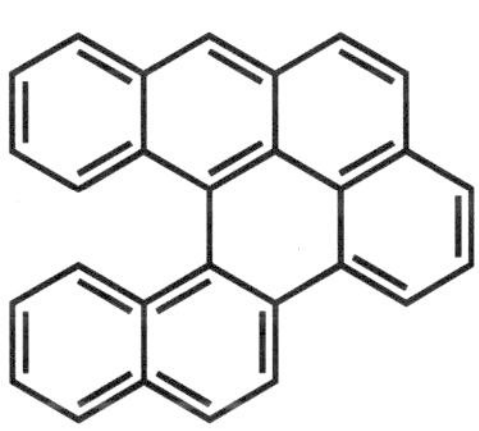

61

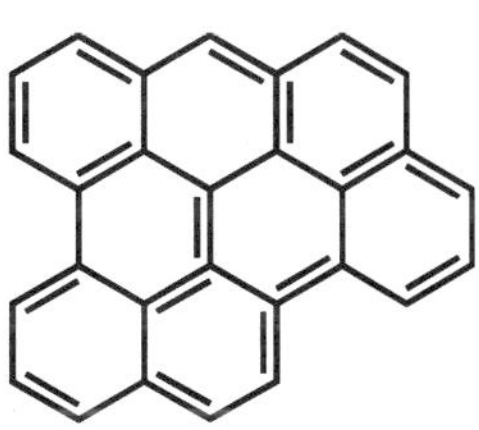

62

Clar and co-workers (25–27) also used this reaction as the final step in the synthesis of many new LPAHs. The reactants were formed in a variety of ways, including the Diels-Alder and Grignard reactions described in the following paragraphs. A few examples are shown in Figure 4.

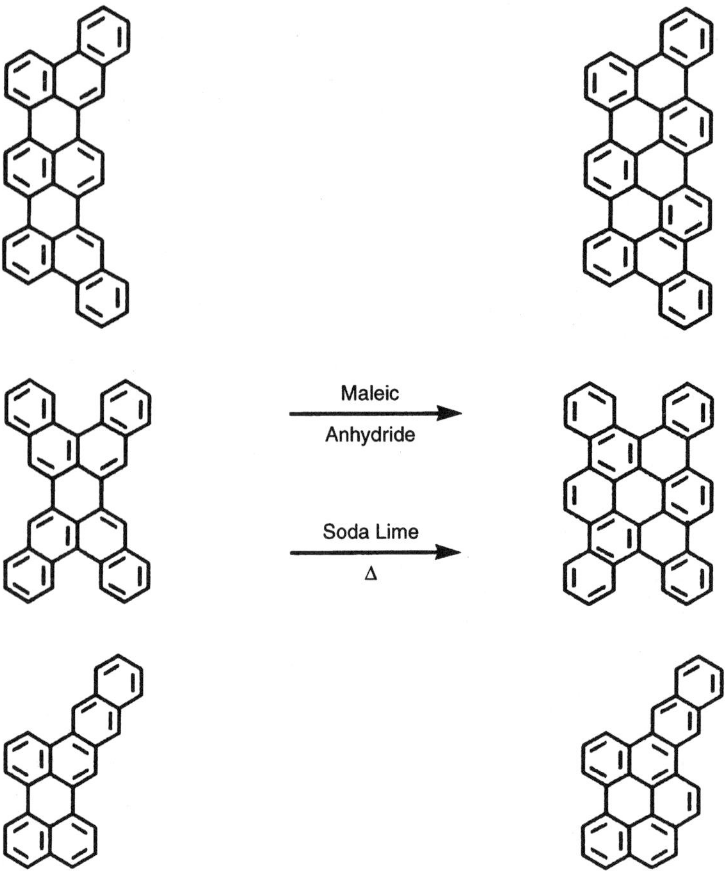

Figure 4. Maleic anhydride addition reactions to add rings onto LPAH structures.

In a German patent application, Clar, Schmidt, and Zauner (28) describe a large number of varied syntheses resulting in LPAHs consisting of a coronene core with fused benzo, naphtho, phenanthro, and other ring groups. The patent lists these as potential materials when doped into resins because of their optical properties and long-lived phosphorescence. The LPAHs are synthesized mainly by a Grignard reaction of a halo-derivative of a PAH and a fused-aromatic substituted quinone (Figs. 5, 6, and 7 are examples).

Lempka, Obenland, and Schmidt (29) fused two coronene molecules by the Friedel-Crafts reaction to form the red compound benzo[1,2,3-*bc*:4,5,6-*b′c′*]dicoronene (*compound 53*). This is a variation of the reaction that is called a Scholl condensation, named after the chemist who first used it for the condensation of two naphthalene molecules to form perylene. Lempka et al. then

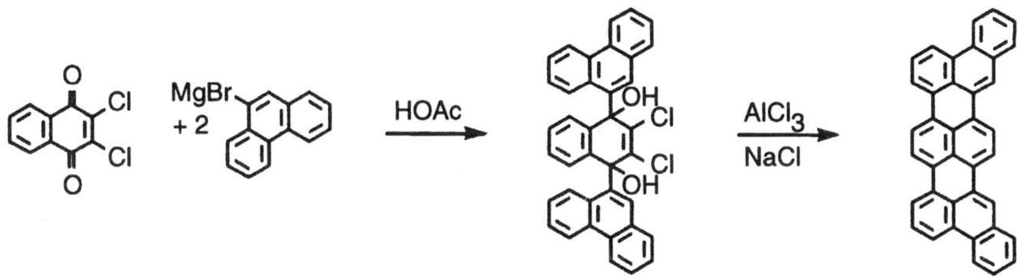

Figure 5. Grignard reaction of magnesium halo-PAH and quinone.

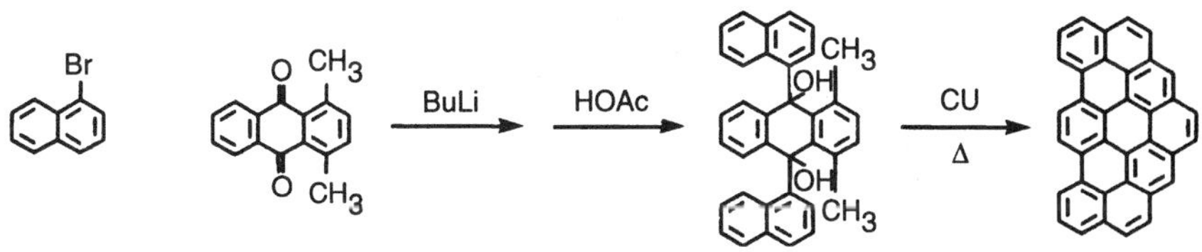

Figure 6. Grignard reaction of lithium halo-PAH and quinone.

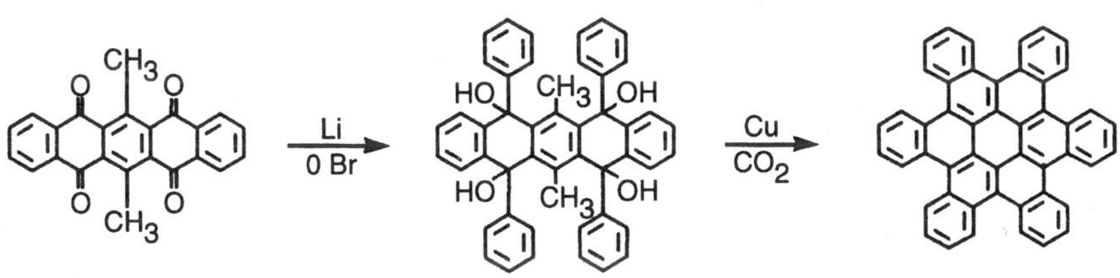

Figure 7. Grignard reaction of lithium halo-PAH and quinone.

proved that this was the same material that Boente had produced in a similar reaction many years earlier. In that early report, Boente (30) also isolated an insoluble black residue. From the proportions of carbon and hydrogen (determined by chemical means), he proposed it to be the condensation product of four coronene molecules, i.e., two of the red dimers. Boente's work had been done during and after World War II to investigate the chemistry of coal conversion to motor fuels. This process had been widely used in Germany during the war as an alternative to petroleum processing. LPAHs, especially coronene, were produced in those processes.

Fetzer (31) also used this reaction to create many new LPAHs by condensation of mixtures of the smaller PAHs pyrene, benzo[*ghi*]perylene, coronene, and ovalene. Many of the LPAHs found were new products of 40-64 carbons. One of the structures is shown as *compound 63*, the only alternant

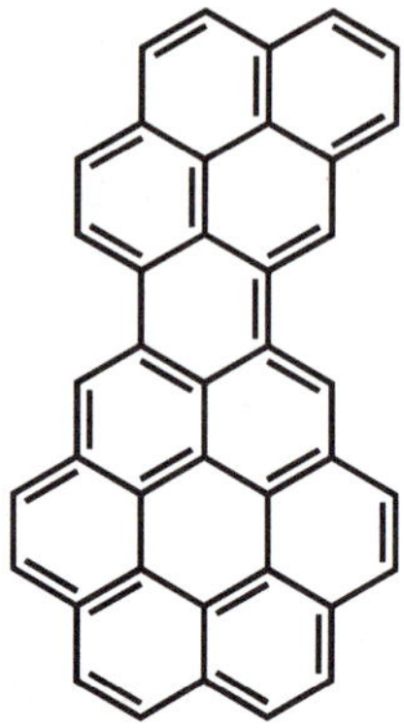

63

product of the pyrene and coronene mixture. The symmetry of the reactants is reflected in the number of possible isomers. The condensation of pyrene and coronene has only one possible product, whereas that of coronene and ovalene results in two possible products. Two ovalene molecules give six possible products through formation of one bridging ring and another with two bridging rings. None of the LPAHs was isolated because they could not be separated by chromatography because of their low solubilities and strong adsorption on any type of column.

The Friedel-Crafts reaction can be very vigorous, leading to gross rearrangements of the reactants. Bluemer, Gunderman, and Zander (32) produced dibenzo[*h,rst*]pentaphene from 5-phenyl-dibenz[*a,j*]anthracene. The only similarity between product and reactant was that they each have 28 aromatic carbon atoms. These gross rearrangements when aluminum(III) chloride is used have resulted in several misidentifications of reaction products when these were later examined by photoelectron spectroscopy.

Other, less vigorous reactions can be used to condense hydrocarbon structures through formation of bridging bonds. Platinum or palladium on activated carbon and photolysis using iodine are two examples.

The simplest reaction that is used to produce new LPAHs through formation of an additional ring is the Diels-Alder reaction route to form an additional ring on certain smaller PAHs. Many peri-condensed PAHs act as the diene, reacting in their bay region (25–27). This part of the PAH can react with maleic anhydride (the dienophile), forming a new peripheral ring by the addition of two carbons. This intermediate is then aromatized by heating in chloranil or soda lime, which removes the anhydride group. Thus tribenzo[*fg,mn,xyz*]-heptaphene adds 2 rings to form the new 12-ring LPAH dibenzo[*fg,mn*]-phenanthro[2,1, 10,9,8,7-*vwxyza′b′*]heptaphene (*compounds 64 and 65*). Biermann and Schmidt (33, 34) reviewed the reactivity of many ortho-fused

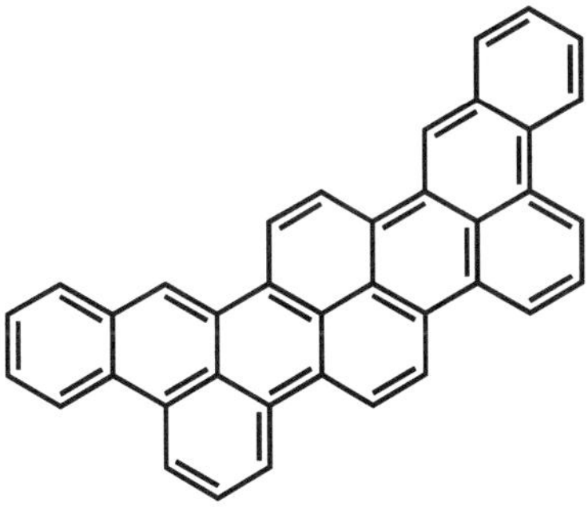

64

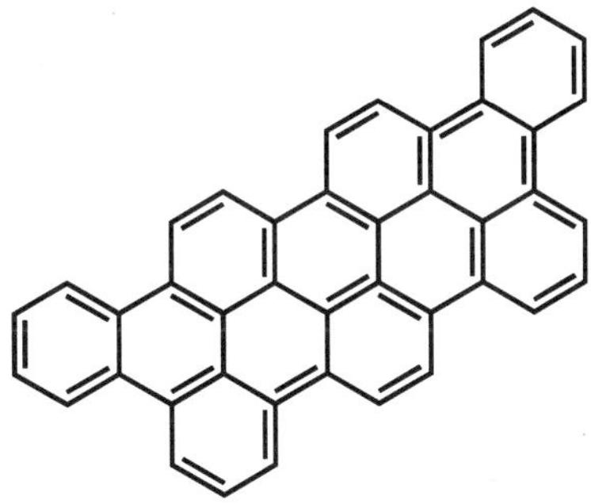

65

LPAHs to Diels-Alder addition. They determined experimental reactivities and rate constants with maleic anhydride and found that there was a correlation with the ionization potentials determined through the photoelectron spectra, the position of absorbance bands in the UV spectra, and to phosphorescence lifetimes.

Zander and Friedrichsen (35, 36) used this reaction on the 15-ring LPAH benzo[1,2,3-*bc*:3,4,5-*b*′*c*′]dicoronene to yield the 16-ring, 50-carbon dinaphtho[2,1,8,7-*defg*:2′,1′,8′,7′-*mnop*]pentapheno[3,4,5,6,7,8,9,10-$uvwxyza_1b_1c_1d_1$]-heptaphene and 17-ring, 52-carbon benzo[13,14] pentapheno[3,4,5,6,7,8,9,10-$uvwxyza_1b_1c_1d_1$]dinaphtho[2,1,8,7-*defg*:2′,1′,8′,7′-*mnop*]heptaphene, shown in Figure 8, through reaction on one or two bay regions. (Note the changes in the order of preference in the fusing group names because of the fusion of the alphabetically earlier benzo group to the pentapheno group.)

If certain larger anhydrides such as phthalic anhydride are used as the dienophile instead of maleic anhydride, then two or more rings can be added to the diene PAH structure. This was the approach of Clar and co-workers (25–27) in synthesizing several very high-carbon number LPAHs.

Schlieke, Schlueter, Hauser, and Heinze (37) used the Diels-Alder reaction to couple several small aromatic units together to form a series of very large and long PAHs. The key reactant is the bifunctional one with an aromatic core structure that has a dienophilic function on either end. This couples to two of the cyclopentadienones. The last compound has 98 carbons in the condensed 29-ring portion and is calculated to be 4.5 nanometers long. The alkyl substitution makes these compounds highly soluble, and their absorbance and fluorescence spectra were collected. Because these structures have several phenyl substituents, their absorbance spectra will be shifted to higher wavelengths (these substituents are not as efficient at extending the π electron conjugation as direct fusion of rings, but they do have a significant effect). The structures are nonalternant, but the large number of aromatic carbon atoms is sufficient to give high

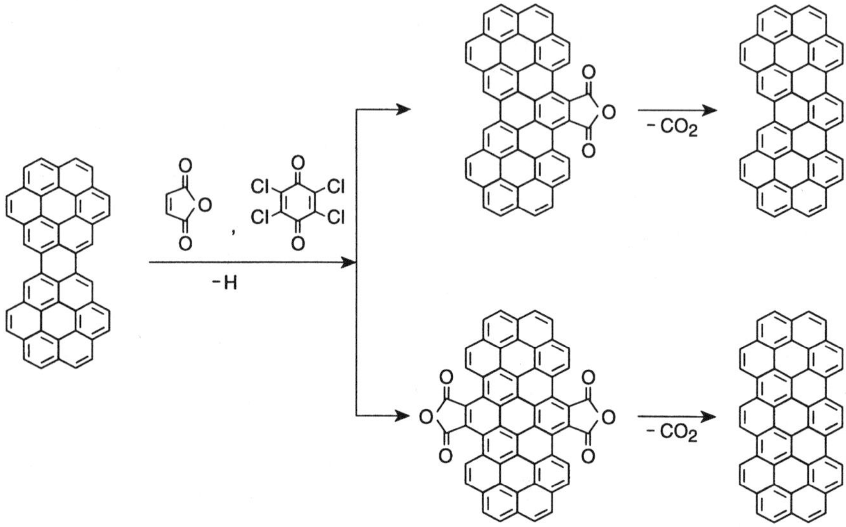

Figure 8. Dicoronylene maleic anhydride reaction used by Zander and Friedrichsen.

wavelength (low energy) bands at 581 and 611 nm. They also show vibronic character (they are somewhat symmetrical in their fluorescence excitation and emission, although the second compound has a more diffuse emission spectrum and a wide Stokes' shift of 30 nm between the excitation and emission band).

Three patents (38–40) describe an analogous reaction that was used to add rings to perylene and benzo[*ghi*]perylene. Benzyne was generated in a solution that contained one or the other PAH. Either naphtho[1,2,3,4-*ghi*]perylene (*compound 66*) or benzo[*a*]coronene was formed. Some of the naphtho[1,2,3,4-*ghi*]perylene was converted to benzo[*a*]coronene through a maleic anhydride Diels-Alder reaction. The benzo[*a*]coronene was claimed to be used in a variety of photo-transparent and photo-sensitive uses, such as in sunglasses and photocopiers. The third patent describes these uses of benzo[*a*]coronene and states that the other LPAHs dibenzo[*a,g*]coronene and tribenzo[*b,n,pqr*]perylene (tribenzo[*fg,ij,rst*]pentaphene) could also be used for these applications.

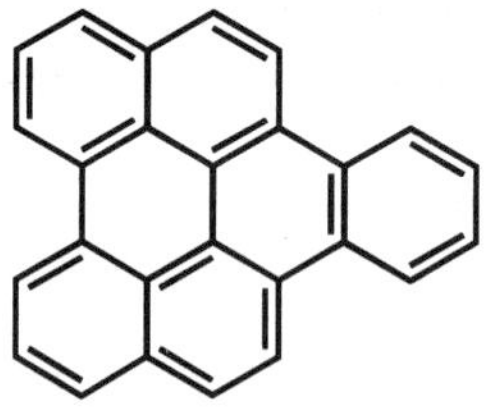

66

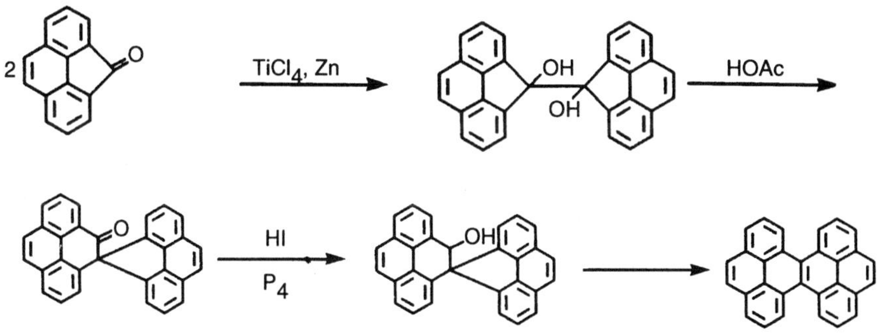

Figure 9. Formation of tetrabenzo[*de*,*hi*,*mn*,*qr*]tetracene through pinacol-pinacolone reaction.

The formation of perylene anion and dianion can be used to produce benzo-[*ghi*]perylene or coronene with much higher yields than the maleic anhydride Diels-Alder route (41). For benzo[*ghi*]perylene, the yield was 79% versus 57%. For coronene, it was 55% versus 41%. The anionic forms of perylene and benzo[*ghi*]perylene were produced by sodium metal in tetrahydrofuran solution. The perylene was made from the commercially-available perylene dicarboxylic dianhydride.

One specific reaction, producing tetrabenzo[*de*,*hi*,*mn*,*qr*]tetracene, was a pinacol-pinacolone rearrangement (42). This reaction mechanism, shown in Figure 9, is so specific that it cannot be used to produce any other LPAH products if similar reactants are used instead.

Fetzer and Biggs (43, 44) used the Clar synthesis to produce over two dozen new LPAHs. This reaction condenses the perinaphthenone-type starting ketones by using a high-temperature melt of zinc dust, zinc chloride, and sodium chloride. The four ketones perinaphthenone, benz[*de*]anthracenone (benzanthrone), dibenzo[*h*,*de*]anthracenone, and benzo[*cd*]pyren-9-one (naphthanthrone) and binary mixtures of them were used as the starting ketone reactants. The structures of these starting ketones and one example of the reactions are shown in Figures 10 and 11. Key to this work was the use of high-performance liquid chromatography (HPLC) to separate the complex mixtures into individual components. As an indication of the power of the separations, a new isomer was isolated from each of the reaction product mixtures of benzanthrone or naphthanthrone when reacted by itself, although Clar and his co-workers had performed the same reactions and discovered numerous products two decades earlier.

Agranat and Suissa (45) examined the many literature references to the Clar reaction of four starting ketones used in the work of Clar and Fetzer. They then used the yield data and molecular modeling to predict the yields of several of the possible LPAH isomers that were not found by either group. They also discussed

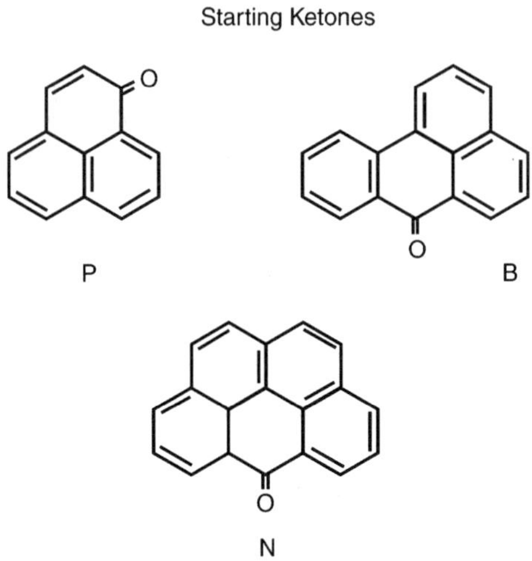

Figure 10. Some of the cyclic ketones used in the Clar reactions investigated by Fetzer and Biggs.

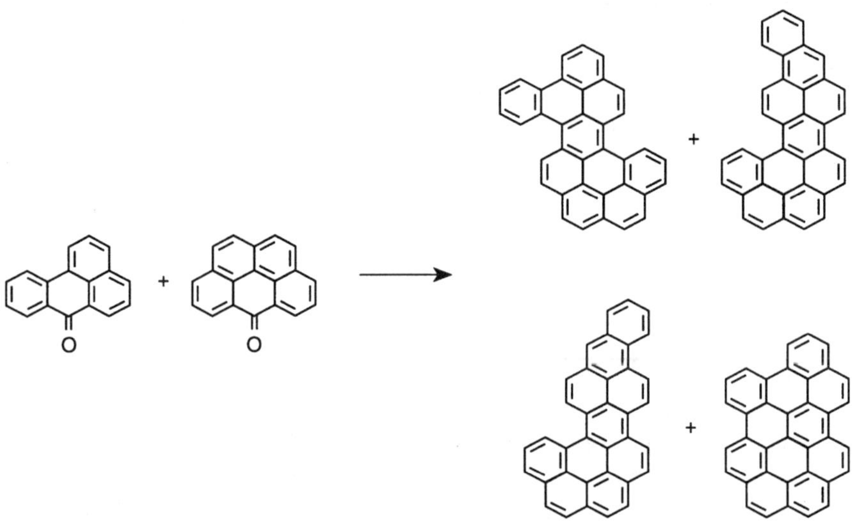

Figure 11. One of the mixed-ketone Clar reactions performed by Fetzer and Biggs.

the probable mechanisms of the condensation and reduction steps involved in these syntheses, and the effects of steric hindrance on isomer yields were examined

Fujisawa et al. (46–51) have extended the group's earlier work into large peropyrene-type LPAHs. This group had synthesized several known LPAHs,

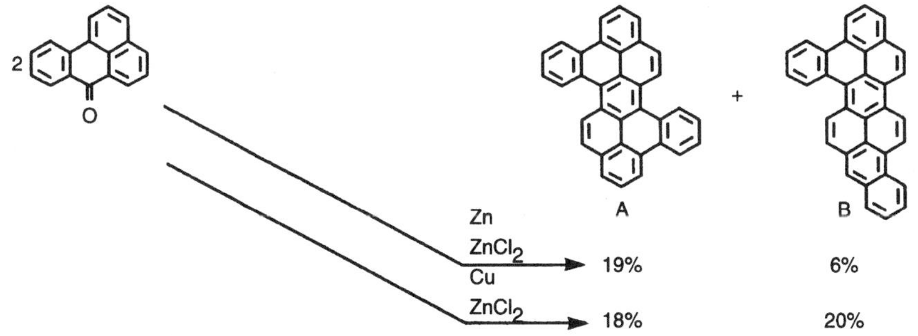

Figure 12. An example of the variation in isomer production in the Clar reaction obtained with the use of different catalysts.

varied the reaction mechanisms, and studied the nonplanarity due to steric effects. They found that both the use of copper metal rather than zinc dust (44) and the use of starting ketones with specific mono-halogen substitution (46, 47, 52) changed the distribution and yield of the LPAH isomers produced (Fig. 12). Li and Gotoh (52) used specific bromo- and iodo-derivatized ketones to find a new, different isomer in the benzanthrone condensation. This group's more recent work has synthesized several new types of the five-ring perinaphthenone starting ketones, which are then used in the Clar synthesis. More variations in reaction products were also made when they utilized mixtures of the starting ketones to make the cross-reaction products (similarly to the earlier work of Fetzer and Biggs). Some of these involved the formation of two bridging rings through the creation three bonds. Two of these new LPAH products are shown as *compounds* 67 *and 68.*

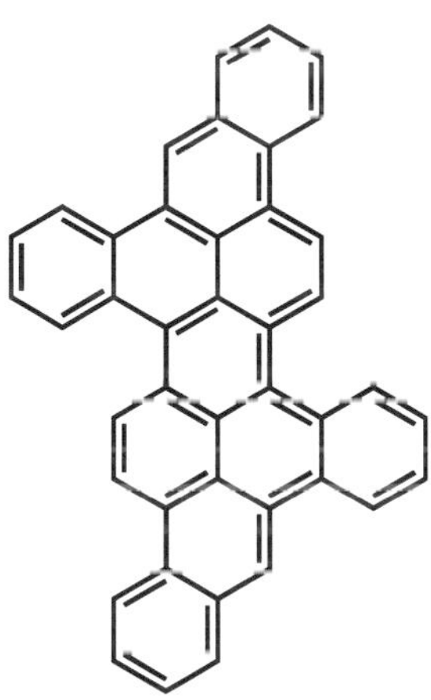

67

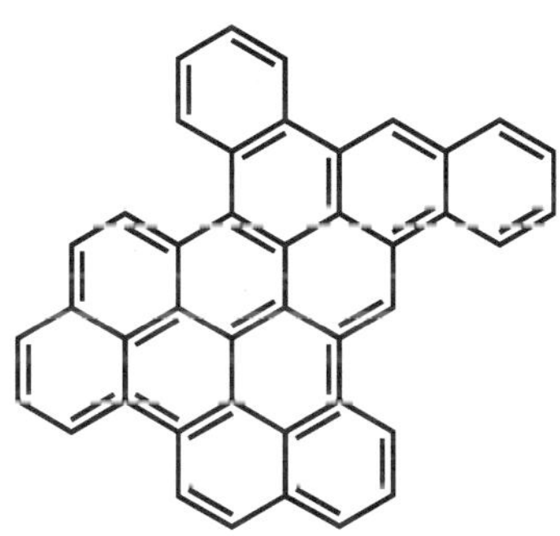

68

A variation of the Clar reaction can be used to reduce large polycyclic aromatic quinones and ketones to the parent LPAHs. A small amount of water can be added to the zinc dust mixture before heating. The three compounds pyranthrone, violanthrone, and isoviolanthrone are the largest polycyclic ketones that are commercially-available. They were once commonly used as dyestuffs. A long-chain alkyl-derivative is patented for commercial use as a marker for lubricating oil or other petroleum products because of its high wavelength and intense fluorescence (53). Pyranthrone, violanthrone, and isoviolanthrone were reduced in this fashion to give the corresponding LPAHs of 30 or 34 ring carbons. These samples were purified and used for studies on their HPLC retention and fluorescence spectroscopy.

Solodar et al. (54, 55) report another variation on the condensation of these types of ketones. They synthesized dibenzo[*cd,lm*]perylene from perinaphthenone when a variety of small alkylamines or hydrazine derivatives in aqueous solution were added and heated. The yields were from 15 % to over 30 %. None of the larger ketones was tried. Perinaphthenone can also be reacted electrochemically to produce dibenzo[*cd,lm*]perylene in 10 % yield. Electron-spin resonance spectroscopy indicated that this reaction progressed through the formation of the phenalene radical.

Metacyclophanes were dehydrogenated with pyridinium hydrobromide perbromate to form intermediate partially hydrogenated structures (56). These were aromatized by using 2,3-dichloro-5,6-dicyanoquinone (DDQ) in boiling xylene to yield dibenzo[*cd,lm*]perylene and the new 10-ring benzo[*rst*]dinaphtho[8,1,2-*cde*:2′,1′,8′-*klm*]pentaphene (*compound 69*). The authors used the trivial name "teropyrene" for this new $C_{36}H_{18}$ LPAH. This was to combine Clar's trivial names of dibenzo[*cd,lm*]perylene (peropyrene) and tribenzo-[*de,hi,rst*]pentaphene (terrylene; *compound 70*), but they misspelled their new

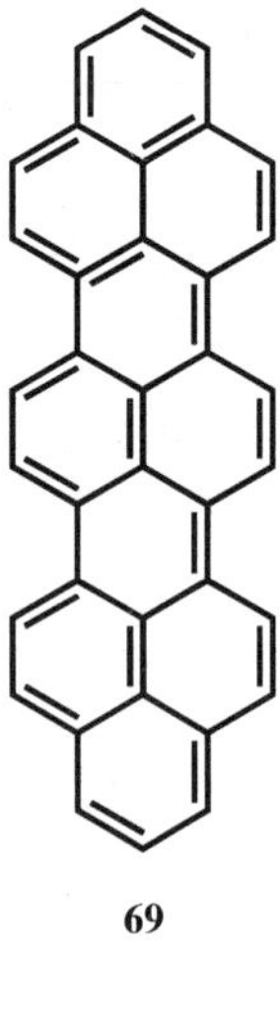

69

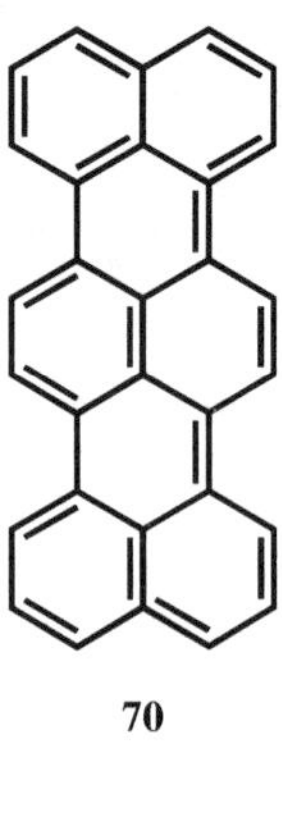

70

name by omitting one of the "r"s. A similar reaction was used by Grutzmacher and Huseman (57) to photolytically form tetrabenzo[*de*,*hi*,*mn*,*qr*]naphthacene from a cyclophane.

There has been a flurry of recent research from many groups throughout the world aimed at the total synthesis of fullerenes (58–75). This has resulted in dozens of new nonplanar LPAHs being made and characterized. Almost all of these new LPAHs are nonalternant, with some containing two, three, four, or more five-member rings.

One of the more prominent ones is flash-vacuum pyrolysis, specifically tailored to produce these compounds. Flash-vacuum pyrolysis relies on the condensation to larger LPAH structures from smaller ones. The starting material in the flash-vacuum pyrolysis are small PAHs having labile substituents in certain locations. The heating causes the facile-leaving group (the substituent) to be removed. The resulting reactive species readily condenses into the desired structures by forming new bonds from the sites produced by the leaving group to spatially adjacent portions of the PAH structure. The first report of a PAH synthesis using this sort of reaction was by Scott, Hashemi, Meyer, and Warren (58). They produced the previously-known $C_{22}H_{10}$ PAH corannulene. Their yield and the simplicity of the route were much better than the older synthesis.

The work of the research groups of Rabideau, Scott, Taylor, Zimmerman, and others has resulted in many new compounds and new synthetic routes. The target LPAH structures are portions of the C-60 fullerene's 3D spherical structure, such as corannulene or pyracyclene or larger LPAHs with these as the core structure. Additionally, through either side reactions or "failed" attempts at these desired structures, many other LPAHs have been made, isolated, and characterized. Some of these structures include several isomers of the extended 30-carbon corannulene-like structures with bowllike shapes ("half-fullerenes" or "hemi-fullerenes" or "Bucky bowls"). Two examples are shown as *compounds 71 and 72.*

Another route from smaller parts to the fullerene structure has been used by Taylor and his research group (76, 77). They have pyrolized naphthalene,

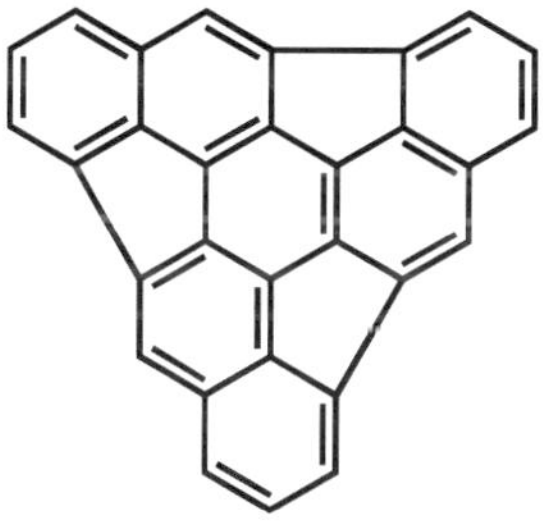

71

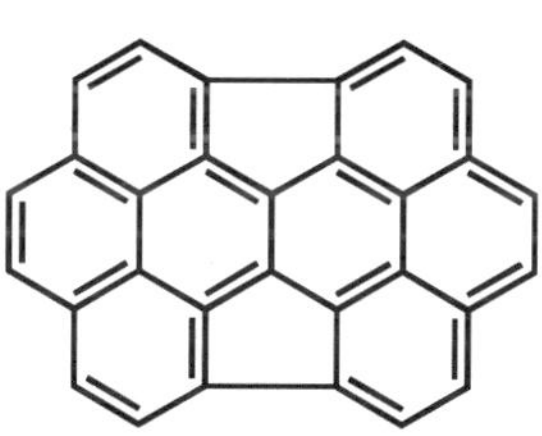

72

benzo[*k*]fluoranthene, and corannulene. C-60 fullerene was found in the product. Some of the PAH products of the naphthalene reaction were identified by GC and HPLC analyses. The intermediate LPAHs have not been isolated and characterized, although mass spectroscopy matched LPAH masses. This group states that with the synthesis of the partial fullerene structures and the other LPAHs resulting from flash pyrolysis reactions, several of these intermediates should be identified.

Sarobe (as a member of Jenneskens' research group) used the reactivity of the cyclopenta-fused smaller PAHs to form decacyclene and larger benzologues by flash vacuum pyrolysis (78). The cyclopenta portion of three of the reacting molecules forms a central bridging ring. Thus acenaphthylene reacts to form decacyclene (*compound 73*). This reaction is a lower-temperature analog to Taylor's work, by fusing small units into the fullerene-type arrangement of five- and six-member rings. Other reactions used cyclopenta[*cd*]fluoranthene, cyclopenta[*cd*]benzo[*ghi*]fluoranthene, and benz[*j*]acephenanthrylene (*compounds 74–76*) as the starting materials. The products of these reactions have 54 or 60 ring carbons. The products are very insoluble and so could not be isolated. Mass spectra, however, showed the molecular ions of the fully condensed products of $C_{54}H_{24}$ (trisindeno[1,2,3-*cd*:1′,2′,3′-*no*:1′,2′,3′-*yz*]decacyclene), $C_{60}H_{24}$, and $C_{60}H_{24}$ (trinaphtho[2,1-*b*:2′,1′-*m*:2′,1′-*x*]decacyclene). Many of these LPAHs should be the same ones that Taylor and co-workers identified in the pyrolyses of the small PAHs.

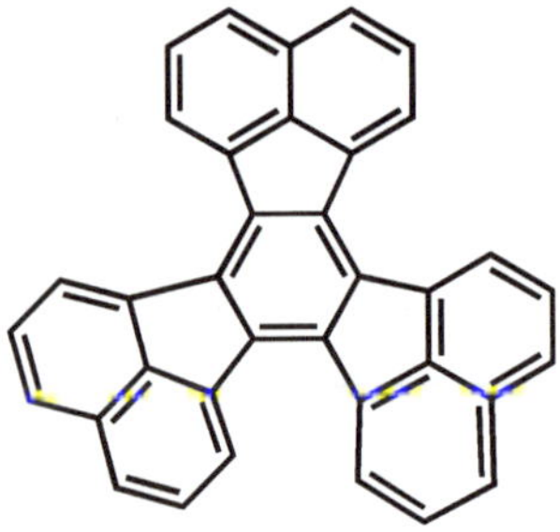

73

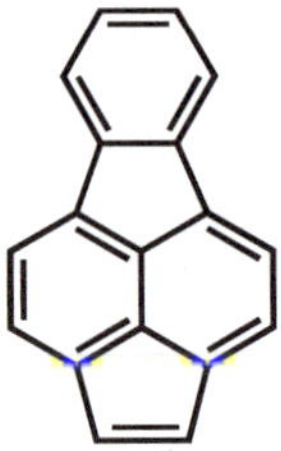

74

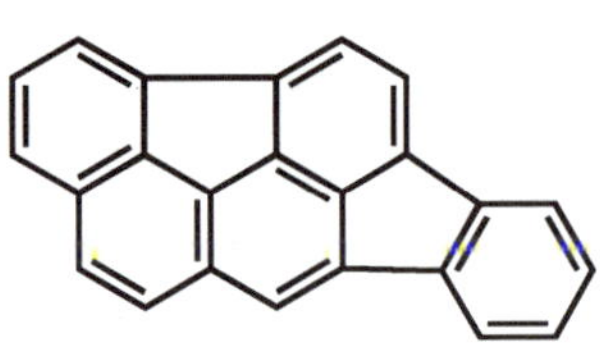

75

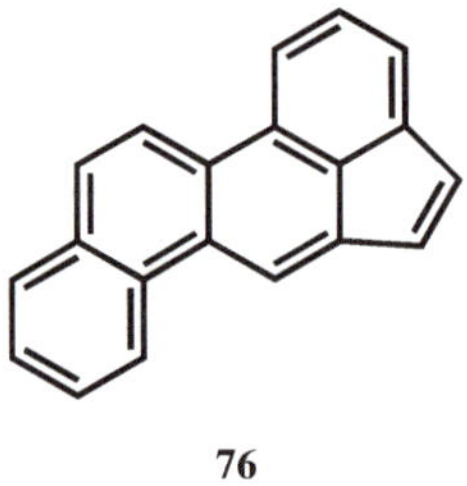

76

Thermal condensation of PAHs produces many of the same reactions as condensation with aluminum(III) chloride. Lang and co-workers (79, 80) condensed naphthalene and found many compounds in the complex product mixture, including perylene and the LPAH tribenzo[*de*,*kl*,*rst*]pentaphene. In a different reaction, they reacted benzene and pyrene. Besides the two alternant pyrene condensation products dinaphtho[2,1,8,7-*defg*:2′,1′,8′,7′-*ijkl*]pentaphene and dinaphtho[2,1,8,7-*defg*:2′,1′,8′,7′-*opqr*]pentacene (*compounds 77 and 78*), the nonalternant mixed reaction product diindeno[1,2,3[*cd*:1′,2′,3′-*jk*]pyrene was found (*compound 79*). Lewis and Edstrom (10) reported that thermal condensation of acenaphthylene yielded decacyclene.

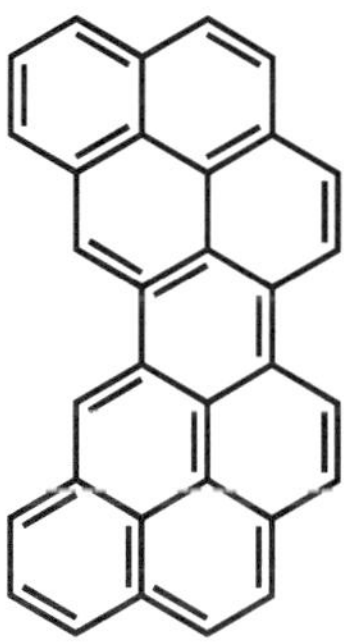

77

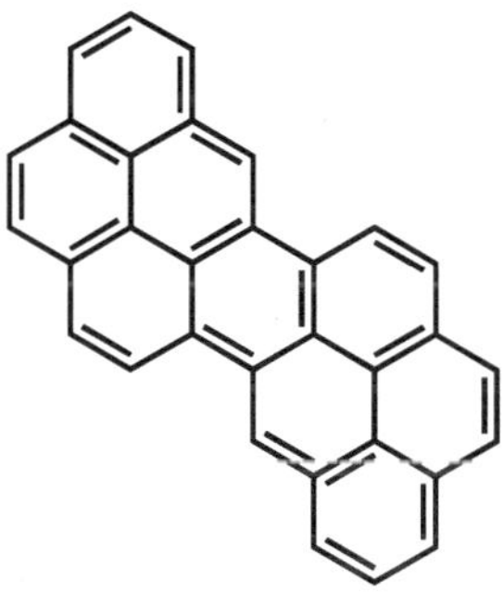

78

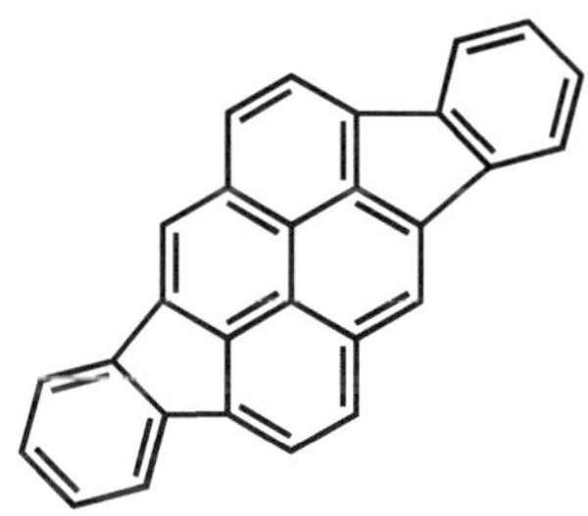

79

Zander, Haase, and Dreeskamp (81) reproduced the work of Zinke and co-workers (82, 83) by thermally pyrolyzing perylene at 450 degrees and isolated the red-orange needles of a 40-carbon alternant product, dibenzo[*jk*,*pq*]di-naphtho[8,1,2-*cde*:2′,1′,8′-*tua*]pentacene (*compound 80*). The structure was determined by high-field nuclear magnetic resonance (NMR) spectroscopy. This

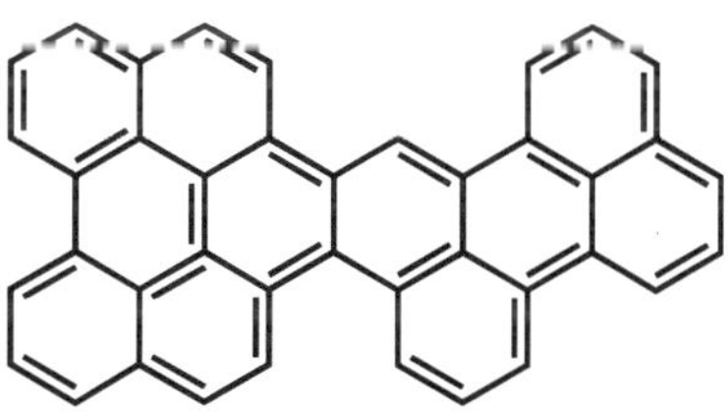

80

product is different from the isomeric products isolated from the Scholl condensation in aluminum(III) chloride of perylene. This is mainly the isomeric alternant greenish benzo[1,2,3-*cd*:4,5,6-*c*′*d*′]diperylene (quaterrylene, *compound 81*) with a much lesser amount of the analogous nonalternant reddish-brown cyclopenta[1,2-*b*:3,4,5-*c*′*d*′]diperylene. These were reported earlier by Zinke and co-workers and Clar, Kelly, and Laird (84).

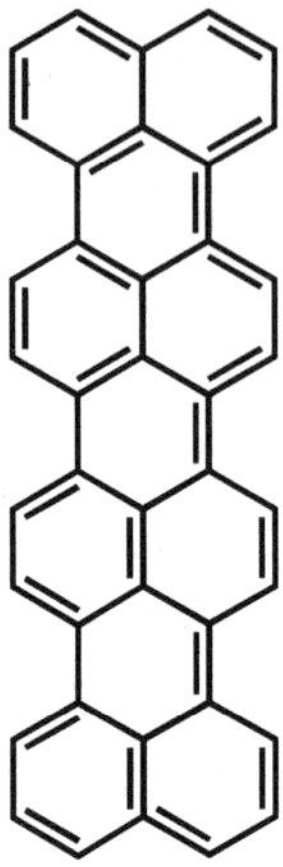

81

In one of the more involved LPAH syntheses, Broeme and Dieterich (85) used the 10-step route shown in Figure 13 to produce the 40-carbon, 13-ring phenanthro[3,4,5,6-*bcdef*]ovalene (which they called by the trivial name of circumanthracene). The overall yield of this series of reactions was only 3.7 %. The UV absorbance spectrum compared favorably to one that had been theoretically calculated. This structure had originally been claimed to be made by Clar, Kelly, Robertson, and Rossman (86), but later Clar, Robertson, Schoegl, and Schmidt (87) used photoelectron spectroscopy (see Chapter 10) to show that this was not the product.

Erickson and Milliken (88) synthesized the very symmetrical ortho-fused benzo[*c*]naphtho[2,1-*p*]chrysene (*compound 82*). They condensed two naphthalene derivatives to gain a substituted benzo[*c*]chrysene. This compound was in turn reacted with triphenylphosphine to form an intermediate olefinic compound. This was photolyzed into the target product.

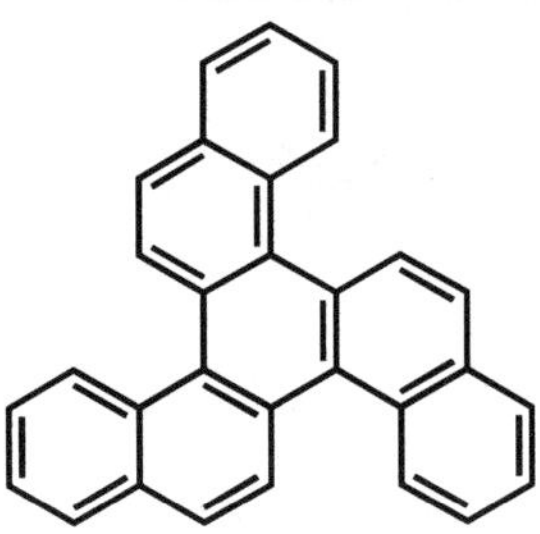

82

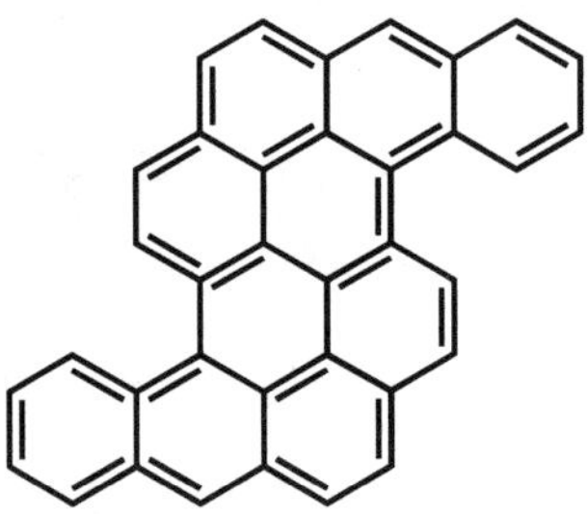

83

Fujisawa, Takekawa, Nakamura, Uchida, Ohshima, and Oonishi (89) synthesized the new $C_{36}H_{18}$ isomer dibenzo[*ghi*,*o*]anthra[2,1,9-*cde*]perylene (*compound 83*) through use of the Ullmann reaction. They synthesized 1-bro-

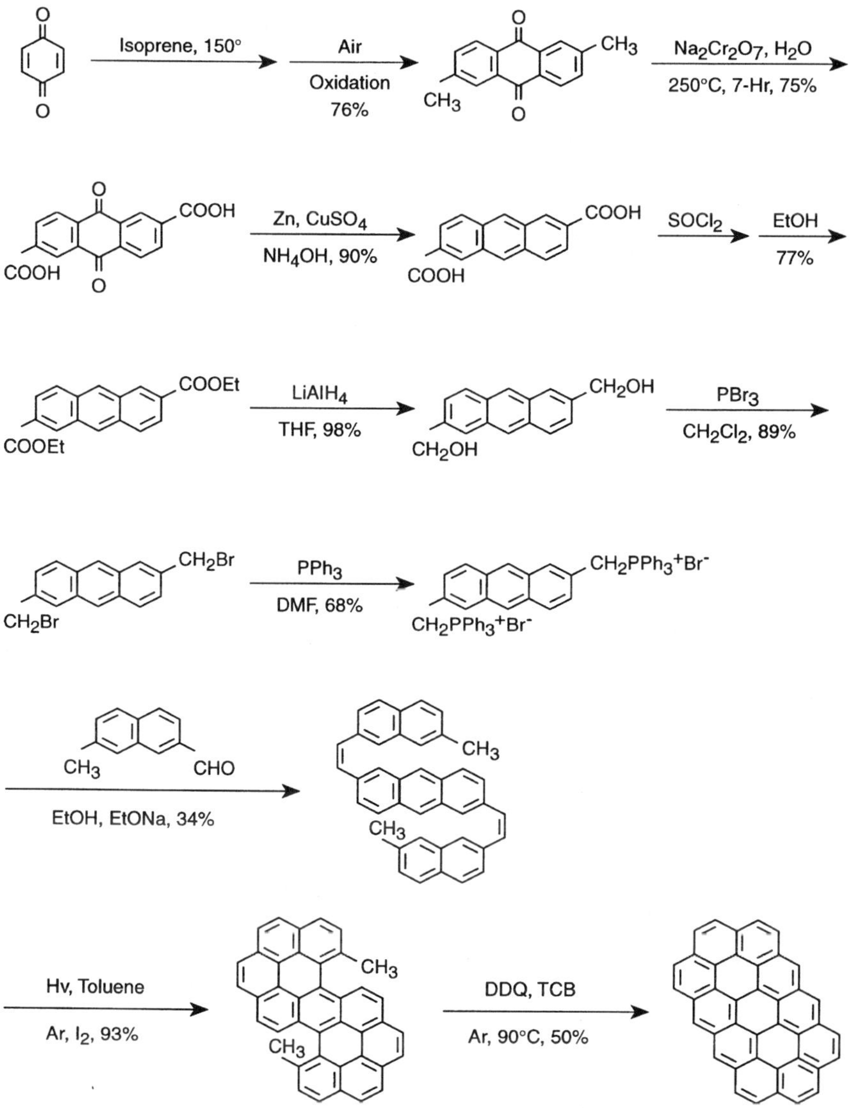

Figure 13. Multi-step synthesis of circumanthracene from R. D. Broeme and F. Dieterich, *Tetrahedron Lett.* **32** 5230 (1991), reproduced with permission from Elsevier Science Publishers BV.

mobenz[*a*]anthraquinone. Heating a slurry of this starting material, activated copper powder, and naphthalene gave 1,1′-dibenzanthrinyl. This was cyclicized by sulfuric acid and zinc dust. The cyclic dione was reduced to the desired product with a melt of zinc dust, zinc chloride, and sodium chloride. They reported an absorption maximum in dichlorobenzene of 492 nm.

Harvey, Yang, and Yang (90) made dibenzo[*h,rst*]pentaphene and naphtho[1,2,3,4-*rst*]pentaphene (*compound 84*). They produced 1,4-bis-(bromomethyl)anthracene and 9,10-bis-(bromomethyl)anthracene. These were reacted with 1-pyrrolidino-1-cyclohexane to yield an intermediate diketone. These were condensed and dehydrogenated to the LPAH with 10% palladium-charcoal in triglyme.

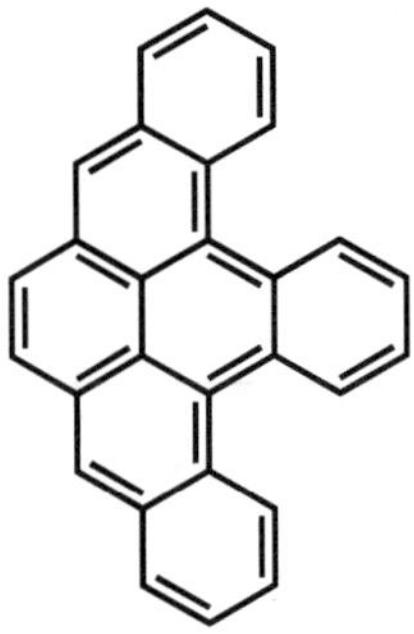

84

In a series of cyclodehydrogenation reactions, Laarhoven and co-workers (91, 92) reacted several "stilbenelike" molecules to form LPAHs. These reactive molecules were induced to condense through photolysis. Through this reaction route, several structures were made that contained the dibenzo[*c,g*]phenanthrene and phenanthro[3,4-*c*]phenanthrene substructures. Some of those with the dibenzo[*c,g*]phenanthrene substructure were successfully condensed to the peri-fused LPAH (the resulting part of the LPAH being a benzo[*ghi*]perylene. In this way, tribenzo[*c,g,p*]chrysene (*compound 85*) was formed and then converted to benzo[*b*]naphtho[1,2,3,4-*pqr*]perylene (*compound 86*). Other reactions gave dibenzo[*c,g*]chrysene and dibenzo[*b,pqr*]perylene, naphtho[1,2-*a*]picene and phenaleno[1,2,3,4,5-*stua*]picene (*compounds 87–90*). The last compound was a newly-discovered, peri-fused LPAH, and a UV absorbance spectrum is given for it (and the many, many ortho-fused LPAHs).

Muellen et al. (93–97) produced other stilbene-type reactants. These were condensed into VLPAHs. This group also similarly condensed large polyphenylenes into LPAHs. Hexaphenylbenzene can be condensed by aluminum(III) chloride to hexabenzo[*bc,ef,hi,kl,no,qr*]coronene. Muellen and co-workers have made several large planar structures using this reaction for condensation of large

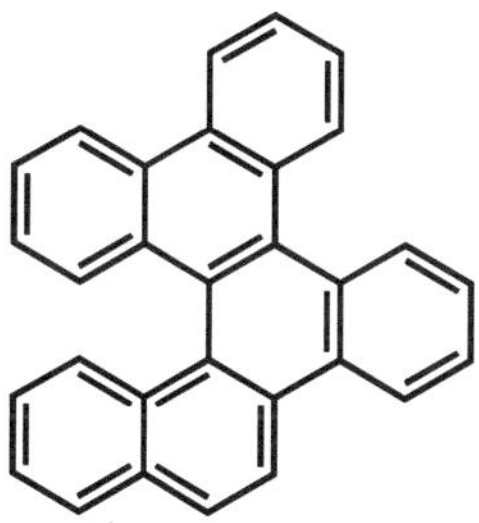

85

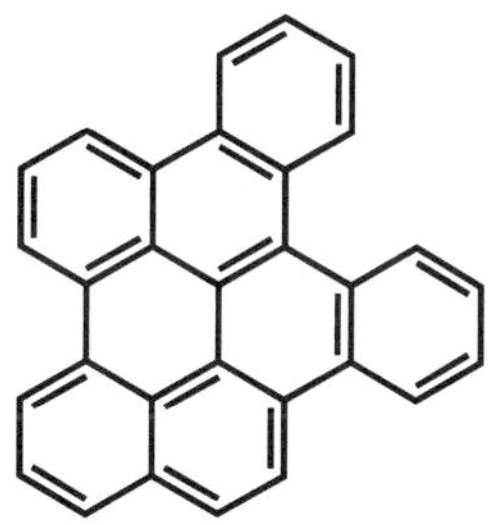

86

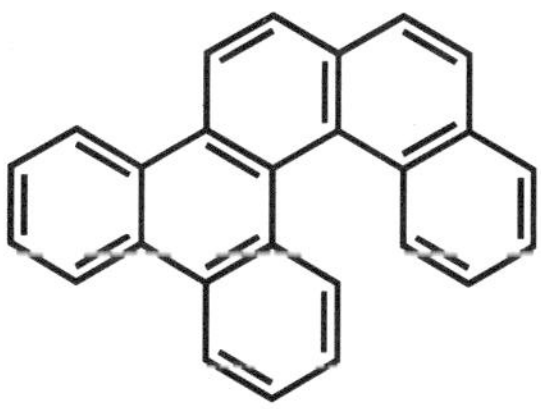

87

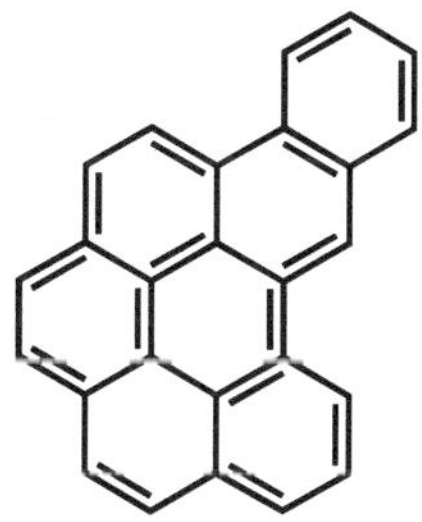

88

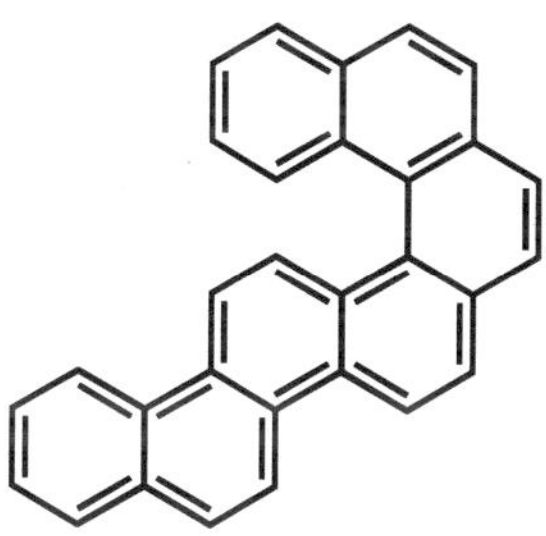

89

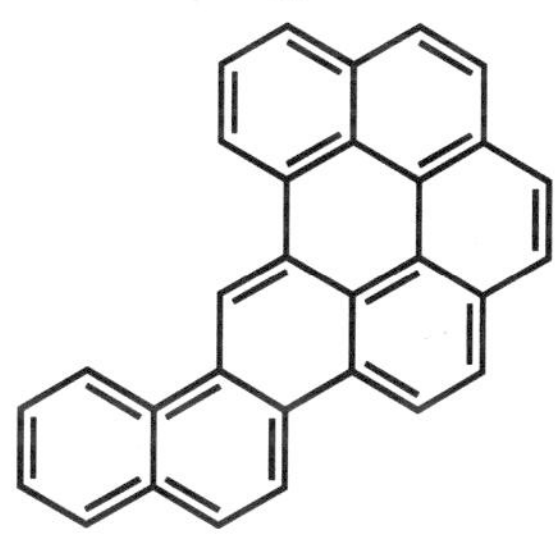

90

symmetrical polyphenyl structures. The polyphenyl compounds were generally formed by Diels-Alder reactions of acetylene derivatives of benzene and a dienone with phenyl substituents. An example is shown in Figure 14, resulting in a 78-carbon LPAH. The cyclodehydrogenations that yield the LPAH occur through the use of a variety of reagents and conditions.

A few of the larger structures synthesized through this manner are also shown as *compounds 91–93*, highlighting the power of this approach in extending the range in carbon number of the LPAHs. This group also has synthesized some alkyl derivatives with higher solubilities.

These large planar structures extend the "record" for largest PAH to well over 200 carbon atoms, with one possible product having the formula $C_{222}H_{42}$ (*compound 94*). Muellen and co-workers were unable to unambiguously isolate

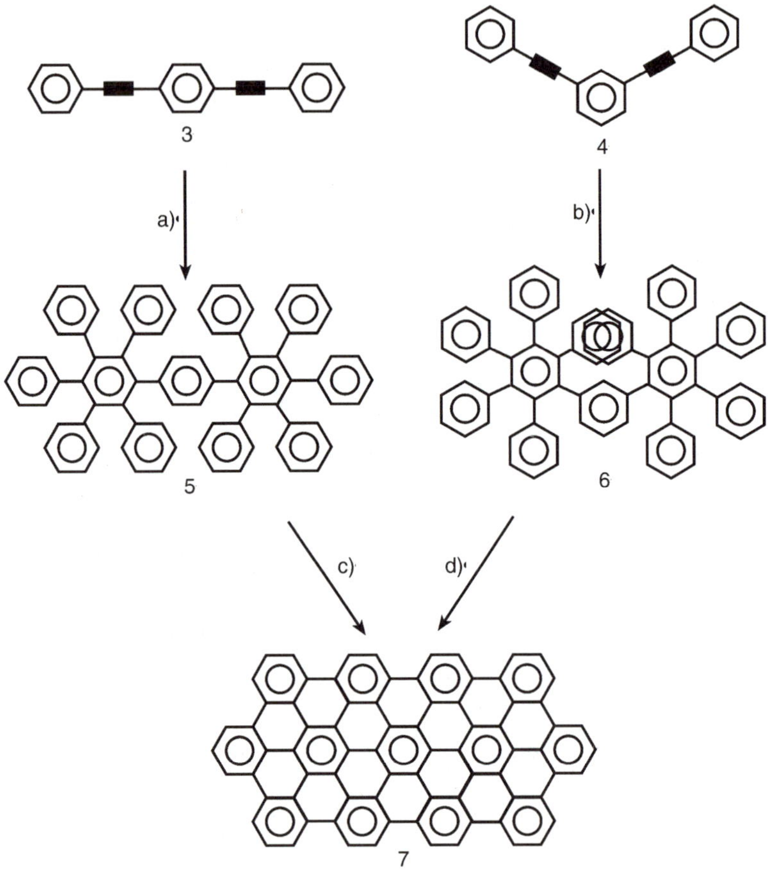

Figure 14. Muellen synthesis of 78-carbon LPAH from the condensation of a polyphenyl from V. S. Iyer, M. Wehmeier, J. D. Brand, M. A. Keegstra, and K. Muellen, *Angew. Chem. Int. Ed. Engl.* **36** 1604 (1997), reproduced with permission from Wiley-VCH.

and characterize the isolated black product as the fully-condensed molecule. The 37-ring polyphenyl was condensed with aluminum(III) chloride and copper(II) triflate in carbon disulfide solution. Mass spectroscopy was inconclusive because there was a broad cluster of peaks around the expected mass for that LPAH. The electrical properties of the material were consistent with full condensation. Dislocations in the aromatic π network, such as those caused by aromatic rings with free edges, would drastically affect the conducting properties of these large molecules.

These structures also raise the question of when PAH behavior decreases and the properties of the structure become similar to those of graphite (98). The

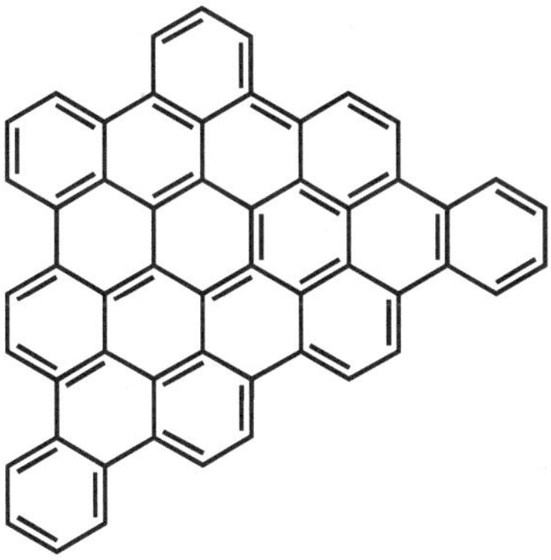

91

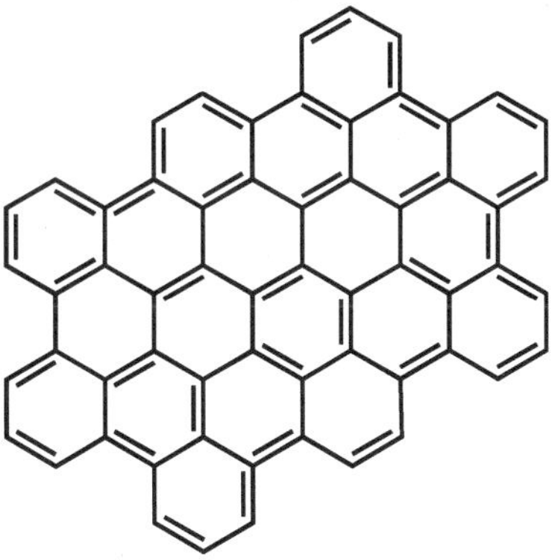

92

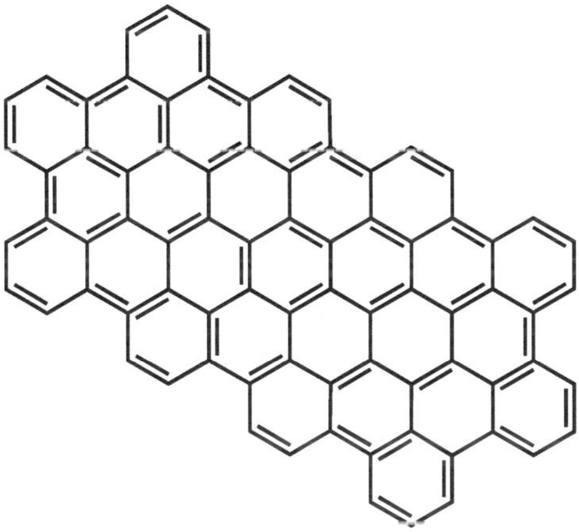

93

extended π structure of graphite gives it identical ionization potential (IP) and electron affinity (EA) values of 4.39 electron volts. As the LPAHs get larger, their IPs decrease and their EAs increase (99). These values have not yet been determined experimentally for these new LPAHs. Some theoretical results (100, 101) indicate that graphite may not be a good model for "infinitely large" LPAHs. The effects of the peripheral hydrogen atoms and "defects" caused by the structure not being both fully-benzenoid and highly symmetrical yield a finite difference between IPs and EAs. In addition, the intrinsic effects of the electron repulsions and the probable semiconductor behavior of very large structures lead to different behavior than that of graphite. Zhang, Bertran, and Lee (102) examined a variety of LPAH structures and calculated the energy gap

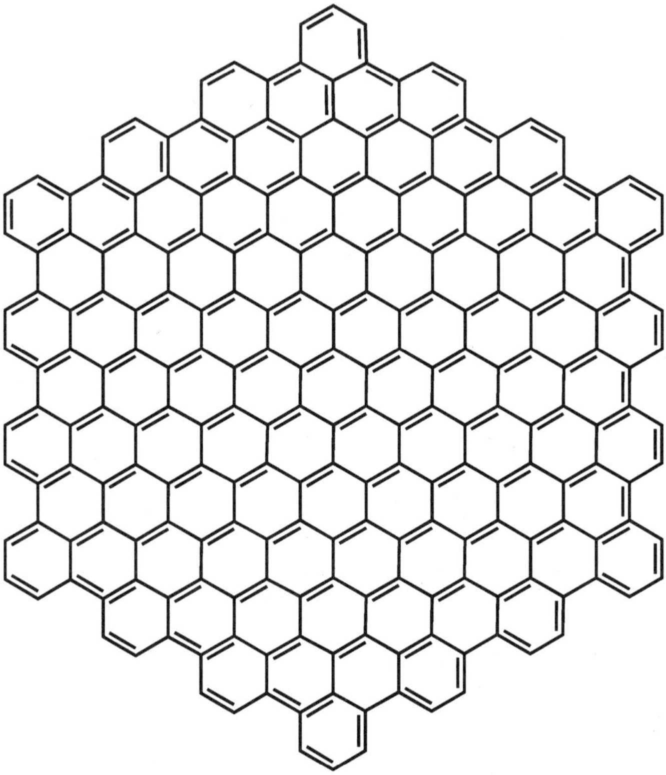

94

of each. The wide variation in these values is then described as a cause of the broad-band luminescence found in many graphitic and carbon materials.

Muellen's research group (103) also synthesized several LPAHs of the fused-naphthalene series through the use of coupling reactions of substituted naphthalene. The first LPAH of these is tribenzo[*de*,*kl*,*rst*]pentaphene, trivially called by Clar "terrylene." The second benzo[1,2,3-*cd*:4,5,6-c′d′]diperylene, trivially called "quaterrylene." can also be made by Scholl condensation of perylene. To study the electrochemical oxidation and reduction of these LPAHs, they also synthesized soluble derivatives with several *tert*-butyl groups. They readily formed the cation, dication, anion, and dianions of perylene. For terrylene, the trication and trianion were observed, and for quaterrylene the tetracation and tetra-anion were found. The increased stability of the higher ionic forms is reflective of the higher conductance of the larger LPAH structure of this "oligorylene" series. Chemical formation of the anion resulted in a blue dianion for terrylene, compared to the purple dianion of perylene. The research group of Naamann, Neumann, and Muench (104) successfully synthesized the fifth member of the series. This compound, bisbenzo[5,10]anthra[9,1,2-*cde*:2′,1′,9′-*klm*]benzo[*rst*]pentaphene (*compound 95*), could be trivially called penterrylene. It is an electrical conductor when it is doped with a strong Lewis acid.

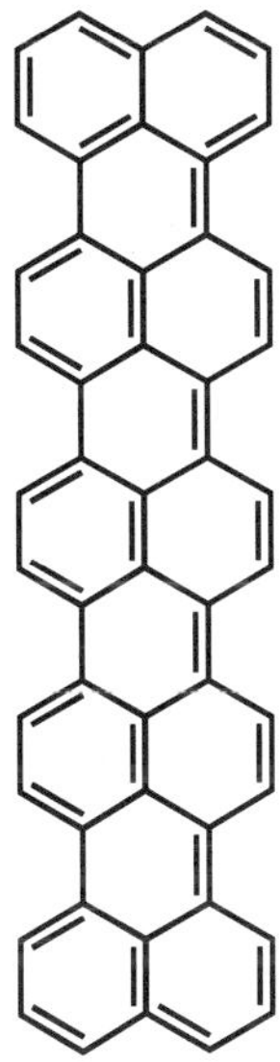

95

The advances from Muellen and his co-workers have accelerated the increase in size of the largest experimentally-produced LPAHs. Only a decade ago, the largest known LPAH was the $C_{52}H_{20}$ benzo[13,14] pentapheno[3,4,5,6,7,8,9,10-*uvwxyza*$_1$*b*$_1$*c*$_1$*d*$_1$]dinaphtho[2,1,8,7-*defg*:2′,1′,8′,7′-*mnop*]heptaphene (*compound 32*). Now, this newer work has succeeded to the point where there are even sets of several isomers of these larger PAHs. Three $C_{54}H_{22}$ isomers, tribenzo[-*hi*,*o*,*uv*]triphenaleno[2,1,12,11-*bcdef*]ovalene, tetrabenzo[*bc*,*de*,*o*,*uv*]phenanthro [2,1,10,9-*hijk*]ovalene, and tribenzo[*ef*,*hi*,*o*]triphenaleno[2,1,12,11-*uvabc*] ovalene, have been synthesized and characterized. Their ring structures are shown in Figure 15.

Obenland and Schmidt (105) used a similar reaction to unambiguously produce the previously-known triphenyleno[3,4,5,6-*fghij*]pentaphene (*compound 96*). This compound had also been made through 1,1′-biphenanthryl and maleic anhydride. After decarboxylation, aluminum(III) chloride was used to condense the intermediate to triphenyleno[3,4,5,6 *fghij*]pentaphene. This LPAH was reacted with maleic anhydride to produce the new, fully-benzenoid LPAH tribenzo[*a*,*g*,*m*]coronene.

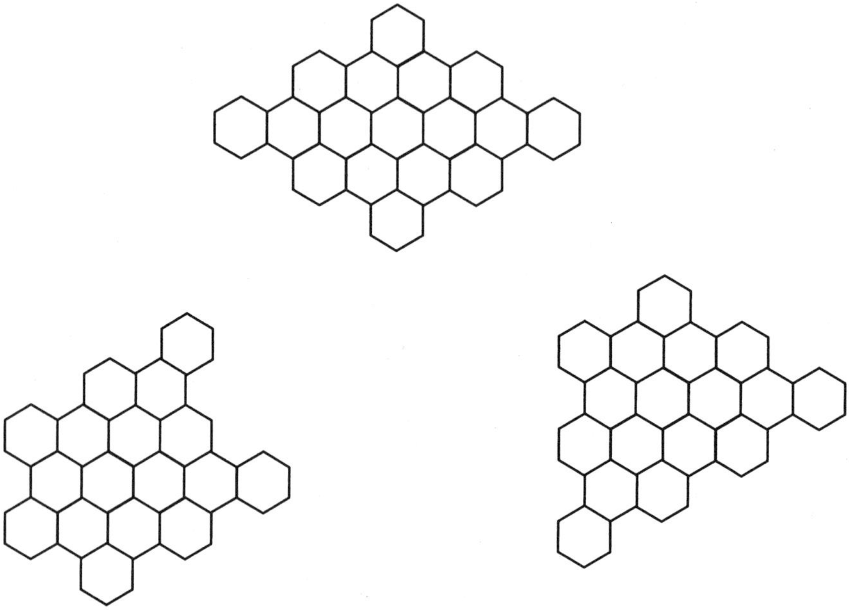

Figure 15. Three 54-carbon LPAH isomers synthesized by Muellen and co-workers.

Barnett, Ho, Baldridge and Pascal (106) unequivocally produced dibenzo[*f,j*]phenanthro[9,10-*s*]picene (*compound 97*) through pyrolytic condensation of phenanthrene-9,10-dicarboxylic anhydride. Four previous claims of this synthesis had either been refuted or proven ambiguous, but this group was able to isolate the compound and make crystals for X-ray crystallography. The structure proved to be highly nonplanar because of the many steric interactions.

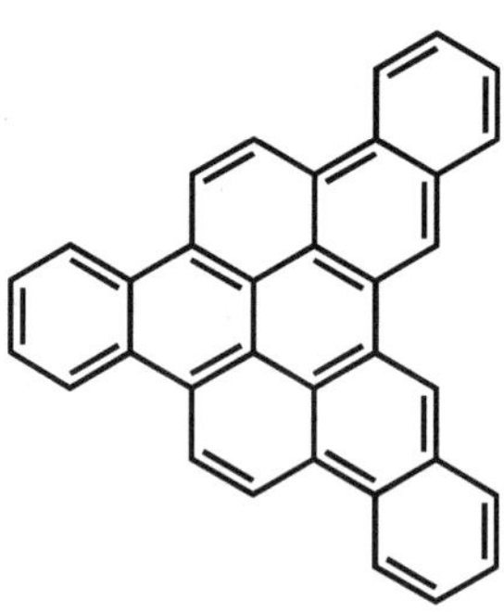

96

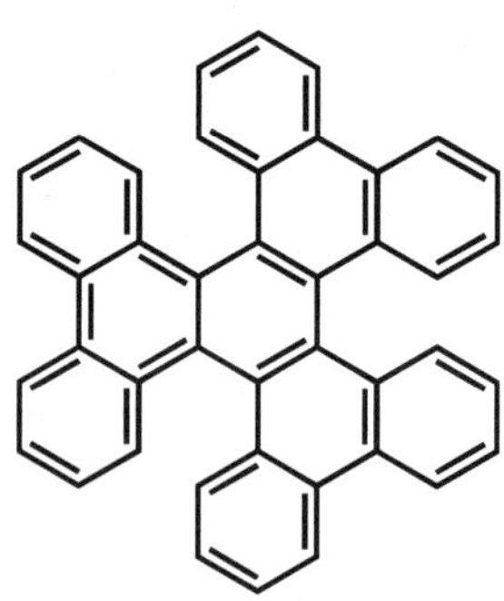

97

Once only the theoretic chemists could discuss these structures and model their properties through computer simulations, but now the validity of those models can be tested through comparison with authentic samples.

There is one issue that can be raised at this point as an aside to the work of the research groups of Schlueter and Muellen and their co-workers that synthesize these extremely large structures. As pointed out earlier, there are IUPAC rules for naming PAHs. The original rules were set by the IUPAC in 1964. There have been no additional aromatic cores added to this list, although there have been several modifications of these rules. The number of both alternant and nonalternant LPAHs has grown both in number of newly synthesized PAHs and in the size of the ring systems.

The nomenclature for these compounds can be quite cumbersome and confusing. The resulting situation is combinations of many, many fusing and fused cores being used to produce the complete names. In some cases, fused structures are fused onto other fused structures. This leads to the use of the symbols "′"and "′" to denote secondary and tertiary fused substructures, i.e., those not directly fused to the core naming structure but to fused groups.

Working with these rules can be very difficult either for a researcher trying to name a structure or for one who has a name and is trying to draw the corresponding structure. There are numerous examples in the literature of the average researcher attempting to name LPAHs and not getting them correct (and it is highly likely that there are many scattered through this book). This leads to confusion over which LPAH is being discussed and whether it is the same one mentioned in another publication, but with a different name. There can even be errors and inconsistencies in Chemical Abstracts, which routinely names new compounds by using the IUPAC rules and is looked upon as the supreme naming entity. It is hoped that as more and more research is done in LPAHs, the nomenclature rules will be changed to accommodate these new compounds.

2.4 THEORETICAL STUDIES OF LPAHs

For many years, the LPAHs have been one of the most favorite classes of compounds of theoretical and computational chemists. In fact, until some of the recent synthetic work just described, this type of work was the only literature reference for most of these compounds. Because of their relatively simple structures, computer programs could be readily written that readily created models of the LPAHs. These models have been used to study trends in many areas of LPAH properties. These include the 3D structures, the resonance energies, the ionization potentials, the stability from or propensity toward reaction, and other properties (107–113).

Randic (114) has calculated the theoretical boiling (or sublimation) points (bps) of planar LPAHs using a model set of 22 PAHs with known bps. The structural features of the 22 PAHs were assessed, and their contributions to the bps were calculated. These values were then extrapolated to larger PAHs with similar structural features.

Simonsick and Hites (115) used the calculated relative stabilities and the resonance energies of LPAH isomers to determine the ionization potentials (IPs) of LPAH isomers. These were then used to assign structures to chromatographic peaks when detected by chemical-ionization mass spectrometry with methane as the reagent gas (116–118). The relative intensity of the molecular ion and the M+1 ion can be correlated to the IP. The distinctiveness of the IPs of some isomers was so high that Simonsick and Hites could assign structures to some GC peaks without comparison to standards. This set of calculations also included structural diagrams of each LPAH with a tabulation of the electron density at each atom in the LPAH (described as "net atomic charges"). These values can be useful in ascertaining reaction mechanisms. Schmidt and co-workers used thermodynamic calculations to assign possible structures to three nonalternant $C_{24}H_{12}$ (300 dalton) LPAHs. They calculated the formation energies and assumed that their three unknown LPAHs were the three most stable nonalternant isomers.

There has been a lot of argument as to how to calculate the number of isomers for a particular carbon number and degree of condensation. One flaw in some of these enumerations is the inclusion of nonaromatic molecules. Having a certain number of six-member rings and π bonds does not determine whether or not a structure is a LPAH (119). Some structures, such as zethene (dibenzo[*de*,*mn*]tetracene; *compound 98*) and its benzologues (such as dibenzo[*de*,*op*]pentacene; *compound 99*), are not fully aromatic. If all the Kekule structures of zethene are drawn, it will be found that for every one of these structures the π bonds in the central bridging rings are located in the same carbon-carbon bonds at the 7,7a

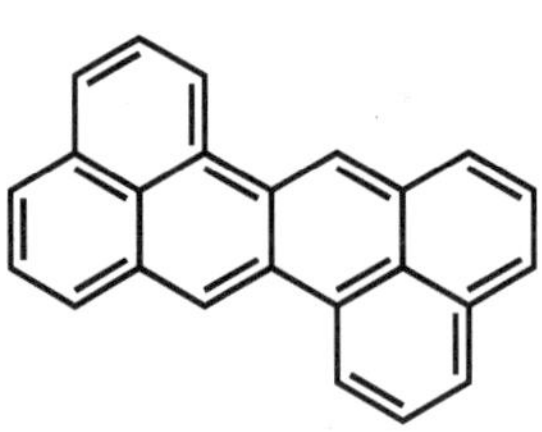

98

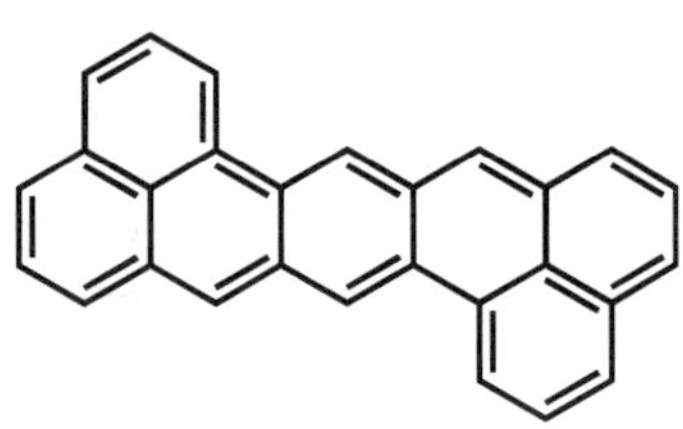

99

and 14,14a positions. There is no delocalization of these bonds, and they are alkene in their character, not aromatic. This theoretical model is proved by the NMR spectrum, which shows localized double-bond character, and by the very high-wavelength absorbance band at 556 nm (in benzene). These types of structures should not be enumerated as LPAHs.

Other structures can be made graphically that are composed of only hexagonal carbon rings, but they cannot exist in reality because the aromatic structure does not have all the electrons in π bonds. Two electrons remain unpaired, and a diradical is the only realistic graphical representation. The stable form of these structures is a partially hydrogenated form, because the diradical cannot generally exist as a stable species. The smallest of these is the PAH (22 carbon) 4H,8H-dibenzo[*cd,mn*]pyrene (*compound 100*). The simplest such LPAH structure is 7H,8H-dibenzo[*de,hi*]tetracene (*compound 101*). One larger "diradical" 34-carbon diradical structure is shown as *compound 102*. The nature of this hydrocarbon was first pointed out by Clar and MacKay (120), when they reported unsuccessful attempts at synthesizing it. All of the π electrons cannot be drawn as double bonding in this structure. Two are always left over after arranging the π bonds. The two "odd" electrons must be in an extremely unstable diradical configuration, with each "odd" electron being located at opposite ends of the molecule. Clar devoted three chapters in his book, *The Aromatic Sextet*, to the hydrocarbons with fixed double bonds or permanent diradical structures (119).

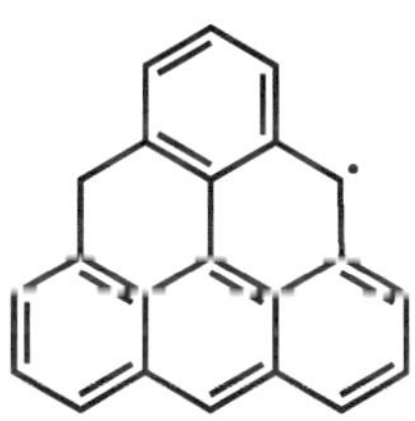

100

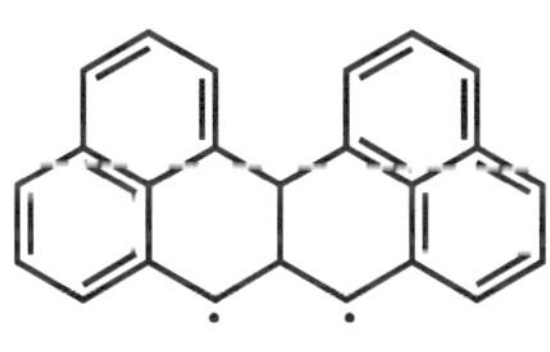

101

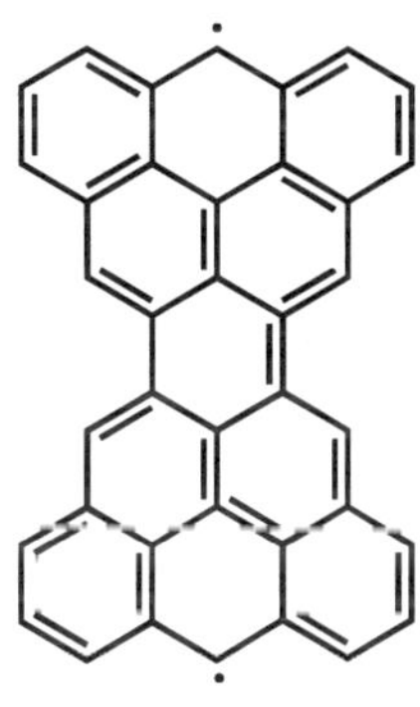

102

These structures also are enumerated as valid isomers of the LPAHs of the same empirical formulae by a simple geometric counting algorithm that only fuses hexagonal rings together, but they are not LPAHs. Herndon has calculated the number of alternant isomers, while taking both of these factors into account.

Herndon and Dias (121–137) have been particularly active in the areas of enumeration of LPAH isomers and their classification into families of LPAHs based on empirical formulae or structural features. Herndon and his co workers have calculated the potential for new LPAH structures to be carcinogenic and correlated their structures to their enthalpies.

The concept of a periodic table for the PAHs was covered by Dias in several papers (125–130). Because this class of compounds is only composed of hydrogen atoms (with only 1 single-bond linkage) and carbon atoms (tetravalent with the added constraint of having π-bonding), it is only natural that there are arrangements by carbon number and degree of condensation.

Dias (137) also calculated thermodynamic stabilities for the fully-benzenoid LPAHs, including one study specifically aimed at those of 60 carbons. He later correlated LPAH structural families by finding those with the same numbers of isomers and common structural features (which he called constant-number isomer sets). Randic, Hosoya, and Nakada (138) described how to determine the Clar structure of LPAHs. They focused on how to determine whether a structure was fully-benzenoid.

Rouvray (139–145) has also developed methods to count PAH isomers. He did the calculations to enumerate the LPAHs of six and seven rings. Because the definition to classify the PAHs was ring number and not carbon number, this meant that seven rings included $C_{22}H_{14}$, $C_{24}H_{12}$, $C_{26}H_{14}$, and $C_{28}H_{16}$. He took into account these diradical structures. He also examined the "all-benzenoid" class of LPAHs, those composed only of interlinked aromatic sextets (in the Clar style structure). In two later collaborations, the calculation of spectral characteristics was gained through the calculation of the molecular orbital energies.

Wenchen and Wenjim (146) developed different algorithms for isomer enumeration. They calculated systems up to nine 6-member rings and compared their results to the work of Dias and others. They did not deal with any sterically strained systems or focus only on stable, nonradical species.

Balaban (147) examined five aspects of LPAH behavior through the use of graph theory. These were to explain the reactivity of maleic anhydride for a Diels-Alder reaction and to predict it for new LPAHs; to predict carcinogenic activities; to assess the possibility of new inner-edge structures analogous to kekulene; to model better the LPAHs to assess whether an internal ring current exists for coronene, ovalene, and similar structures; and to easily assess whether a structure is an LPAH, a localized zethene-type hydrocarbon, or a stable free radical.

Computer models do a very good job of assessing steric strain and determining whether a LPAH structure will be planar or nonplanar and how nonplanar the structure may be. For the planar LPAHs without bay, cove, or fjord regions to induce steric strain and nonplanarity, molecular modeling was relatively straightforward. The effects of steric strain on aromaticity and delocalization were accounted for as successive generations of computer modeling programs were developed (148). For example, models yielded two conformational structures for the newly synthesized LPAH tetrabenzo[*a*,*cd*,*f*,*lm*]perylene (*compound 103*) (149). These conformers differed by less than 4 J, indicating that the two were both likely to occur in equilibrium with

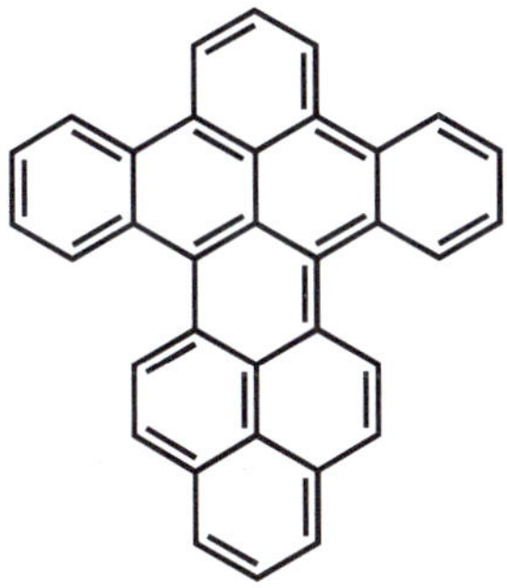

103

each other at room temperature. One conformer reduced the steric strain by twisting around the bridging ring while the second (the slightly less-favored conformer) flexed along the long central axis. The conversion of these forms was the explanation for its anomalous UV absorbance spectrum. Other examples include the matches of computer models with crystallographic data for several other new LPAHs synthesized by Fujisawa and co-workers (150, 151).

Alberty, Chung, and Reif (152) calculated the thermodynamic properties of a large number of peri-fused LPAH isomer groups. The entropy, enthalpy, Gibbs free energies of formation, and heat capacities are calculated for $C_{16}H_{10}$, $C_{22}H_{12}$, and $C_{24}H_{12}$, as well as the first four benzologous series (each formula plus an increment of an ortho-fused ring, C_4H_2). Temperature-dependent values could be calculated using the tabulated coefficients. A few individual isomeric values were calculated, highlighting both that there is some difference between isomers and that these differences are small relative to the absolute values of each property. Somayajulu and Zwolinski (153) calculated the heats of sublimation for several LPAHs. They used a triatomic additivity algorithm, which partitions the energy of the whole molecule into sums of three atom bits (formed through the carbon-carbon and carbon-hydrogen bonds)

Stein (154) has calculated which LPAHs are the most thermodynamically favored for each empirical formula. The result was a series of structures possible in a mechanism with a series through simple one-ring buildup. He also calculated the enthalpies of formation, entropies, and heat capacities of many LPAHs. He then used these to calculate the chemical equilibria and relative stabilities. Two reaction schemes were highlighted. First, a very high temperature equilibrium between the LPAHs and free reactive acetylene units results in a series starting with benzene and continuing through naphthalene, phenanthrene, pyrene, benzo[*e*]pyrene, benzo[*ghi*]perylene, and finally coronene. Each of these is the most stable PAH of its particular carbon number. This was exactly the sequence found by Fetzer and co-workers (155–157) in the catalytic hydrocracking of petroleum and in deep-sea hydrothermal vents.

An alternate buildup route for lower temperatures, also proposed by Stein in that paper, involves sequential additions of benzene rings through formation of a single bond and subsequent fusion to form the next PAH. This series involves benzene, biphenyl, *o*-terphenyl, triphenylene, 4-phenyl-triphenylene, dibenzo-[*fg*,*op*]tetracene (*compound 104*), 4-phenyl-dibenzo[*fg*,*op*]tetracene, and tribenzo[*fg*, *ij*,*rst*]pentaphene (*compound 105*). These calculations showed that there is a very large driving force toward the LPAHs through these production routes because of the very rapidly increasing stability and resonance energies of the larger structures. It is interesting that this second route involves the identical series of most stable LPAH structures that result from Clar's conceptually model of LPAH stabilities that involve counting the fully aromatic sextets (158).

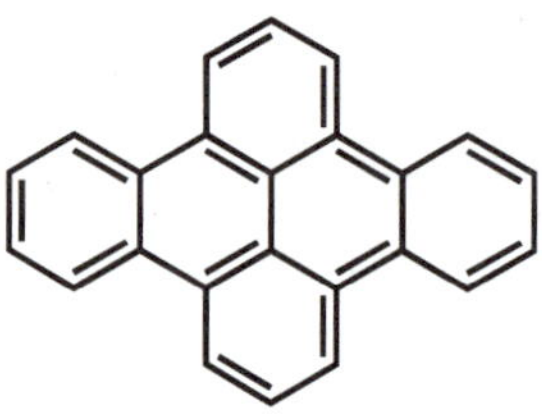

104

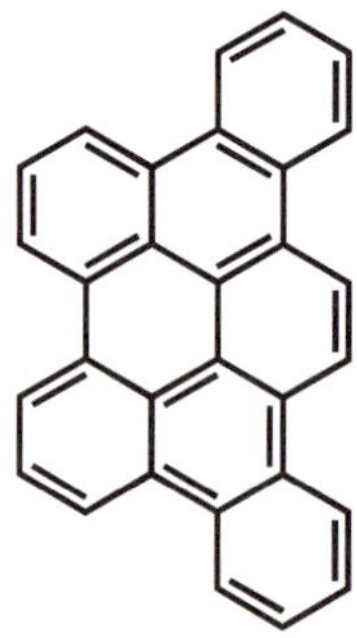

105

Acevedo, Ranaudo, Gutierrez, and Escobar (159) used perturbation molecular orbital calculations to assess the energies of formation of many PAHs. They then used these results to model the asphaltenes in petroleum (these are materials insoluble in saturated hydrocarbon solvents). They concluded that pyrene is more likely than chrysene and that thermodynamics is not the cause for the observed lack of LPAHs in appreciable quantities in petroleum.

As mentioned earlier, the theoretical question of LPAH behavior relative to that of graphite is of interest. Stein and Brown (160) have calculated that there is essentially no difference between the two as far as edge effects and the chemistry and behavior of the peripheral hydrogen atoms on either LPAHs or graphite. In earlier work (161), they had shown that the carbons on the periphery were similar to those in phenanthrene or polyacene structures. The internal carbons were like those in coronene. Reactivities were not dependent on molecule size or the energies of the HOMO.

Cyvin, John, Gutman, Brunvoll, and Cyvin (162) examined the buildup of larger LPAH structures by algorithms that added C_2, C_3H, and C_4H_2 units (the peripheral ring structures that can be fused to core structures). It is interesting to note the parallelism between the graphical structure building, the calculated thermodynamic stabilities and real-world production in petroleum processing and on deep-sea hydrothermal vents.

Agranat and Suissa (163) used topological descriptions and the experimental yields from known reactions to predict the stabilities and yields of several "peropyrene-type" LPAHs. The products of the Clar reaction of phenalenone and other similar ketones were assessed. Four undiscovered isomers were described, with one being very probable as produced in the condensation of benzanthrone. The authors contended that this isomer is unknown because it has not yet been isolated from the complex reaction product mixture.

One of the more active areas in modeling of LPAHs has been the estimation of carcinogenicity. In a pair of papers, Sung (164, 165) examined the K and L regions of a large number of LPAHs. These two regions were defined as being analogous to the 9,10 carbon-carbon bonds in phenanthrene and anthracene. These are the areas of highest electron localization (and therefore highest reactivities) in each of these two compounds. Therefore, for a K region to be defined, the ring structure should have a ring with two adjacent hydrogen atoms (ortho to each other) with high electron density across that bond. An L region is thus a ring with two opposite hydrogen atoms (in the para positions) and high electron density. Sung found no correlation between known PAH carcinogens and those with K or L regions.

In contrast, Zhang, Sannes, Shusterman, and Hansch (166) examined a much larger data set of PAHs and hetero-atom polycyclic compounds (239 compounds vs. ~60). They found some correlation between activity and the presence of K and L regions. They also found that there were correlations with the logarithm of the octanol-water partition coefficient (log K_{ow}) and with the energy of the HOMO. It should be noted that the correlation with the log K_{ow} values probably is only coincidental because of the specific compounds used in the data set. Isomeric PAHs like perylene, benzo[*e*]pyrene, and benzo[*a*]pyrene have similar log K_{ow} values but very different levels of carcinogenicity.

Subach (167) found a correlation between carcinogenic activity and the calculated enthalpies of formation. A data set of 22 randomly-selected PAHs was used, some of which were LPAHs. Of this set, 11 had carcinogenic activity. All of these had relatively high energy values, from 64 to 83 kcal/ mol. Of the noncarcinogenic PAHs, only four had values in this range. Methyl substitution was also found to raise the value of enthalpy of formation, indicating less stability. These values correlated with the known biological activities for the methyl-derivatives of chrysene and benz[*a*]anthracene (tetraphene).

Several molecular descriptors, based on shape and size, have been correlated with PAH carcinogenicity. Miyashita, Okuyama, Yamaura, and Jinno (168) used a combination of bay region and K region to estimate the carcinogenicity of the LPAHs found in diesel particulate. Stigliani, Bonnafous, and Grassy (169) used a set of five structural descriptors to estimate carcinogenicity of 51 PAHs, including a large number of 24- and 26-carbon PAHs. Wang and Wang (170) calculated the molecular surface area and found a correlation. Rachin (171, 172) calculated the levels of carcinogenic activity for several LPAHs with a pyrene substructure and for the LPAHs in diesel exhaust.

Flesher, Horn, and Lehner (173, 174) calculate the mutagenicities of several LPAHs and their benzylic carbonium ions. These calculations used both geometric rules, including the topologies for bay regions, and electron density calculations to estimate electrophilic substitution.

Yan (175) calculated the electron densities of the carbonium ions of 50 PAHs. The delocalization (or localization) was found to correlate to carcinogenic activity.

This wealth of theoretical work is rather surprising because very little actual experimental work on the biological activities of the LPAHs has been performed. The validity has not been shown of extrapolating to the biological activities of LPAHs by using models based on the smaller PAHs. Differences in solubilities, metabolic reactivities, permeation through cell membranes and other tissues, the fit of metabolites into DNA or RNA to form adducts, and other factors may be important and unaccounted for.

REFERENCES

1. J. C. Fetzer, in *Chemical Analysis of Polycyclic Aromatic Compounds*, T. Vo-Dinh, Ed., John Wiley and Sons, New York, 1989, p. 59–109.
2. M. Zander, *Polycycliche Aromaten*, Teubner, Stuttgart, Germany, 1995.
3. J. C. Fetzer and W. R. Biggs, *Org. Prep. Proceed. Int.* **20** 223–230 (1988).
4. W. Schmidt, G. Grimmer, J. Jacob, and G. Dettbarn, *Toxicol. Environ. Chem.* **13** 1–16 (1986).
5. W. Schmidt, G. Grimmer, J. Jacob, G. Dettbarn, and K. W. Naujack, *Fresenius Z. Anal. Chem.* **326** 401–413 (1987).
6. S. A. Wise, B. A. Benner, H. Liu, G. D. Byrd, and A. Colmsjo, *Anal. Chem.* **60** 630–637 (1988).
7. J. C. Fetzer, B. R. T. Simoneit, P. Garrigues, and H. Budzinski, *Polycyclic Arom. Compds.* **9** 109–120 (1996).
8. F. Mikes, G. Boshart, and E. Gil-Av, *J. Chem. Soc. Chem. Commun.* 99–100 (1976).
9. Y. H. Kim, A. Tishbee, and E. Gil-Av, *J. Am. Chem. Soc.* **102** 5915–5917 (1980).
10. I. C. Lewis and T. Edstrom, *J. Org. Chem.* **28** 2050–2057 (1963).
11. S. Obenland and W. Schmidt, in *Polycyclic Aromatic Hydrocarbons and Astrophysics*, A. Leger, L. d'Hendecourt, and B. Boccara, Eds., Reidel, Dordrecht, The Netherlands, 1987, p. 165–172.
12. M. Zander, *Fuel* **74** 305–306 (1995).
13. J. D. Dunitz and A. Gavezzotti, *Acc. Chem. Res.* **32** 677–684 (1999).

14. N. Wakayawa and H. Inokuchi, Bull. *Chem. Soc. Jpn.* **40** 2267–2271 (1967).
15. A. Gavezzotti and G. R. Desiraju, *Acta Cryst. B* **44** 427–434 (1988).
16. S. Iwashima, H. Honda, J. Aoki, and H. Inokuchi, *Mol. Cryst. Liq. Cryst.* **59** 207–219 (1980).
17. K. K. Laali, M. Tanaka, and J. C. Fetzer, *J. Chem. Soc. Perkin Trans.* **2** 1315–1318 (1997).
18. M. Suenaga, Y. Miyahara, N. Shimizu, and T. Inazu, *Angew. Chem. Int. Ed. Engl.* **37** 90–91 (1998).
19. E. Clar, *Polycyclic Hydrocarbons*, John Wiley and Sons, New York, 1964, Vols. 1 and 2.
20. J. R. Dias and K. K. Koehler. *Polycyclic Arom. Cmpds.* **5** 79–86 (1994).
21. D. M. Jewett, in *Chromatography in Petroleum Analysis*, K. H. Altgelt and T. H. Gouw, Eds., Dekker, New York, 1979, p. 281–284.
22. R. G. Harvey, *Polycyclic Aromatic Hydrocarbons*, John Wiley and Sons, New York, 1996.
23. J. C. Fetzer, *Org. Prep. Proced. Int.* **21** 47–65 (1989).
24. F. Vingiello and L. Ojakaar, *Tetrahedron* **22** 847–860 (1966).
25. E. Clar, C. T. Ironside, and M. Zander, *Tetrahedron* **22** 3527–3533 (1966).
26. E. Clar and A. Mullen, *Tetrahedron* **27** 5239–5245 (1971).
27. E. Clar and B. McAndrew, *Tetrahedron* **28** 1137–1142 (1972).
28. German patent application P2914994.8, E. Clar, W. Schmidt, and T. Zauner, April 12, 1979 (to BASF A.-G., Munich, Fed. Rep. Germany).
29. H. J. Lempka, S. Obenland, and W. Schmidt, *J. Chem. Phys.* **96** 349–360 (1985).
30. L. Boente, *Brennstoff Chem.* **36** 210–214 (1955).
31. J. C. Fetzer, *Polycyclic Aromat. Compd.* **11** 317–324 (1996).
32. G. Bluemer, K. Gunderman, and M. Zander, *Chem. Ber.* **109** 1991–2000 (1976).
33. D. Biermann and W. Schmidt, *J. Am. Chem. Soc.* **102** 3163–3173 (1980).
34. D. Biermann and W. Schmidt, *J. Am. Chem. Soc.* **102** 3173–3181 (1980).
35. M. Zander and W. Friedrichsen, *Chemiker-Ztg.* **115** 360–361 (1991).
36. M. Zander, in *Polycyclic Aromatic Compounds: Synthesis, Properties, Analytical Measurements, Occurrence, and Biological Effects*, P. Garrigues and M. Lamotte, Eds., Gordon and Breach Science, Yverdon, Switzerland, 1993, p. 321–330.
37. B. Schlieke, A.-D. Schueter, P. Hauser, and J. Heinze, *Angew. Chem. Int. Ed. Eng.* **36** 1996–1997 (1997).
38. U.S. patent 3,364,274, Jan. 16, 1968, G. Stork and K. Matsuda, (to American Cyanimid).
39. U.S. patent 3,364,275, Jan. 16, 1968, G. Stork and K. Matsuda, (to American Cyanimid).
40. Netherlands patent 7,301,782, Aug. 12, 1974 (to American Cyanamid).
41. J. T. M. van Dijk, A. Hartwijk, A. C. Bleeker, J. Lugtenburg, and J. Cornelisse, *J. Org. Chem.* **61** 1136–1139 (1996).

42. T. Kimura, M. Minabe, and K. Suzuki, *J. Org. Chem.* **43** 1247–1248 (1978).
43. J. C. Fetzer, in *Polynuclear Aromatic Hydrocarbons*, L. Ebert, Ed., Advances in Chemistry 217, American Chemical Society, Washington, DC, 1988, p. 309–331.
44. J. C. Fetzer and W. R. Biggs, *Org. Prep. Proc. Int.* **20** 223–230 (1988).
45. I. Agranat and M. R. Suissa, *Polycyclic Arom. Cmpds.* **3** 51–61 (1992).
46. J. Aoki, M. Takekawa, S. Fujisawa, and S. Iwashima, *Bull. Chem. Soc. Jpn.* **50** 1017–1018 (1977).
47. J. Aoki, M. Takekawa, S. Fujisawa, and S. Iwashima, *J. Org. Chem.* **46** 3922–3923 (1981).
48. S. Fujisawa, I. Oonishi, J. Aoki, Y. Ohashi, and Y. Samada, *Bull. Chem. Soc. Jpn.* **55** 3424–3426 (1982).
49. I. Oonishi, S. Fujisawa, J. Aoki, and Y. Ohashi, *Bull. Chem. Soc. Jpn.* **59** 2233–2236 (1986).
50. I. Oonishi, S. Ohshima, S. Fujisawa, J. Aoki, Y. Ohashi, and T. M. Krygowski, *J. Mol. Struct.* **265** 283–292 (1992).
51. S. Fujisawa, S. Ohshima, A. Uchida, and I. Oonishi, *Polycyclic Arom. Cmpds.* **11** 267–273 (1996).
52. J. H. Li and N. Gotoh, *Yuki Gosei Kagaku Kyokai Shi* **32** 718–722 (1978), in Japanese.
53. U.S. patent 4,278,444, July 14, 1981, R. E. Beyer and J. P. Malgoulas (to Mobil Oil).
54. V. L. Solodor, I. D. Razgonyaeva, and E. Y. Khmel'nitskaya, *Zh. Org. Khim.* **22** 882–883 (1986).
55. S. L. Solodar, I. D. Razgonyaeva, Y. Y., Uman, S. G. Alekseeva, and N. T. Ioffe, *Zh. Obshch. Khim.* **55** 2108–2117 (1985).
56. S. Umemoto, T. Kawashima, Y. Sakata, and S. Misumi, *Tetrahedron Lett.* **12** 1005–1006 (1975).
57. H. Grutzmacher and W. Huseman, *Tetrahedron Lett.* **26** 2431–2434 (1985).
58. L. T. Scott, M. M. Hashemi, D. T. Meyer, and H. B. Warren, *J. Am. Chem. Soc.* **112** 7082–7084 (1991).
59. P. W. Rabideau, A. H. Abdourazak, H. E. Folsom, Z. Marcinow, A. Sygula, and R. Sygula, *J. Am. Chem. Soc.* **116** 7891–7892 (1994).
60. S. Hagen, U. Nuechter, M. Nuechter, and G. Zimmermann, *Tetrahedron Lett.* **35** 7013–7014 (1994).
61. R. Faust, *Angew. Chem. Int. Ed. Eng.* **34** 1429–1432 (1995).
62. S. Hagen, H. Christoph, and G. Zimmerman, *Tetrahedron* **51** 6961–6970 (1995).
63. S. Hagen, U. Neuchter, M. Neuchter, and G. Zimmermann, *Polycyclic Arom. Cmpds.* **4** 209–217 (1995).
64. A. H. Abdourazak, Z. Marcinow, A. Sygula, R. Sygula, and P. W. Rabideau, *J. Am. Chem. Soc.* **117** 6410–6411 (1995).
65. C. Marcinow, F. R. Fronczek, Y.-H. Liu, and P. W. Rabideau, *J. Org. Chem.* **60** 7015–7016 (1995).

66. G. Mehta, K. V. Rao, and K. Ravikumar, *J. Chem. Soc. Perkin Trans.* **1995** 1787–1788.
67. L. T. Scott and A. Necula, *J. Org. Chem.* **61** 386–388 (1996).
68. L. T. Scott, M. S. Bratcher, and S. Hagen, *J. Am. Chem. Soc.* **118** 8743–8744 (1996).
69. S. Hagen and L. T. Scott, *J. Org. Chem.* **61** 7198–7199 (1996).
70. P. W. Rabideau and A. Sygula, *Acc. Chem. Res.* **29** 235–242 (1996).
71. M. D. Clayton and P. W. Rabideau, *Tetrahedron Lett.* **38** 741–744 (1997).
72. S. Hagen, M. S. Bratcher, M. S. Erickson, G. Zimmermann, and L. T. Scott, *Angew. Chem. Int. Ed. Engl.* **56** 406–408 (1997).
73. A. Sygula and P. W. Rabideau, *J. Am. Chem. Soc.* **120** 12666–12667 (1998).
74. M. Sarobe, R. W. A. Havenith, and L. W. Jenneskens, *Chem. Commun.* 1021–1022 (1999).
75. G. Mehta, G. Panda, and P. V. V. Sarma, *Tetrahedron Lett.* **39** 5835–5836 (1999).
76. R. Taylor, G. J. Langley, H. W. Kroto, and D. R. M. Walton, *Nature* **366** 728–731 (1994).
77. C. Crowley, H. W. Kroto, R. Taylor, D. R. M. Walton, M. S. Bratcher, P.-C. Cheng, and L. T. Scott, *Tetrahedron Lett.* **36** 9215–9218 (1995).
78. M. Sarobe, PhD. Thesis, Universiteit Utrecht, 1998.
79. K. F. Lang, H. Buffler, and J. Kalowy, *Monatsh. Chem.* **90** 2888–2893 (1957).
80. K. F. Lang and H. Buffler, *Monatsh. Chem.* **90** 2894–2898 (1957).
81. M. Zander, J. Haase, and H. Dreeskamp, *Erdol Kohle Erdgas Petroch.* **35** 65–69 (1982).
82. A. Zinke, H. Nudmueller, and R. Ott, *Monatsh. Chem.* **86** 853 (1955).
83. R. Ott, E. Zeschko, and A. Zinke, *Monatsh. Chem.* **94** 51–62 (1963).
84. E.Clar, W. Kelly, and R. M. Laird, *Monatsh. Chem.* **87** 391 (1956).
85. R. D. Broeme and F. Dieterich, *Tetrahedron Lett.* **32** 5227–5230 (1991).
86. E. Clar, W. Kelly, J. M. Robertson, and M. G. Rossman, *J. Chem. Soc.* **1956** 3878–3881.
87. E. Clar, J. M. Robertson, R. Schoegl, and W. Schmidt, *J. Am. Chem. Soc.* **103** 1320–1328 (1981).
88. M. S. Erickson and J. Milliken, *Polycyclic Arom. Cmpds.* **8** 1–7 (1996).
89. S. Fujisawa, M. Takekawa, Y. Nakamura, A. Uchida, S. Ohshima, and I. Oonishi, *Proceedings of the Sixteenth International Symposium on Polycyclic Aromatic Compounds*, Charlotte, NC, October 1997.
90. R. G. Harvey, D. T. C. Yang, and C. Yang, *Polycyclic Arom. Cmpds.* **4** 127–133 (1994).
91. W. H. Laarhoven, T. J. H. M. Cuppen, and R. J. F. Nivard, *Tetrahedron* **26** 4865–4881 (1970).
92. A. H. A. Tinnemans and W. H. Laarhoven, *J. Am. Chem. Soc.* **96** 4617–4622 (1974).

93. M. Mueller, H. Mauermann-Duell, M. Wagner, V. Endelmann, and K. Muellen, *Chem. Eur. J.* **1** 1583–1586 (1995).
94. F. Morgenroth, C. Kuebel, and K. Muellen, *J. Mater. Chem.* **1** 1207–1211 (1997).
95. V. S. Iyer, M. Wehmeier, J. D. Brand, M. A. Keegstra, and K. Muellen, *Angew. Chem. Int. Ed. Engl.* **36** 1604–1607 (1997).
96. M. Mueller, V. S. Iyer, C. Kuebel, V. Enkelmann, and K. Muellen, *Angew. Chem. Int. Ed. Engl.* **36** 1607–1610 (1997).
97. M. Mueller, C. Kuebel, F. Morgenrith, V. S. Iyer, and K. Muellen, *Carbon* **36** 827–831 (1998).
98. L. B. Ebert and J. C. Scanlon, in *Polynuclear Aromatic Hydrocarbons*, L. Ebert, Ed., Advances in Chemistry 217, American Chemical Society, Washington, DC, 1988, 1988, p. 367–382.
99. O. W. Adams and R. L. Miller, *J. Am. Chem. Soc.* **88** 404–412 (1966).
100. N. Tyityukov, G. Madjaruva, F. Dietz, and K. Miellen, *J. Phys. Chem. B* **102** 10183–10189 (1998).
101. F. Dietz, N. Tyutyukov, S. Karabunarliev, J. Hristov, and K. Muellen, *Polycyclic Arom. Cmpds.* **13** 241–259 (1999).
102. R. Q. Zhang, E. Bertran, and S.-T. Lee, *Diamond Relat. Mater.* **7** 1663–1668 (1998).
103. A. Bohnen, K.-H. Koch, W. Luettke, and K. Muellen, *Angew. Chem. Int. Ed. Engl.* **29** 525–527 (1990).
104. German patent 3226780, H. Naamann, P. Neumann, and V. Muench, (to BASF A.-G. Munich, Fed. Rep. Germany).
105. S. Obenland and W. Schmidt, in *Polycyclic Aromatic Hydrocarbons and Astrophysics*, A. Leger, L. d'Hendecourt, and B. Boccara, Eds., Reidel, Dordrecht, The Netherlands, 1987, p. 165–172.
106. L. Barnett, D. M. Ho, K. K. Baldridge, and R. A. Pascal, *J. Am. Chem. Soc.* **121** 727–733 (1999).
107. J. R. Dias, in *Handbook of Polycyclic Hydrocarbons, Part A Benzenoid Hydrocarbons*, Elsevier, Amsterdam, 1987.
108. J. R. Dias, in *Handbook of Polycyclic Hydrocarbons Part B Polycyclic Isomers and Heteroatom Analogs of Benzenoid Hydrocarbons*, Elsevier, Amsterdam, 1988.
109. *Advances in the Theory of Benzenoid Hydrocarbons II*, I Gutman, Ed., Springer-Verlag, Berlin, 1992.
110. M. Randic, *Int. J. Quantum Chem.* **17** 549–586 (1980).
111. W. C. Herndon, *Tetrahedron* **40** 3987–3995 (1984).
112. M. Randic, V. Solomon, S. C. Grossman, D. J. Klein, and N. Trinajstic, *Int. J. Quantum Chem.* **32** 35–59 (1987).
113. W. C. Herndon, P. C. Nowak, D. A. Connor, and P. Lin, *J. Am. Chem. Soc.* **114** 41–47 (1992).
114. M. Randic, *New J. Chem.* **20** 1001–1009 (1996).

115. W. Simonsick and R. Hites, in *Calculated Molecular Properties of Polycyclic Aromatic Hydrocarbons*, Elsevier, Amsterdam, 1987.
116. W. Simonsick and R. Hites, in *Polynuclear Aromatic Hydrocarbons: Eighth International Symposium on Mechanisms, Methods, and Metabolism*, M. W. Cooks and A. J. Dennis, Eds., Batelle, Columbus, OH, 1985, p. 1227–1237.
117. W. Simonsick and R. Hites, *Anal. Chem.* **56** 2749–2754 (1984).
118. W. Simonsick and R. Hites, *Anal. Chem.*, **58** 2114 (1986).
119. E. Clar, *The Aromatic Sextet*, Wiley-Interscience, New York, 1972, p. 103–119.
120. E. Clar and C. C. MacKay, *Tetrahedron* **28** 6041–6047 (1972).
121. W. C. Herndon, *J. Am. Chem. Soc.* **112** 4546–4547 (1990).
122. W. C. Herndon and L. V. Szentpaly, *J. Mol. Struct.* **148** 141–152 (1986).
123. W. C. Herndon, D. A. Conner, and P. Lin, *Pure Appl. Chem.* **62** 435–444 (1990).
124. J. R. Dias, *J. Chem. Inf. Comput. Sci.* **24** 124–135 (1984).
125. J. R. Dias, *Can. J. Chem.* **62** 2914–2922 (1984).
126. J. R. Dias, *Acc. Chem. Res.* **18** 241–248 (1985).
127. J. R. Dias, *J. Mol. Struct.* **137** 9–29 (1986).
128. J. R. Dias, *Thermochim. Acta* **122** 313–337 (1987).
129. J. R. Dias, *J. Mol. Struct.* **185** 57–81 (1989).
130. J. R. Dias, *J. Chem. Inf. Comput. Sci.* **30** 251–256 (1990).
131. J. R. Dias, *J. Phys. Org. Chem.* **3** 765–783 (1990).
132. J. R. Dias, *Z. Naturforsch. A* **45** 1335–1341 (1990).
133. J. R. Dias, *J. Chem. Inf. Comput. Sci.* **31** 89–96 (1991).
134. J. R. Dias, *Polycyclic Arom. Cmpds.* **4** 87–106 (1994).
135. J. R. Dias, *Chem. Phys. Lett.* **176** 559–562 (1991).
136. J. R. Dias, *Struct. Chem.* **3** 389–398 (1992).
137. J. R. Dias, *J. Chem. Inf. Comput. Sci.* **33** (1993).
138. M. Randic, H. Hosoya, and K. Nakada, *Polycyclic Arom. Cmpds.* **4** 249–269 (1995).
139. D. H. Rouvray, *S. Afr. J. Sci.* **71** 106–110 (1975).
140. W. E. Harris, S. L. Thomas, and D. H. Rouvray, *S. Afr. J. Sci.* **71** 309 (1975).
141. D. H. Rouvray, *S. Afr. Chem. Chron.* **4** 39–49 (1975).
142. D. H. Rouvray, *S. Afr. Chem. Chron.* **5** 191–197 (1976).
143. O. E. Polansky and D. H. Rouvray, *MATCH* **2** 91–106 (1976).
144. I. Gutman and D. H. Rouvray, *Chem. Phys. Lett.* **62** 384–388 (1979).
145. D. H. Rouvray and S. El-Basil, *THEOCHEM* **42** 9–20 (1988).
146. H. Wenchen and H. Wenjim, *Tetrahedron* **42** 3291–3299 (1986).
147. A. Balaban, *Pure Appl. Chem.* **54** 1075–1096 (1982).
148. W. C. Herndon, in *Polynuclear Aromatic Hydrocarbons*, L. Ebert, Ed., Advances in Chemistry 217, American Chemical Society, Washington, DC, 1988, p. 1–12.
149. J. C. Fetzer and W. Schmidt, *Spectrochim. Acta A* **45A** 503–505 (1989).

150. I. Oonishi, S. Fujisawa, J. Aoki, and T. Danno, *Bull. Chem. Soc. Jpn.* **51** 2256–2260 (1978).
151. S. Fujisawa, I. Oonishi, J. Aoki, and Y. Ohashi, *Acta Crystallogr.* **C42** 1390–1392 (1986).
152. R. Alberty, M. B. Chung, and A. K. Reif, *J. Phys. Chem. Ref. Data* **18** 77–109 (1989).
153. G. R. Somayajulu and B. J. Zwolinski, *J. Chem. Soc. Faraday Trans.* **2** 70 1928–1941 (1974).
154. S. Stein, *J. Phys. Chem.* **82** 866–871 (1978).
155. R. F. Sullivan, M. M. Boduszynski, and J. C. Fetzer, *Energy Fuels* **3** 603–612 (1989).
156. J. C. Fetzer, in *Catalysts in Petroleum and Petrochemical Processing 1995*, M. Absi-Halabi, J. Besharah, H. Qabazard, and A. Stanislaus, Eds., Elsevier, 1996, p. 263–271.
157. B. R. T. Simoneit and J. C. Fetzer, in *Organic Geochemistry: Developments and Applications to Energy, Climate, Environment and Human History*, J. O. Grimalt and C. Dorronsoro, Eds., A. I. G. O. A., Donostia, San Sebastian, Spain, 1995, p. 414–417.
158. E. Clar, *The Aromatic Sextet*, Wiley-Interscience, New York, 1972.
159. S. Acevedo, M. A. Ranaudo, L. B. Gutierrez, and G. Escobar, *Fuel* **75** 1139–1144 (1996).
160. S. E. Stein and R. L. Brown, *J. Am. Chem. Soc.* **113** 787–793 (1991).
161. S. E. Stein and R. L. Brown, *J. Am. Chem. Soc.* **109** 3721–3729 (1987).
162. S. J. Cyvin, P. E. John, I. Gutman, J. Brunvoll, and B. N. Cyvin, *J. Mol. Struct.* **334** 243–247 (1995).
163. I. Agranat and M. R. Suissa, *Polycyclic Arom. Cmpds.* **3** 51–61 (1992).
164. S.-S. Sung, *C. R. Acad. Sci. Ser. D* **273** 1247–1250 (1971).
165. S.-S. Sung, *C. R. Acad. Sci., Ser. D* **274** 1597–1600 (1972).
166. L. Zhang, K. Sannes, A. J. Shusterman, and C. Hansch, *Che.-Biol. Interact.* **81** 149–180 (1992).
167. D. J. Subach, *Health Phys.* **27** 403–404 (1974).
168. Y. Miyashita, T. Okuyama, K. Yamaura, and K. Jinno, *Anal. Chim. Acta* **202** 237–240 (1987).
169. J. L. Stigliani, M. Bonnafous, and G. Grassy, *Eur. J. Med. Chem.* **25** 157–170 (1990), in French.
170. L. Wang and X. Wang, in *Recent Advances in Chemistry and Molecular Biology for Cancer Research*, Q. Dai, M.-A. Armour, and Q. Zheng, Eds., Science Press, Beijing, 1993, p. 271–275.
171. E. I. Rachin, *Bulg. Chem. Commun.* **28** 344–352 (1995).
172. E. I. Rachin, *Bulg. Chem. Commun.* **28** 353–357 (1995).
173. J. W. Flesher, J. Horn, and A. F. Lehner, *THEOCHEM* **362** 29–49 (1996).
174. A. F. Lehner, J. Horn, and J. W. Flesher, *THEOCHEM* **366** 203–217 (1996).
175. L. S. Yan, *Carcinogenesis (London)* **6** 1–6 (1985).

CHAPTER

3

ULTRAVIOLET/VISIBLE ABSORPTION SPECTROSCOPY

3.1 GENERAL CHARACTERISTICS OF LPAH ABSORBANCE SPECTRA

One of the most striking properties of the PAHs that sets them apart from other classes of organic compounds is the nature of their UV/visible absorbance spectra. When the absorbance spectrum of a solution of any PAH is examined, several observations are made (1, 2).

First, the spectral response is extremely high. The probability of incident photons interacting with the PAH's π electrons is very high. This results in very large molar absorptivity values. LPAHs with molar absorptivities of greater than 1 million have been reported.

The various transitions in the UV absorbance spectrum of a LPAH can be classified by the electronic transition that causes them. For this discussion, the terminology of Clar will be used. He identified three main types of transitions and designated them the α, β, and ρ bands. The α band is the highest wavelength of the three and corresponds to the transition from the second highest occupied molecular orbital to the lowest unoccupied molecular orbital (LUMO). This is also known as the 0-0 transition. The β band is the lowest wavelength of the three and corresponds to a transition from the highest occupied molecular orbital (HOMO) to the second lowest unoccupied orbital. The ρ band is of intermediate wavelength and corresponds to a transition from the HOMO to the LUMO. The fine structure in many of the spectra of LPAHs, shown by the clusters of adjacent bands of regularly decreasing wavelengths and intensities, arises from transitions involving the different vibrational states of the ground and excited states of the molecules.

Second, there are usually numerous bands in the absorbance spectrum of a LPAH. These bands generally occur in groups. For example, the two spectra of the 11-ring LPAH isomers dinaphtho[8,1,2-*abc*:8′,1′,2′-*jkl*]coronene and dinaphtho[8,1,2-*abc*:2′,1′,8′-*jkm*]coronene (*compounds 106* and *107*) are shown in Figure 16. Each group of bands corresponds to a specific set of π electron energy transitions. The transitions can be ascribed to certain structural features

Large (C>=24) Polycyclic Aromatic Hydrocarbons: Chemistry and Analysis, by John C. Fetzer.
ISBN 0-471-36354-5

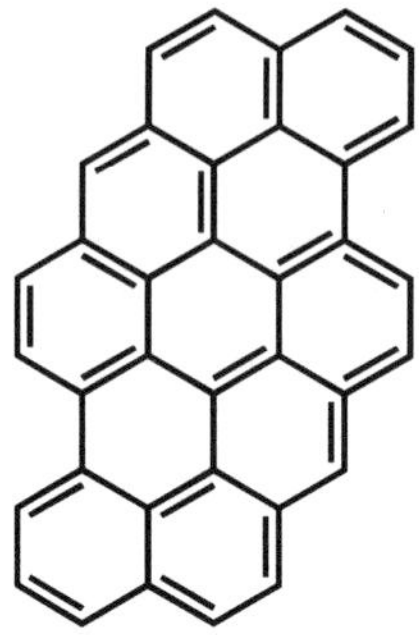

106

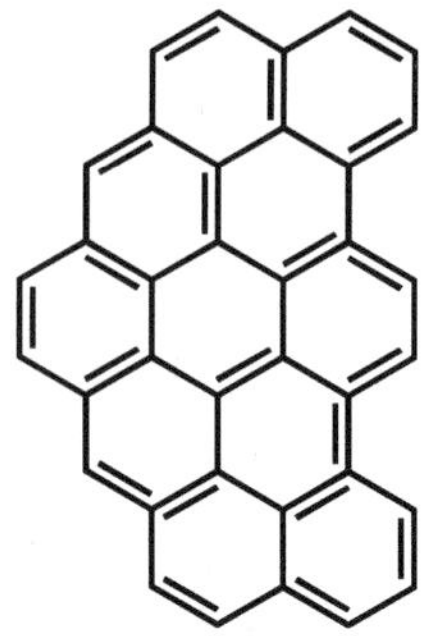

107

of the overall LPAH structure or to substructures within it. The energy of the transition, and thus the wavelength of the absorbance band, is determined by the localization of the π electrons in the LPAH.

The spectra of these two LPAHs also show another very important facet of PAH absorbance spectra. When a set of isomers are compared, drastic differences are seen, even for somewhat similar structures. The spectrum for each isomer is different from any of the others, and in many cases, very drastically so. The arrangement of π electrons in a PAH is dominated by the overall shape of the molecule (an approximate analogy is that the electrons are the "particles in a box" of very differently shaped boxes). Isomeric PAHs, by definition, will have different shapes. This difference in structure will lead to very dissimilar spectra for isomers that structurally appear to be almost identical. A discussion on annellation theory and the use of UV spectra in structural determination is given in the chapter on determining LPAH structures.

Ortho-fused PAHs yield spectra characterized by a very intense low-wavelength band with several smaller bands at higher wavelengths. The low-wavelength band results from absorption of energy due to π electrons localized in only one ring of the structure. Because an ortho-fused LPAH has many of these resonance structures and no others, this band is intense. The higher-wavelength bands result from interactions with resonance structures with conjugation of isolated double bonds and an aromatic ring.

Peri-fused PAHs have extremely rich spectra characterized by clusters of bands spread throughout a wide spectral range. In many instances, a series of bands is seen. These clusters of bands are evenly spaced and increase in intensity with higher wavelengths. These clusters arise from absorption of the incident light by ground-state molecules being at thermal equilibrium (and thus at slightly different energy levels). The transition to the first excited state, therefore, can be several discrete energies. The most probable is the lowest-energy transition or highest-wavelength absorbance band.

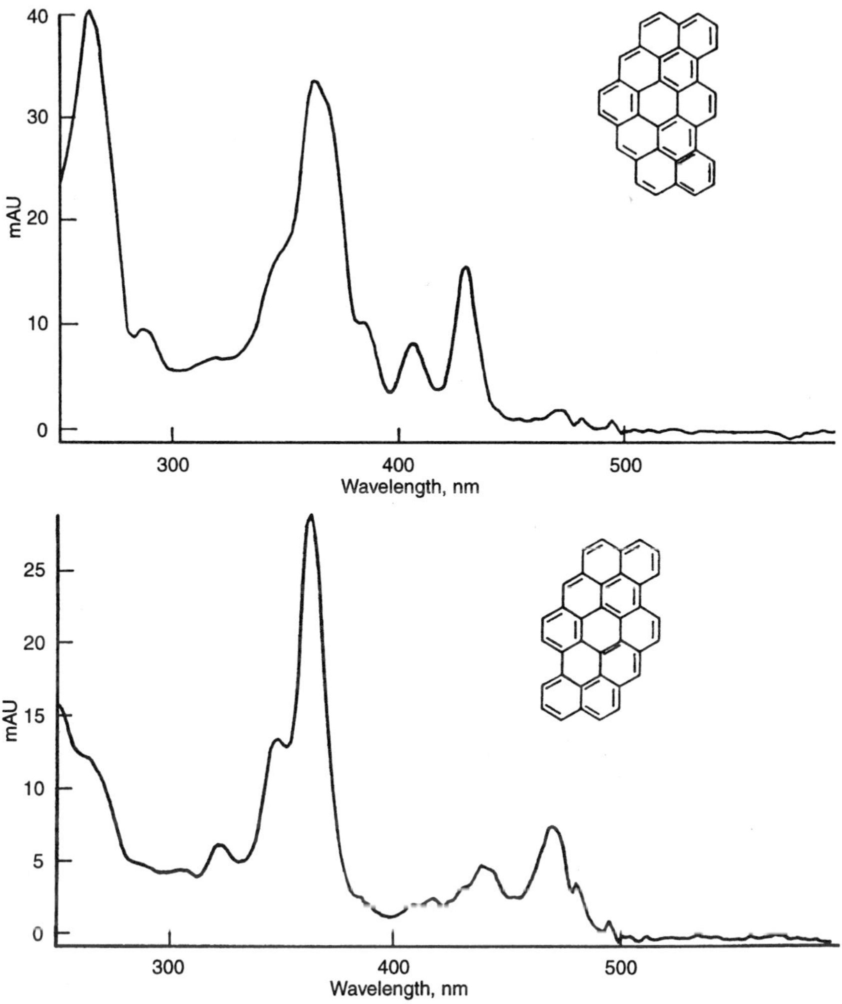

Figure 16. Example of LPAH UV absorbance spectra of the 11-ring LPAH isomers dinaphtho[8,1,2-*abc*:8′*m*1′,2′-*jkl*]coronene and dinaphtho[8,1,2-*abc*:8′,1′,2′-*ijk*]coronene.

The band positions of alternant LPAHs are shifted slightly by changes in solvent. If the spectra in weak solvents, such as the alkanes, are compared with stronger solvents, there is a red shift when going from weak to strong. This usually amounts to only about 5 or fewer nanometers. An example for the 31 carbon dibenzo[*a*,*rst*]naphtho[8,1,2-*cde*]pentaphene is shown in Figure 17. The band shape differences are caused by solvent-induced small changes in the nonplanarity of this molecule. For compounds with weak 0-0 transitions, i.e.,

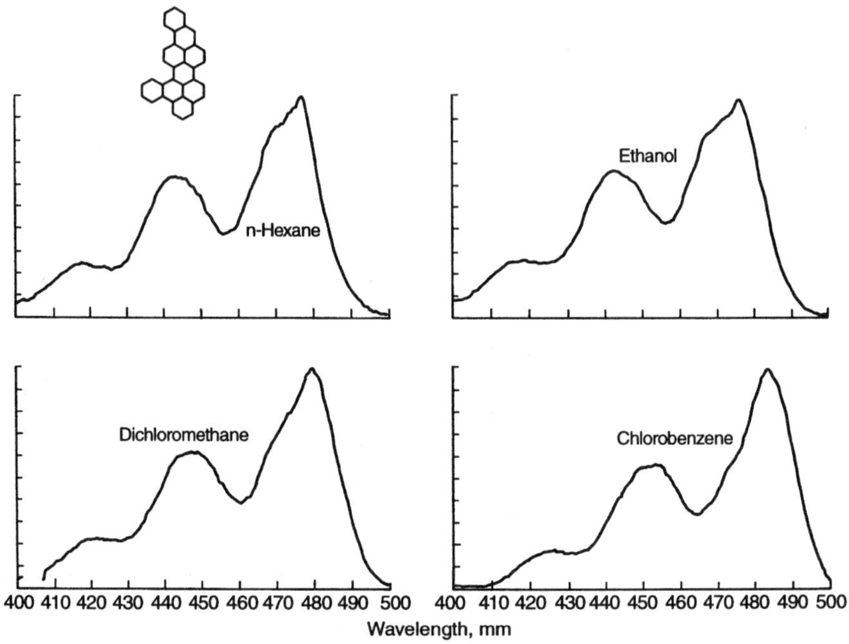

Figure 17. The solvent-induced shifts in location and band shape in the spectra of dibenzo[*a,rst*]-naphtho[8,1,2-*cde*]pentaphene.

an α band intensity of <1000, changes in solvent can have drastic changes in intensity of this band. Obenland and Schmidt reported that the α band of tribenzo[*a,g,m*]coronene increased from a molar absorptivity of 80 in 1,4-dioxane and of 160 in 1,2,4 trichloroenzene, Its location also shifts from 427 nm to 430 nm (3). The β and ρ bands shift ~4 nm but do not change appreciably.

Another type of red shift is commonly observed for the LPAHs. Alkyl substitution shifts the spectra slightly to higher wavelengths. The relative band positions and intensities do not change very much, leaving the same spectral pattern for an LPAH and its alkyl derivatives. An alkyl group attachment, such as a methyl, ethyl, or longer-chain substituent, usually causes a 1- or 2-nm shift. The effect is somewhat additive. The position of substitution can also cause minor differences in the magnitude of the shift. Saturated-ring substituents have a much greater effect. Each saturated ring moves the spectrum upward between 5 and 10 nm. An example of this effect, with the PAH core pyrene, is shown in Figure 18. This series was chosen because no long-chain or saturated-ring substituted standards are available for any LPAH. Fromhers, Thaler, and Wolf report an extreme example of this shift by saturated rings (4). In their report on

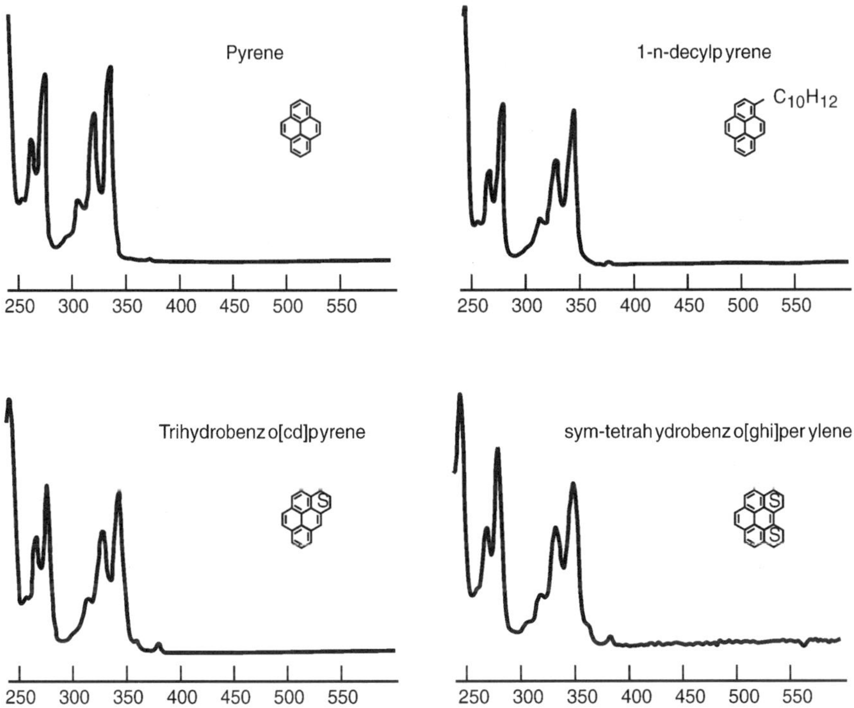

Figure 18. The UV absorbance spectral similarity for the pyrene family of PAHs.

the synthesis of dinaphtho[8,1,2-*abc*:2′,1′,8′-*klm*]coronene, they also reported the results of hydrogenating this LPAH. Many of the products were partially hydrogenated species. The decahydro derivative has a coronene aromatic structure. The UV spectrum of this compound has the same pattern as that of coronene but is shifted ~12–13 nm to the red. The octadecahydro derivative, with a pyrene core, was shifted 25–26 nm higher than pyrene but retained the spectral pattern.

Snyder and Saunders report a method to quantitate alkyl coronenes in processed petroleum oils (5). It uses an adsorption chromatography fractionation. The coronene fraction is collected in 12% (v/v) dichloromethane in n-pentane. Absorbances at 301, 307, and 313 nm determine the various types of coronene and an absorbance at 385 nm corrects for any interferences due to benzo[*ghi*]perylene.

Nonalternant PAHs generally have less defined spectra, with broad bands. When a five-member ring is found in a LPAH structure, not as many Kekule structures can be drawn. Therefore, the π electrons are much more localized. A comparison example is shown from the work of Zander and Friedrichsen of the spectra of the two 15-ring structures that are formed through condensation of two coronene molecules, benzo[1,2,3-*bc*:4,5,6-*b′c′*]dicoronene and cyclopenta[1,2-*a*:3,4,5-*b′c′*]dicoronene (6). Each is formed from two coronene molecules being joined by formation of a bridging ring. The alternant species (*compound 53*) has a highest-wavelength absorbance band at 504 nm in 1,2,4-trichlorobenzene. The nonalternant version, bridged with a five-member ring (*compound 108*), has a maximum at 494 nm, and its spectral bands are much less

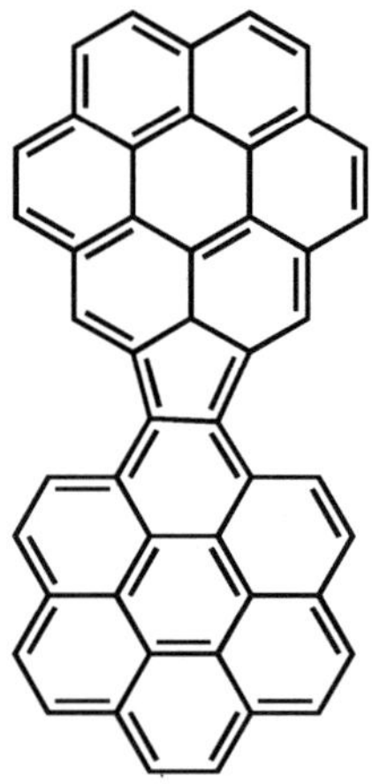

108

intense for the series of higher-wavelength bands with molar absorptivity values of less than one-fourth as strong (the estimated fluorescence yields for the two are 0.81 and 0.13). The spectra of these two structures are shown in Figure 19.

The spectra of nonplanar, alternant, peri-fused LPAHs show a characteristic feature. The basic structure is dibenzo[*cd,lm*]perylene (peropyrene). The successive addition of fused rings to the side faces of this core structure yield the spectra seen in Figures 20–22. The valleys between the spectral bands of planar LPAHs are very deep, so that the bands themselves are very narrow. The valleys for nonplanar structures are broader and the valleys in between are shallow, so that the bands appear to be rounder and less distinct. The nonplanar structures are not as rigid as the planar structures. This increased flexibility leads to a wider distribution of energies for the ground state and a spreading of the absorbance bands (15–17). Another example of the contrast between isomers is found in the second LPAH in this series, tribenzo[*a,cd,lm*]perylene, and its isomer benzo[*rst*]naphtho[8,1,2-*cde*]pentaphene (originally produced in the same Clar condensation reaction. Figure 23 shows the spectra of this isomer, which is totally different than the upper spectrum in Figure 20.

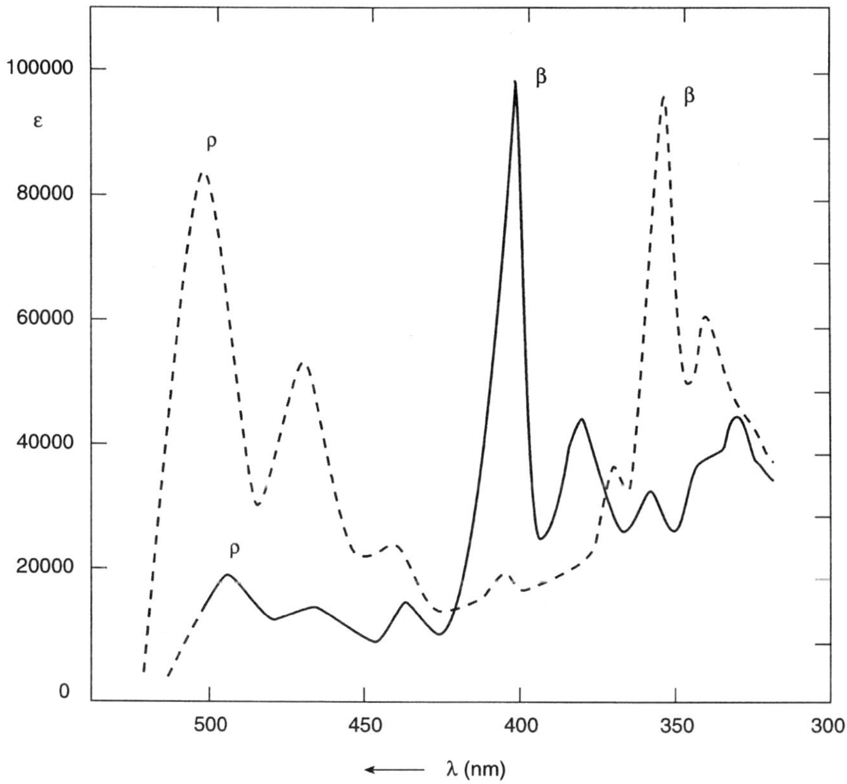

Figure 19. Comparison of UV absorbance spectra for the nonalternant (solid line) and alternant (dashed line) bridged dicoronene LPAHs. From M. Zander and W. Friedrichsen, *Z. Naturforsc. A Phys. Sci.* **47** 1315, 1316 (1992).

Another area made possible by Clar's huge collection of LPAHs and absorbance spectra was the creation of a theory connecting the two. Clar called this the "annellation theory." It described the relationship of structural features and other similarities in LPAH structure and how they were reflected as patterns in the absorbance spectra. Clar even extended the theory to predict the absor bance spectra of some unknown LPAHs that his research group later synthesized. Annellation theory is discussed in greater detail in the chapter on methods for LPAH structure determination.

One interesting consequence of the extended conjugation of the π electron system in LPAHs is their color. Many LPAHs are yellow, orange, or red because of their absorbance at high wavelengths. If the absorbance bands have higher wavelengths, the LPAHs may even be bright blue or green. The benzologous

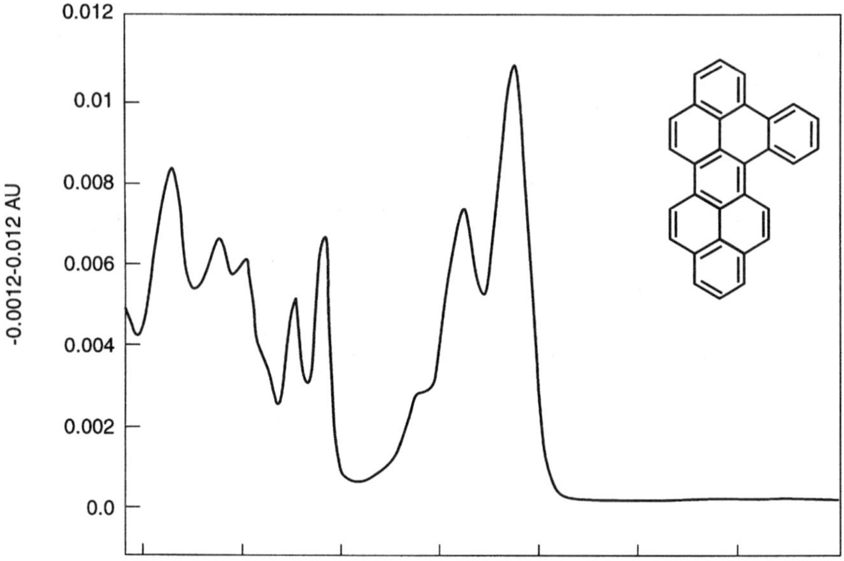

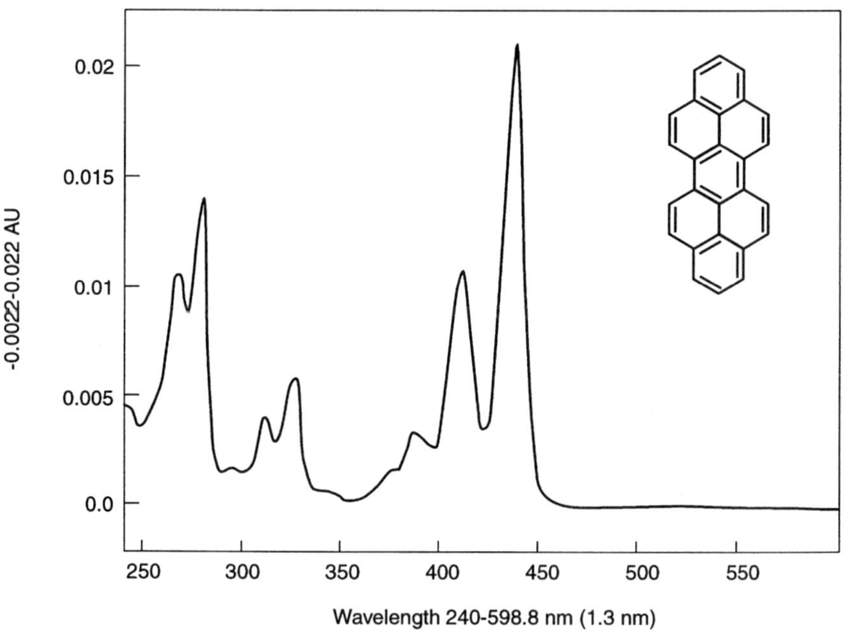

Figures 20, 21, and 22. The series of UV spectra of dibenzo[*cd*,*lm*]perylene (peropyrene) and some of its benzologues. Note the shallower valleys between the bands due to increased steric hindrance and more nonplanarity.

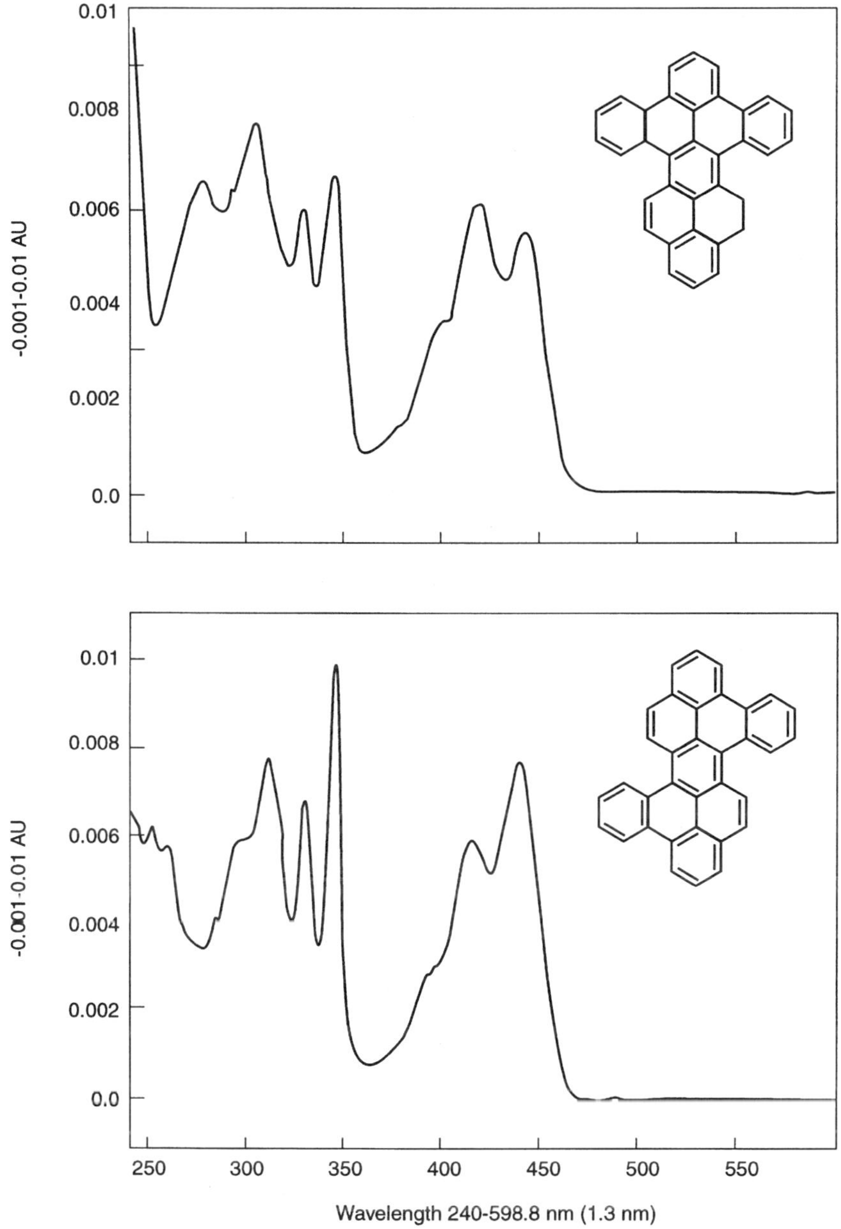
0.01
0.008
0.006
0.004
0.002
0.0
-0.001-0.01 AU
0.01
0.008
0.006
0.004
0.002
0.0
-0.001-0.01 AU
250
300
350
400
450
500
550
Wavelength 240-598.8 nm (1.3 nm)

Figures 21

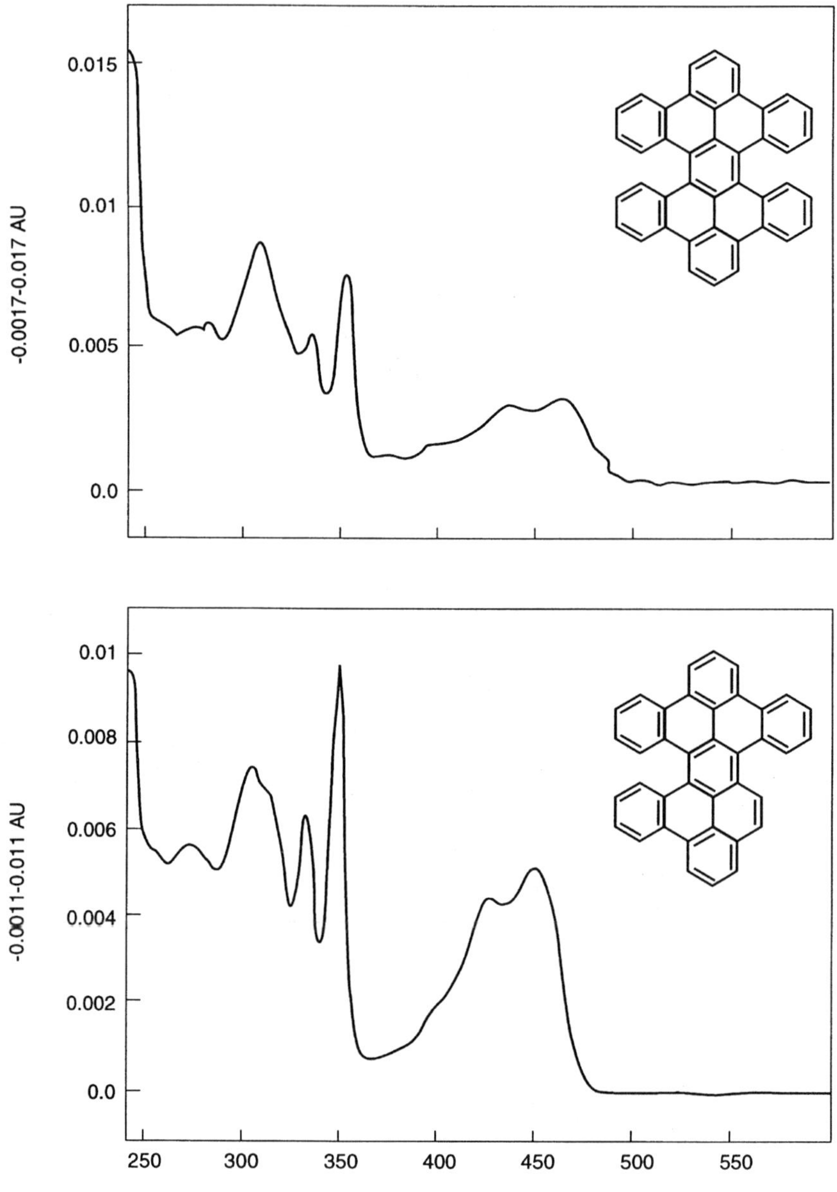
0.015
0.01
0.005
0.0
-0.0017-0.017 AU
0.01
0.008
0.006
0.004
0.002
0.0
-0.0011-0.011 AU
250
300
350
400
450
500
550
Wavelength 240-598.8 nm (1.3 nm)

Figures 22

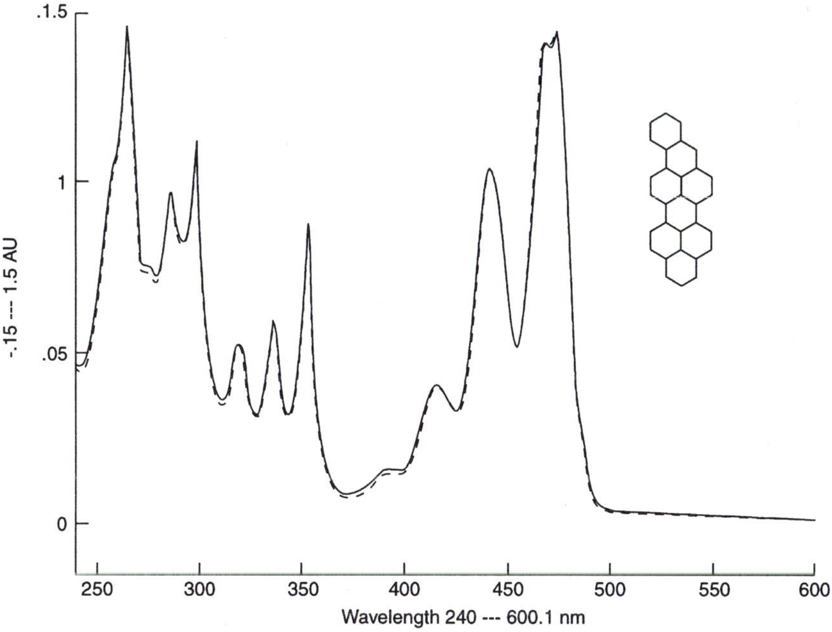

Figure 23. The UV absorbance spectrum of benzo[*rst*]naphtho[8,1,2-*cde*]pentaphene.

series of tribenzo[*de*,*kl*.*rst*]pentaphene (*compound 70*), dibenzo[*de*,*kl*]naphtho-[1,2,3,4-*rst*]pentaphene (*compound 109*), and tribenzo[*de*,*h*,*kl*]naphtho[1,2,3,4-*rst*]pentaphene (*compound 110*) are red, green, and blue. These colors result from absorbance bands of > 600 nm. The corresponding fluorescence emissions from these bands are at even higher wavelengths and can be orange, yellow or red.

109

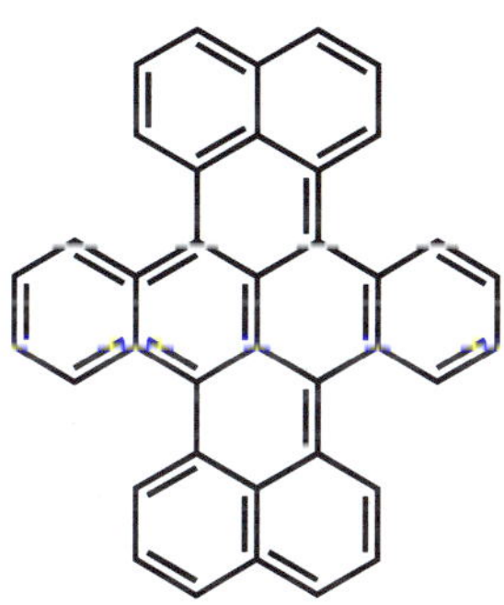

110

The best spectral atlas of the absorbance spectra of LPAHs is still Eric Clar's two-volume set *Polycyclic Hydrocarbons*, even though it was published well over three decades ago. Its continuing utility is due to the fact that it contains an extremely large number of LPAH spectra that could be collected because of Clar's extensive synthesis efforts. His huge collection of LPAHs made spectral collection possible. Some newer spectral atlases contain spectra of certain LPAHs not found in Clar, but no newer comprehensive atlas has ever been published.

In addition to all the LPAH spectra found in Clar's volumes, many individual papers have reported spectra of a variety of newly-synthesized LPAHs (7–18). The scarcity of these standard compounds, and the difficulties in reproducing the syntheses and isolations necessary to obtain pure samples, has made it virtually impossible for anyone today to collect a large set of spectra in the fashion that Clar did. Table 2 includes references that report UV absorbance spectra of more recently discovered peri-fused LPAHs. The spectra of many ortho-fused LPAHs have also been published, but these were not included because their spectra are less distinct and less definitive.

TABLE 2. New UV Spectra

dibenzo[*def,p*]chrysene $C_{24}H_{14}$

Spectral Atlas of Polycyclic Aromatic Compounds, W. Karcher, R. J. Fordham, J. J. Dubois, P. D. J. M. Glaude, and J. A. M. Ligthart, Eds., Reidel, Dordrecht, The Netherlands, 1983, Vol. 1, p. 778–779., The spectrum for this structure in Clar is really that of dibenzo[*b,f*]fluoranthene, see E. Clar, J. M. Robertson, M. Schloegl, and W. Schmidt, *J. Am. Chem. Soc.* **103** 1320–1328 (1981).

tribenzo[*b,def,p*]chrysene $C_{28}H_{16}$

G.-P. Blumer, K.-D. Gundermann, and M. Zander, *Chem. Berichte* **109** 1881–2000 (1976).

phenaleno[1,2,3,4,5-*stua*]picene $C_{28}H_{14}$

A. H. A. Tinnemans and W. H. Laarhoven, *J. Am. Chem. Soc.* **96** 4617–4622 (1974).

tribenzo[*a,g,m*]coronene $C_{36}H_{18}$

S. Obenland and W. Schmidt, in *Polycyclic Aromatic Hydrocarbons and Astrophysics*, A. Leger, L. d'Hendecourt, and B. Boccara, Eds. Reidel, Dordrecht, The Netherlands, 1987, p. 165–172.

naphtho[8,1,2-*abc*]coronene $C_{30}H_{14}$

tribenzo[*a,cd,lm*]perylene $C_{30}H_{16}$

TABLE 2. (*Continued*)

benzo[*rst*]naphtho[8,1,2-*cde*]pentaphene $C_{30}H_{16}$; misidentified in this paper as phenaleno[1,9-*ab*]perylene, but later shown to be this structure, J. C. Fetzer, *Polycyclic Arom. Cmpds* (in press). Figure 23 in this chapter

dibenzo[*j,lm*]naphtho[1,8-*ab*]perylene $C_{34}H_{18}$

J. C. Fetzer and W. R. Biggs, *J. Chromatogr.* **322** 275–286 (1985).

phenanthro[5,4,3,2-*efghi*]perylene $C_{28}H_{14}$

J. C. Fetzer and W. R. Biggs, *Polycyclic Arom. Cmpds.* **5** 197 (1994).

dibenzo[*cd,k*]naphtho[8,1,2,3-*fghi*]perylene $C_{32}H_{16}$

J. C. Fetzer and W. R. Biggs, *J. Chromatogr.* **642** 319–327 (1993).

benzo[*e*]indeno[1,2,3-*hi*]acephenanthrylene $C_{26}H_{14}$

benz[*g*]indeno[1,2,3,4-*mnop*]chrysene $C_{26}H_{14}$

diindeno[1,2,3,4-*defg*:1′,2′,3′,4′-*mnop*]chrysene $C_{26}H_{12}$

S. Hagen, U. Neuchter, M. Neuchter, and G. Zimmermann, *Polycyclic Arom. Cmpds.* **4** 209–217 (1995).

tetrabenzo[*a,cd,f,lm*]perylene $C_{34}H_{18}$

J. C. Fetzer and W. Schmidt, *Spectrochim. Acta A* **45A**, 503–505 (1989).

benzo[*rst*]dinaphtho[8,1,2-*cde*:2′,1′,8′-*klm*]pentaphene $C_{36}H_{18}$

S. Umemoto, T. Kawashima, Y. Sakata, and S. Misumi, *Tetrahedron Lett.* **12** 1005–1006 (1975).

naphtho[8,1,2-*hij*]hexaphene $C_{32}H_{18}$

M. Zander and W. Franke, *Chem. Ber.* **94** 446–450 (1961).

dibenzo[*b,pqr*]perylene $C_{26}H_{14}$

benzo[*uv*]naphtho[2,1,8,7-*defg*]pentacene $C_{30}H_{16}$

M. Zander and W. H. Franke, *Chem. Ber.* **99** 1275–1279 (1966).

phenanthro[3,4,5,6-*bcdef*]ovalene $C_{40}H_{16}$

R. D. Broeme and F. Dieterich, *Tetrahedron Lett.* **32** 5227–5230 (1991).

benzo[*lm*]phenanthro[1,8-*ab*]perylene $C_{34}H_{18}$

pentabenzo[*a,cd,f,j,lm*]perylene $C_{38}H_{20}$

tribenzo[*fg,ij,m*]naphtho[8,1,2-*opq*]pentaphene $C_{38}H_{20}$

TABLE 2. (*Continued*)

dibenzo[*a,rst*]phenanthro[1.10.9-*cde*]pentaphene $C_{38}H_{20}$

dibenzo[*fg,ij*]benz[3,4]phenanthro[2,1,10,9,8,7-*pqrstuv*]pentaphene $C_{38}H_{18}$

dibenzo[*fg,ij*]benz[1.9]anthra[2,3,4,10,5-*opqrst*]pentaphene $C_{38}H_{18}$

tribenzo[*j,lm,o*]phenanthro[5,4,3-*abcd*]perylene $C_{40}H_{20}$

benzo[*a*]naphtho[2′,1′,8′,7′,6′-*pqrst*]phenanthro[1,10,9-*cde*]pentaphene $C_{40}H_{20}$

benzo[11,12]naphthaceno[4,5,6-*abcd*]coronene $C_{40}H_{18}$

dibenz[*fg,ij*]anthra[1,9,8,7-*qrstu*]hexaphene $C_{40}H_{20}$

benzo[*kl*]dinaphtho[1,2,3-*rs*:3′,2′,1′,8′,7′-*vwxyz*]hexaphene $C_{40}H_{20}$

tribenzo[*fg,ij,mn*]naphtho[2,1,8,7-*rstu*]hexaphene $C_{40}H_{20}$

J. C. Fetzer, in *Polynuclear Aromatic Hydrocarbons*, L. Ebert, Ed., Advances in Chemistry 217, American Chemical Society, Washington, DC, 1988, p. 309–331.

benzo[*lm*]phenanthro[4,5,6-*abcd*]perylene $C_{32}H_{16}$

dibenzo[*j,lm*]phenanthro[4,5,6-*abcd*]perylene $C_{36}H_{18}$

dibenzo[*rs,vwx*]naphtho[2,1,8,7-*klmn*]hexaphene $C_{36}H_{18}$

benzo[*rst*]pyreno[3,4,5-*cde*]pentaphene $C_{36}H_{18}$

tribenzo[*jk,qr,uv*]naphtho[2,1,8,7-*defg*]pentacene $C_{36}H_{18}$

benz[2,10]anthra[1,9,8-*abcd*]coronene $C_{36}H_{16}$

naph[7′,8′,1′,2′,3′-*vwxyz*]anthra[2,1,9,8-*klmno*]hexaphene $C_{38}H_{18}$

J. C. Fetzer and W. R. Biggs, *Org. Prep. Proc. Int.* **20** 223–230 (1988).

cyclopenta[1,2-*a*:3,4,5-*b′c′*]dicoronene $C_{48}H_{20}$

benzo[1,2,3-*bc*:4,5,6-*b′c′*]dicoronene

M. Zander and W. Friedrichsen, *Z. Naturforsc. A Phys. Sci.* **47** 1314–1318 (1992).

tetrabenzo[*de,hi,mn,qr*]tetracene $C_{30}H_{16}$

benzo[*p*]naphtho[2,1,8-*ghi*]chrysene $C_{28}H_{16}$

dibenzo[*a,k*]naphtho[1,2,3,4-*ghi*]perylene $C_{34}H_{18}$

dibenzo[*de, xy*]naphtho[8,1,2-*ghi*]hexaphene $C_{38}H_{20}$

E. Clar, J. F. Guye-Vuilleme, and J. F. Stephen, *Tetrahedron* **20** 2107–2117 (1964).

hexabenzo[*a,d,g,j,m,p*]coronene $C_{48}H_{24}$

E, Clar and J. F. Stephen, *Tetrahedron* **21** 467–470 (1965).

TABLE 2. (*Continued*)

benzo[*st*]naphtho[2,1,8,7-*defg*]pentacene $C_{30}H_{16}$

benzo[*wx*]naphtho[2,1,8,7-*defg*]pentacene $C_{30}H_{16}$

dibenzo[*fg*,*wx*]pentacene $C_{28}H_{16}$

dibenzo[*hi*,*wx*]heptacene $C_{36}H_{20}$

benzo[*yz*]naphtho[2,1,8,7-*hijk*]heptacene $C_{38}H_{20}$

E. Clar, C. T. Ironside, and M. Zander, *Tetrahedron* **22** 3527–3533 (1966).

benzo[*h*]naphtho[1,2,3,4-*rst*]pentaphene $C_{32}H_{18}$

benzo[*rst*]naphtho[2,3-*h*]pentaphene $C_{32}H_{18}$

anthra[1,2,3,4-*rst*]pentaphene $C_{32}H_{18}$

dinaphtho[3,2,1-*fg*:1,2,3-*st*]pentacene $C_{36}H_{20}$

E. Clar, B. A. McAndrew and J. F. Stephen, *Tetrahedron* **26** 5465–5478 (1970).

tribenzo[*fg*,*mn*,*xyz*]heptaphene $C_{34}H_{18}$

dibenzo[*fg*,*mn*]naphtho[1,2,3,4-*xyz*]heptaphene $C_{38}H_{20}$

dibenzo[*fg*,*mn*]phenanthro[1,10,9,8-*vwxyza*$_1$*b*$_1$]heptaphene $C_{42}H_{20}$

trinaphtho[2,1,8,7-*fghi*:2′,1′,8′,7′-*klmn*:1′,2′,3′,4′-*xyz*]heptaphene $C_{46}H_{22}$

E. Clar and A. Mullen, *Tetrahedron* **27** 5239–5245 (1971).

tetrabenzo[*a*,*d*,*j*.*m*]coronene $C_{40}H_{20}$

E. Clar and B. A. McAndrew, *Tetrahedron* **28** 1137–1142 (1972).

dianthra[1,9,8,7-*cdefg*:1′,9′,8′,7′-*nopqr*]pentacene $C_{38}H_{18}$

naphth[2′,1′,8′,7′:4,10]anthra[1,9,8]*abcd*]coronene $C_{38}H_{16}$

E. Clar and C. C. MacKay, *Tetrahedron* **28** 6041–6047 (1972).

benzo[*ghi*]cyclopenta[*cd*]perylene $C_{24}H_{12}$

cyclopenta[*bc*]coronene $C_{26}H_{12}$

K. Taghizadeh, P. F. Nelson, C.-Z. Li, A. Necula, and L. T. Scott, *Twenty-Seventh International Symposium on Combustion*, The Combustion Institute, Boulder, Colorado, 1998, p. 1677–1686.

This is a listing of literature references that contain UV absorbance spectra of new LPAHs that have been published since E. Clar's two-volume set, *Polycyclic Hydrocarbons*.

3.2 LOW-TEMPERATURE ABSORBANCE SPECTROSCOPY

When in solution and at room temperature, LPAH absorbance spectra have bands that are many nanometers wide. This is because of the smearing of ground-state and excited-state molecules by interactions and collisions with the solvent. If the LPAHs are dissolved in a suitably inert solvent, such as an *n*-alkane, and rapidly cooled, the absorbance spectra are quite different (19, 20). The *n*-alkane solvent freezes into an orderly array in which the molecules are oriented in parallel to each other. The LPAHs are trapped in this inert orderly array. The effects of solvent interactions and collisions also become negligible. Thus the LPAH electronic transitions become much more defined. The spectral linewidths narrow to <1 nanometer. What once appeared to be a single broad band in the room-temperature solution spectrum splits into many narrow ones. This is known as the Shpol'skii effect. These spectra are so detailed and the bands are so specifically located that they are literally fingerprints for the specific LPAH structure. Therefore, very similar LPAHs like coronene and 1-methylcoronene have very distinct and dissimilar spectra. Because low-temperature spectra are most often collected as fluorescence spectra, experimental examples are given in the chapter on fluorescence spectra.

REFERENCES

1. E. Clar, *The Aromatic Sextet*, Wiley-Interscience, New York, 1972.
2. E. Clar, *Polycyclic Hydrocarbons*, John Wiley and Sons, New York, 1964, Vols. 1 and 2.
3. S. Obenland and W. Schmidt, in *Polycyclic Aromatic Hydrocarbons and Astrophysics*, A. Leger, L. d'Hendecourt, and B. Boccara, Eds., Reidel, Dordrecht, The Netherlands, 1987, p. 165–172.
4. H. Fromhers, L. Thaler, and G. Wolf, *Z. Elektrochem.* **49** 387–400 (1943), in German.
5. L. R. Snyder and D. L. Saunders, in *Chromatography in Petroleum Analysis*, K. H. Altgelt and T. H. Gouw, Eds., Dekker, New York, 1979, p. 268.
6. M. Zander and W. Friedrichsen, *Z. Naturforsc. A Phys. Sci.* **47** 1314–1318 (1992).
7. S. Fujisawa, I. Oonishi, J. Aoki, Y. Ohashi, and Y. Samada, *Bull. Chem. Soc. Jpn.* **55** 3424–3426 (1982).
8. I. Oonishi, S. Fujisawa, J. Aoki, and Y. Ohashi, *Bull. Chem. Soc. Jpn.* **59** 2233–2236 (1986).
9. I. Oonishi, S. Ohshima, S. Fujisawa, and J. Aoki, *J. Mol. Struct.* **265** 283–292 (1992).
10. S. Fujisawa, S. Ohshima, A. Uchida, and I. Oonishi, *Polycyclic Arom. Cmpds.* **11** 267–273 (1996).
11. S. Fujisawa, M. Takekawa, Y. Nakamura, A. Uchida, S. Ohshima, and I. Oonishi, *Proceedings of the Sixteenth International Symposium on Polycyclic Aromatic Compounds, Charlotte, NC, October 1997.*

12. F. A. Vingiello and L. Ojakaar, *Tetrahedron* **22** 847–860 (1966).

13. E. Clar, J. M. Robertson, R. Schloegl, and W. Schmidt, *J. Am. Chem. Soc.* **103** 1320–1328 (1981).

14. T. Umemoto, T. Kawashima, Y. Sakata, and S. Misumi, *Tetrahedron Lett.* **12** 1005–1006 (1975).

15. J. C. Fetzer, in *Polynuclear Aromatic Hydrocarbons*, L. Ebert, Ed., Advances in Chemistry 217, American Chemical Society, Washington, DC, 1988, p. 309–331.

16. J. C. Fetzer and W. R. Biggs, *Org. Prep. Proc. Int.* **20** 223–230 (1988).

17. J. C. Fetzer and W. Schmidt, *Spectrochim. Acta A* **45A** 503–505 (1989).

18. J. C. Fetzer and W. R. Biggs, *Polycyclic Arom. Cmpds.* **5** 193–199 (1994).

19. F. J. Morgan, R. Lamotte, P. Garrigues, and J. C. Fetzer, *Polycyclic Aromat. Compds.* **2** 141–153 (1991).

20. L. Nakhimovsky, M. Lamotte, and J. Joussot-Dubien, *Handbook of Low Temperature Electronic Spectra of Polycyclic Aromatic Hydrocarbons*, Elsevier, Amsterdam, 1989.

CHAPTER

4

FLUORESCENCE SPECTROSCOPY

4.1 INTRODUCTION

Another characterizing feature of almost all LPAHs is their strong fluorescence. If solutions of a wide variety of LPAHs are prepared and placed under a UV light, almost all will exhibit intense fluorescence emission. The diversity of this set of LPAH solutions would be characterized by the wide variety of fluorescent colors seen, from the low-wavelength violet to the very high-wavelength orange and red.

When the fluorescence spectra of these solutions are collected on a spectrometer, the excitation spectra are usually found to be very similar to the compounds' UV/visible absorbance spectra. The abundance of bands and their relative intensities are generally superimposable. This is not surprising because the same photophysical changes in the molecules are caused by the incident light. The incident light in UV/visible absorbance spectra result from the absorption of the light energies; the fluorescence excitation spectra result from these energies sending the molecules into their excited states. The loss of excitation state energy by emission of a photon yields the fluorescence emission spectrum. The fluorescence emission spectra are also spectrally rich (1, 2). Although there are many references describing the electronic transitions that determine PAH fluorescence spectra, References 1 and 2 give fundamental descriptions of many of the topics discussed in this introduction. The basic principles of the techniques described in this chapter are cited in those references.

Although UV absorption and fluorescence excitation are similar processes, the latter is very different from an analytical standpoint. Fluorescence excitation is not directly observed, but it is seen by changes in the secondary process of emission. The electronic transitions seen in fluorescence are fewer in number than in UV absorbance. In UV absorbance, higher-energy transitions involving the second highest occupied molecular orbital and the second lowest unoccupied molecular orbital are observed. These are often not seen in the published fluorescence spectra of the LPAHs, primarily because the best solvents are aromatic ones like toluene, chlorobenzene, or 1-methylnaphthalene.

Large (C>=24) Polycyclic Aromatic Hydrocarbons: Chemistry and Analysis, by John C. Fetzer.
ISBN 0-471-36354-5

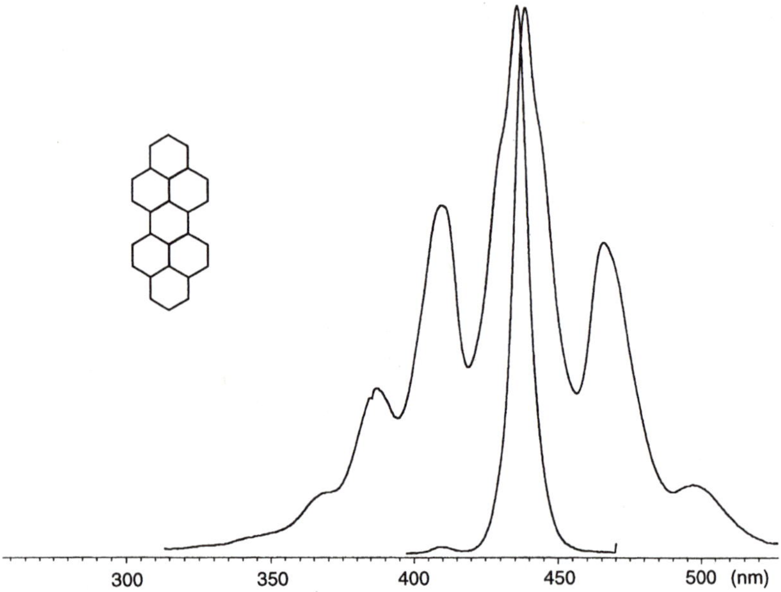

Figure 24. Mirror-image fluorescence spectra of dibenzo[*cd,lm*]perylene.

Often, the emission spectrum of a LPAH will be an apparent mirror image of the higher wavelength portions of the excitation spectrum. Figure 24 shows such a pair of spectra for dibenzo[*cd,lm*]perylene. The group of highest-wavelength excitation bands appear with similar spacings but increasing intensity. This is the same as the pattern of the UV absorbance bands for these PAHs. The emission bands are similarly spaced and have decreasing intensities of approximately the same ratios in height as the corresponding intensities in the excitation spectrum. This pattern is caused by the nature of the electronic states of the molecule. They reflect the most probable transitions between the ground and excited states.

If the potential energy surface of the ground state of the LPAH (designated S_0 because it is a singlet state) is similar in its shapes and the positions of the minima to that of the first excited state (designated S_1 because it also is a singlet state), then this pattern will result. The transitions upward through absorption (and the molecule going from S_0 to S_1) are similar to the transitions downward through emission of a photon. The transition from S_0 to S_1 is known as the 0-0 transition in absorbance spectroscopy.

Generally, the transition occurs through the two most important of these energy levels, the highest occupied molecular orbital (or HOMO) and the lowest unoccupied molecular orbital (or LUMO). The energy difference between these is called the HOMO-LUMO gap. In PAHs with symmetrical fluorescence spectra, this gap is small. The emission of a photon from the excited state readily occurs, and this is the most probable transition. The probabilities of each transition decrease because the probability of the analogous excitation occurring to these higher levels of the excited state is less likely. The other, smaller bands result from the less probable transitions from the occupied orbital below the HOMO to the LUMO, from the HOMO to the second lowest unoccupied orbital, etc.

The wavelength offset between the highest-excitation and the lowest-wavelength emission maxima is called the Stokes shift. The Stokes shift for these mirror-image band structures is usually small, from 5 to 10 nm.

For certain LPAHs, the pair of fluorescence spectra are not mirror images but are asymmetrical. The band heights that reflect the probabilities of the electronic transitions are not the patterns of monotonically increasing intensities in the excitation spectrum or decreasing intensities in the emission spectrum that are due to vibronic effects.

The most common cause of an asymmetrical pair of fluorescence spectra is the different transition probabilities in each mechanism because the 0-0 transition is not favored. This changes the intensity of the other individual bands greatly (i.e., it changes the probability distribution of the various electronic transitions relative to each other). The causes of a weak 0-0 transition are numerous. For the PAH pyrene, it is because of orbital coupling between S_1 and the next higher excited state S_2. Michl and Thulstrup describe this in great detail in their books on the use of polarized spectroscopy (1, 2), and the reader is directed there for this detailed explanation. The symmetry of the PAH and the geometric orientations of the resulting electronic transitions are important factors. Similar behavior is seen for benzo[*ghi*]perylene and the LPAHs coronene, benzo[*a*]coronene, ovalene, and others. The pair of spectra for coronene are shown in Figure 25.

Nonalternant structures generally do not exhibit symmetrical fluorescence spectra. Their less-conjugated structure and the lessened ability of the five-member ring to extend the conjugated π electrons makes it more likely that there would be energy losses or changes in geometry between the ground and excited states.

The fluorescence spectra and behavior of two very similar LPAHs highlights some of the differences between alternant and nonalternant structures (3). The two isomers benzo[1,2,3 *bc*:4,5,6 *b*′*c*′]dicoronene and cyclopenta[1,2-*a*:3,4,5-*b*′*c*′]dicoronene (*compounds 53 and 108*) have quantum efficiencies of 0.81 and 0.13. (The quantum efficiency is a determination of the fraction of molecules that fluoresce and is found by measuring the emitted intensity for a known molar concentration and comparing this to a fluorescence intensity standard like quinine sulfate.) Their highest excitation bands are at 504 and 494 nm [shown in the UV absorbance spectra in Fig. 19; the corresponding emission bands are at 514 and 532 nm (Figure 26). These differences are caused by the six-member

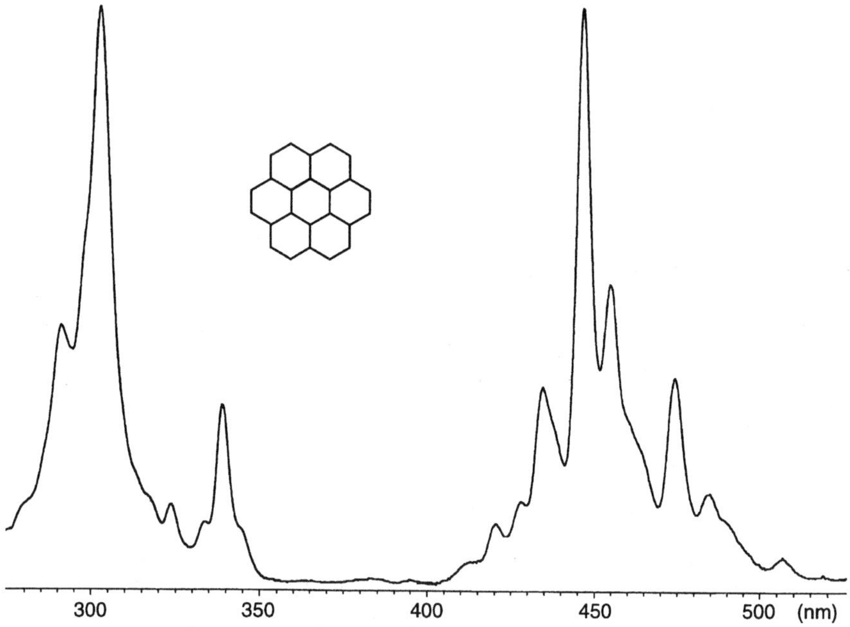

Figure 25. The fluorescence spectra of coronene, showing the non-mirror-image nature of the emission spectrum (on right-hand side).

ring in the first isomer contributing to electron delocalization and the presence of more Kekule structures than the bridging five-member ring in the second isomer.

Bohnen, Koch, Luettke, and Muellen (4) determined the quantum efficiencies for the series of fused naphthalenes starting with perylene. This PAH had a value of 0.94. The value of the second PAH, the LPAH tribenzo[*de*,kl,*rst*]pentaphene (trivially called by Clar "terrylene"), was found to be 0.70, and the third of the series, benzo[1,2,3-*cd*:4,5,6-*c′d′*]diperylene (trivially called "quaterrylene"), had the very low value of 0.05.

The fluorescence spectra of nonplanar, alternant, peri-fused LPAHs show another feature. The valleys between the spectral bands of planar LPAHs are very deep, so that the bands are very narrow. The valleys for nonplanar structures are more broad and shallow, so that the bands also appear to be rounder and less distinct. The nonplanar structures are not as rigid as the planar structures. This increased flexibility leads to a wider distribution of energies for each ground- or excited-state configuration (the "excess" energy in the excited state may be slightly dissipated by transformation into bond movements). The emission spectrum of a nonplanar LPAH of this sort often has even more rounded spectral bands with shallower valleys than its excitation spectrum, indicating that the effect is also augmented by the flexibility of the excited-state molecules.

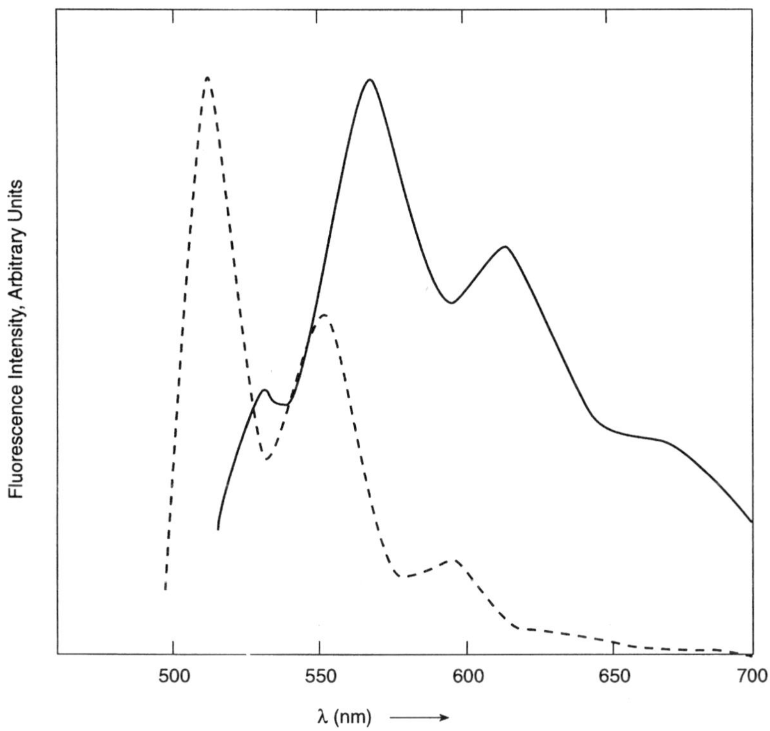

Figure 26. Alternant and nonalternant fused coronene dimers From M. Zander and W. Friedrichsen, *Z. Naturforsc. A Phys. Sci.* **47** 1315, 1316 (1992).

For most PAHs, changes in the solvent usually result in only slight, uniform shifts in the wavelengths of the spectra. These shifts are caused by the differences in solvation strength that affects the energy levels in the ground and excited states. Stronger solvation shifts the spectra to slightly higher wavelengths because the solvent interactions with the π electrons perturb them slightly and the electrons are more spread out in the stronger-solvated LPAH. For example, the magnitude of the shift with a different stronger solvent than one used as a reference spectrum is typically only 1–2 nm. The most extreme case, with a shift of < 10 nm, occurs when changing from nonpolar alkane solvents to very polar solvents such as methanol or acetonitrile.

There are certain LPAHs, however, that show very gross changes in their emission spectra when dissolved in different solvents. LPAHs such as coronene and ovalene show additional bands and great changes in the relative intensities of bands. Figure 27 shows the emission spectra of benzo[*a*]coronene in the nonpolar solvent *n*-hexane and the polar solvent ethanol. Interestingly, many of

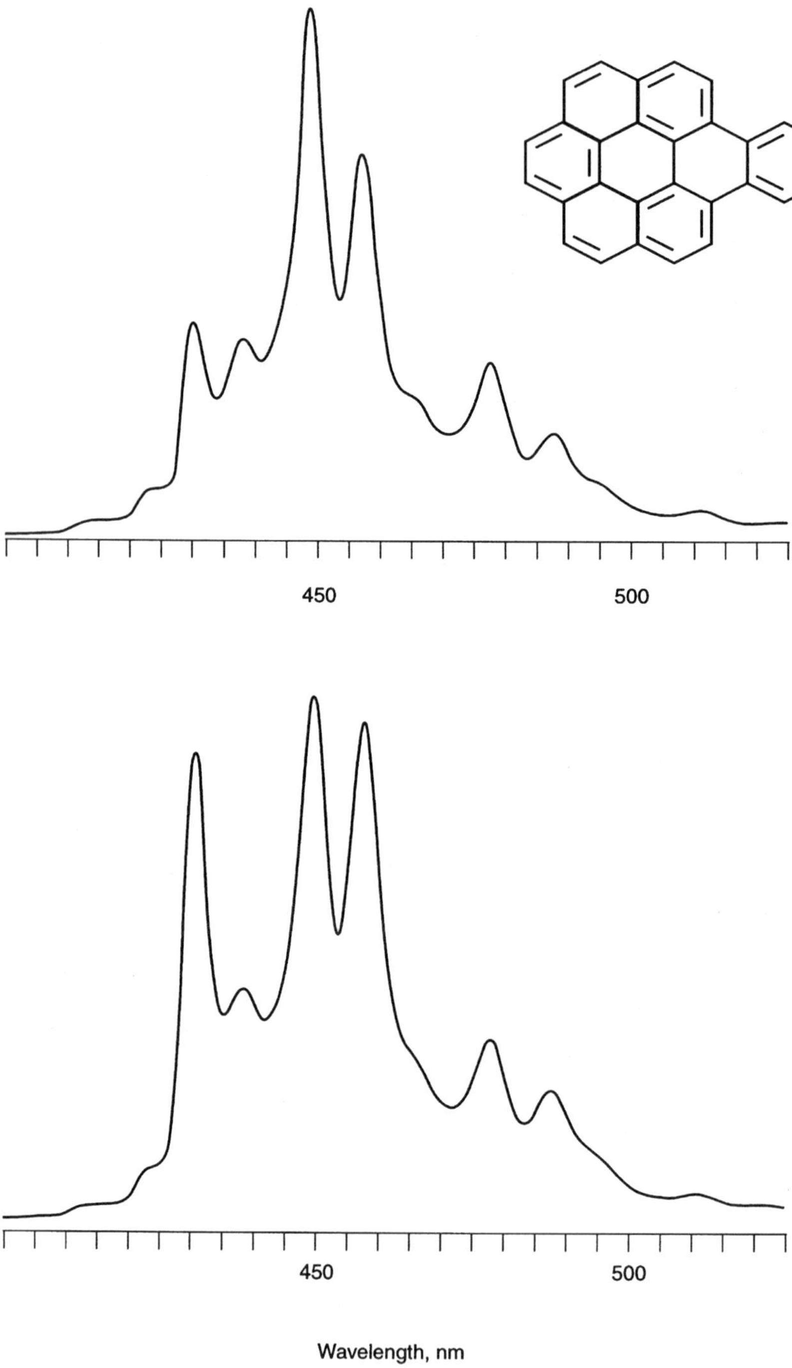
450
500
450
500
Wavelength, nm

these are the same LPAHs that show the non-mirror-image spectra discussed earlier.

A possible explanation of this effect is found if the results of polarized spectroscopy are considered, as also discussed by Michl and Thulstrup (1, 2); then solvation selectively effects some transitions more than others. This augments or decreases the band intensities relative to each other. The higher polarizabilities of certain of the transitions are more likely the cause rather than a selectively-stronger solvent interaction with specific regions of the LPAH caused by the solvent cage. This is because the solvent size or other structural features, which would change the solvent cage geometry, do not seem to have much of an effect.

Acree et al. (5–10), in a long series of papers, studied a wide range of LPAHs to assess structural features that might contribute to this change in spectral patterns. They found that many of the LPAHs that showed this type of behavior were very condensed structures, such as coronene and ovalene.

In fact, all of the PAHs from pyrene and upward in Stein's series of most condensed structures showed this type of behavior. Of the other LPAHs, a few also showed differing emission spectra whereas the vast majority did not. These few generally contained a coronene substructure, as in 1-methylcoronene, naphtho[8,1,2-*abc*]coronene (*compound 29*), dibenzo[*a, j*]coronene (*compound 58*), naphtho[1,2-*a*]coronene (*compound 57*), anthra[1,2-*a*]coronene (*compound 56*), and ovalene (*compound 5*). An understanding of the photophysics of the solvated LPAH can explain these differences in spectral behavior of the LPAHs.

For many of these LPAHs, the changes in spectral features can be correlated to the polarity of the solvent. Pyrene was the first PAH to show this behavior, and later studies found that benzo[*ghi*]perylene showed this behavior as well. The cause of this behavior has not yet been explained either theoretically or experimentally. The idea that the strength of solvation can effect the conformation of LPAHs and their excited states is plausible, because it has been seen in at least one instance. That solvation strength and solvent polarity correlate highly is not surprising, either. This may also explain some of the exceptions to the patterns found by Acree et al. Some solvents may be very polar but not solvate well with a particular LPAH and its excited state. This leads to anomalous behavior when polarity of the solvent is the measure, rather than strength of interaction. The behavior of pyrene and these other PAHs has led to their wide use as probes of the microchemical environments in a wide variety of systems. Those inside micelles or in the "cup" of cyclodetrins or calixarenes are examples.

Figure 27. Changes in the fluorescence emission spectrum of benzo[*a*]coronene. Top spectrum collected on an *n*-hexane solution; bottom spectrum collected on an ethanol solution. Original spectra from A. Colmsjo, unpublished results.

The advantage that some of the LPAHs have over the smaller PAHs usually used as polarity probes, pyrene or benzo[*ghi*]perylene, in this application is that their excitation and emission wavelengths are much higher. For some solvents or systems in which the polarity is being studied the absorption at lower wavelengths prevents the measurement of polarity by using pyrene because of the inner-system filtering that arises (see below). Thus ovalene could be used to determine the polarity of liquid organic salt media. Pyrene and other "solvent polarity probe" molecules with low fluorescence wavelengths could not be used because their spectral wavelengths were lower than those of the salts.

Fluorescence spectrometry also is very popular for LPAHs because of the very low detection limits found for them. LPAHs have very high quantum yields. Because fluorescence spectrometry also has very low spectral backgrounds, detection limits of picograms per liter are possible for analyses in which the spectrometer uses the xenon lamp as the excitation source. Laser-based excitation has even more sensitivity. The one problem that is predominant in the use of fluorescence for PAHs is that the spectra for mixtures can be exceedingly complex. The complexity arises because the spectrum is the sum of all fluorescence of each LPAH at the particular excitation and emission wavelength combination being measured. This necessitates the very time-consuming task of separation into individual components, with isolation of peaks and subsequent off-line spectral collection (the use of the similar on-line detection will be discussed in Chapter 8).

One consequence of the complex spectra of LPAH mixtures is of great concern in fluorescence analysis. The overlap of the excitation spectrum of one LPAH with the emission spectrum of another leads to what is called inner-system filtering. Two types of inner-system filtering can occur. Primary inner-system filtering is the reduction of the intensity of the incident beam by absorption. If the other components have a total absorbance of >0.01 absorbance units (AUs) at the excitation wavelength used, the intensity will be sufficiently reduced to effect the overall fluorescence intensity. Secondary inner-system filtering results when the other components have a similar level of absorbance at the emission wavelength being used.

Mathematical corrections for the effects of inner-system filtering can be used to correct for these matrix absorbances (11). The first step in applying these corrections is measuring the absorbances of the sample at the excitation and emission wavelengths. In doing this, the spectrometer slit width must be the same as each of those that are used in the fluorescence measurement.

Matsushita, Esmi, Suzuki, and Handa et al. (12) used a two-dimensional thin-layer chromatographic (TLC) separation. The spots were scraped off of the silica plates and extracted with chloroform, and the fluorescence spectra were collected. McKay and Latham (13) used TLC and HPLC to isolate LPAH fractions of heavy petroleum distillation cuts. Subsequent fluorescence identified several LPAHs. Paeden, Lee, Hirata, and Novotny (14) used reversed-phase HPLC with off-line fluorescence to isolate and characterize approximately a dozen LPAHs from a carbon black extract. In all three examples, only a few of the LPAHs were unequivocally identified because of a lack of available standards in each research group.

Alternatively to the reliance on separation methods before the spectroscopy, there are methods that rely on spectral approaches to observe the individual LPAHs in mixtures. The following sections will deal with some of these approaches that accomplish this.

4.2 SELECTIVE QUENCHING TECHNIQUES

For many of the alternant and nonalternant LPAHs, each class shows differing fluorescence behavior. Not only do their structures fundamentally differ, but they act as two distinct classes when their spectra are collected in solvent mixtures containing either an electron-donating or an electron-withdrawing reagent. These reagents selectively quench one class and not the other. The most studied reagents have been nitrobenzene and nitromethane as the electron-withdrawing reagents and the trimethoxybenzene isomers as the electron-donating reagents.

Dreeskamp, Koch, Zander, and various other co-workers (15–18) were the first group to study the selective aspects of LPAH quenching. They prepared synthetic mixtures of LPAHs to evaluate the analytical potential of addition of quenchers. The first mixture contained the alternant PAHs anthracene, perylene, dibenzo[*def,mno*]chrysene (anthranthrene), and dibenzo[*b,k*]chrysene and the nonalternant acenaphtheno[1,2-*k*]fluoranthene. The spectrum of the mixture was complex, but on addition of nitromethane only the nonalternant LPAH was seen. A second mixture contained the nonalternant LPAHs acenaphtheno[1,2-*k*]-fluoranthene, dibenz[*a,e*]aceanthrylene, rubicene, and decacyclene and the alternant LPAH dibenzo[*a,j*]pentacene. When 1,2,4-trimethoxybenzene was added, the complex spectrum collapsed into only that of the last compound (Fig. 28).

This class-specific quenching simplifies the fluorescence spectrum of mixtures, allowing some compositional information to be gathered without separation. Acree and co-workers (19–27) have studied the use of nitromethane to selectively quench the alternant LPAHs. They studied the effect of solvent polarity on quenching by using acetonitrile and mixtures of acetonitrile and toluene or acetonitrile and ethyl acetate. It was found that there were a very small number of exceptions. Only dibenzo[*hi,wx*]heptacene (*compound 111*) of the 40 alternant PAHs examined was not quenched and only rubicene (*compound 112*) of the 17 nonalternant LPAHs was quenched in at least one of the three solvents. These studies also showed that two categories of five-member ring compounds did not behave as nonalternant PAHs. These were those

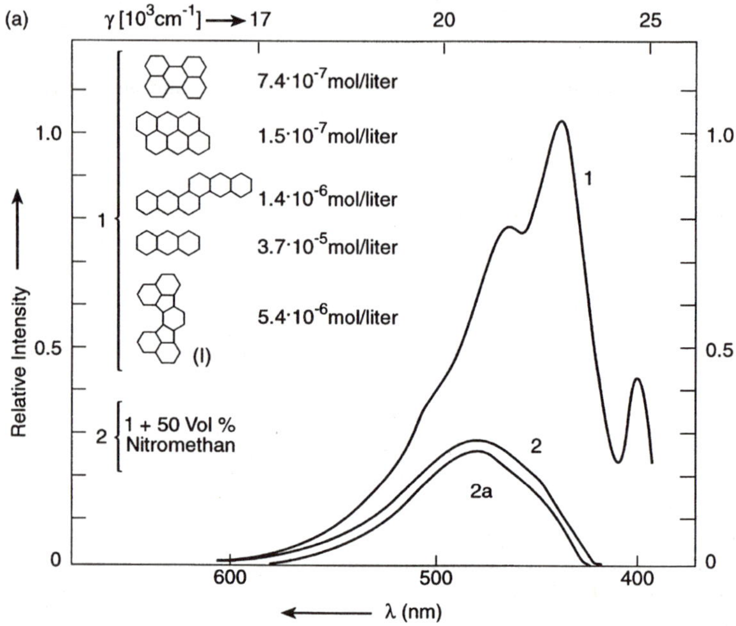
(a)
γ [10³cm⁻¹] → 17
20
25
7.4·10⁻⁷mol/liter
1.5·10⁻⁷mol/liter
1.4·10⁻⁶mol/liter
3.7·10⁻⁵mol/liter
5.4·10⁻⁶mol/liter
(I)
1
2
1 + 50 Vol %
Nitromethan
2a
Relative Intensity
1.0
0.5
0
600
500
400
λ (nm)

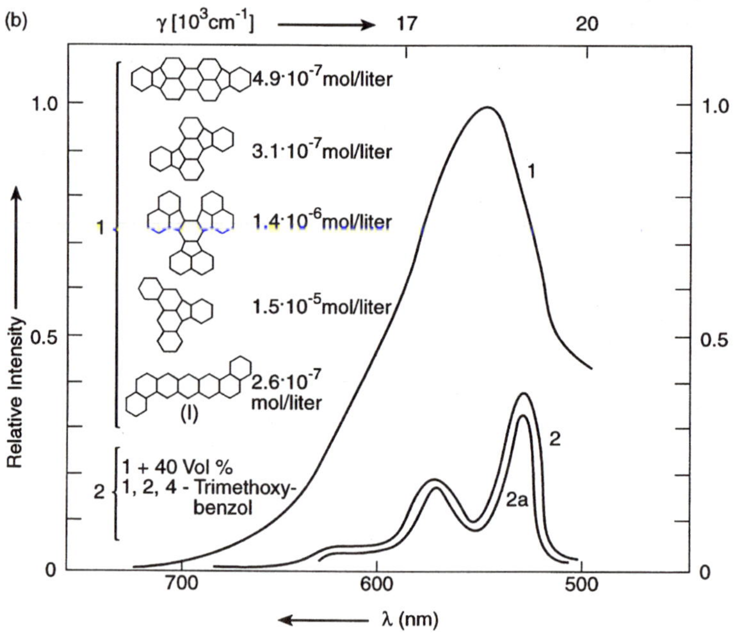
(b)
γ [10³cm⁻¹] → 17
20
4.9·10⁻⁷mol/liter
3.1·10⁻⁷mol/liter
1.4·10⁻⁶mol/liter
1.5·10⁻⁵mol/liter
2.6·10⁻⁷
mol/liter
(I)
1
2
1 + 40 Vol %
1, 2, 4 - Trimethoxy-
benzol
2a
Relative Intensity
1.0
0.5
0
700
600
500
λ (nm)

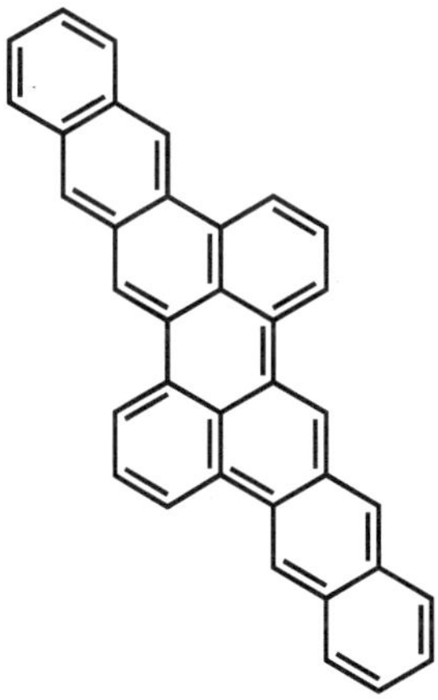

111

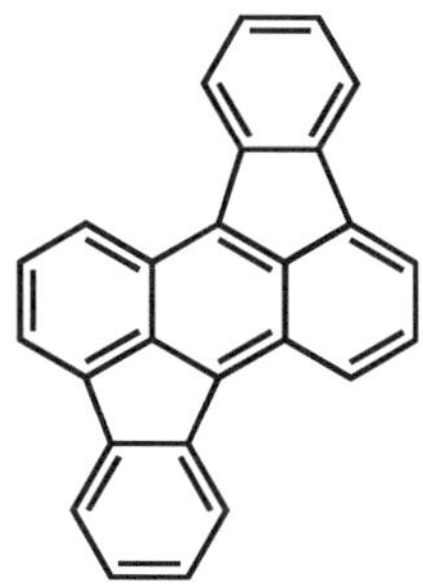

112

containing a fluorene-type bridging five-member ring, such as in benzo[*def*]cyclopenta[*mno*]chrysene (*compound 113*), or those containing one on the periphery, as in benzo[*def*]cyclopenta[*hi*]chrysene (*compound 114*).

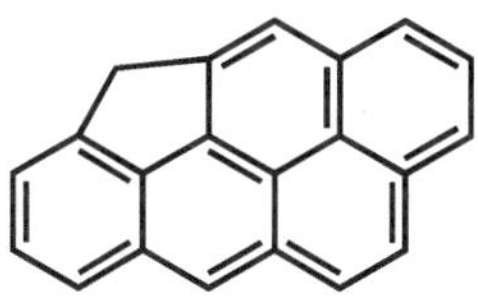

113

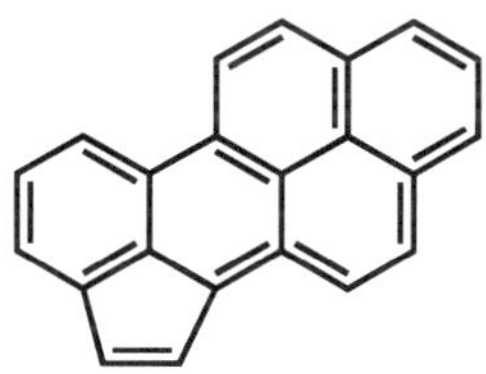

114

Wenzel and Loehmannsroeben (28) studied the fluorescence spectroscopy of dibenzo[*cd,lm*]perylene, including the effect of many quenching molecules in toluene and acetonitrile solutions.

4.3 NARROW LINE-WIDTH SPECTRA: LOW-TEMPERATURE (SHPOL'SKII) AND OTHER MATRIX METHODS

The width of the spectral bands for a fluorescence molecule arise from three sources. At normal temperatures, the molecules populate a range of energy

---

Figure 28. Examples of the selective quenching of alternant vs. nonalternant PAHs upon addition of either nitromethane or 1,2,4-trimethoxybenzene. From U. Breymann, H. Dreeskamp, E. Koch, and M. Zander, *Fresen. Z. Anal. Chem.* **293** 208 (1978), reproduced with permission of Dr. M. Zander and Springer-Verlag GmbH.

states. Also, collisions of the fluorescent molecules with each other and the solvent molecules occur, and finally, there are interactions as well with the solvent molecules themselves. All of these broaden the bands.

To reduce these sources of spectral broadening, a solution of the fluorescent molecule can be cooled quickly. This results in a low-temperature, crystalline matrix. A solvent that has little interaction with the solute, such as an alkane, is preferred. When this is done for PAHs, the broad bands seen in room-temperature spectra collapse into spectra with exceedingly narrow bandwidths. This is known as the Shpol'skii effect. The prevalent theory is that the size of the solvent matrix molecules must be similar to those of the analyte, so that the analyte can readily fit into the regular crystal array of the frozen solvent without a large number of dislocations. There are some variations in the spectra with the *n*-alkane used because of how well or poorly the analyte molecules fit into the crystal structure.

Little work on Shpol'skii spectra of LPAHs has been done because the spectrometric set-ups are rather sophisticated and standards are rare. There have only been a few researchers with access to both. Shpol'skii spectroscopy, however, has been used in these few instances with dramatic results that highlight its strength to easily differentiate the many LPAH structures from each other.

Colmsjo (29, 30) collected low-temperature spectra in frozen *n*-hexane matrices of 16 $C_{26}H_{16}$ isomers. Only six isomers showed the narrow and intense bands indicative of the Shpol'skii effect (the majority of the rest had spectra of less resolution and 3 did not fluoresce). The six isomers were dibenzo[*a*,*c*]-tetracene, benzo[*c*]pentaphene, benzo[*h*]pentaphene, dibenzo[*c*,*k*]tetraphene, dibenzo[*f*,*k*]tetraphene, and dibenzo[*g*,*p*]chrysene (*compounds 115–120*). All of these were noted to have either dibenz[*a*,*c*]anthracene or dibenz[*a*,*h*]anthracene as substructures.

Wise, Benner, Liu, Byrd, and Colmsjo (31) separated a coal tar using a combination of normal- and reversed-phase HPLC methods. Isolated fractions were dissolved in *n*-hexane and rapidly frozen in liquid nitrogen. Many 24-carbon LPAHs were identified.

Morgan, Lamotte, Garrigues, and Fetzer (32) used Shpol'skii spectra to examine the fluorescence behavior of the 34-carbon tetrabenzo[*a*,*cd*,*f*,*lm*]perylene (*compound 103*). This compound was intriguing because its UV absorbance spectrum did not match its fluorescence excitation spectrum (Figure 29), in contrast to the expected similarity seen in Figure 30 for the similar isomer tetrabenzo[*a*,*cd*,*j*,*lm*]perylene (*compound 121*). They found that rapid freezing of solutions of this compound resulted in trapping of two conformers that had different fluorescence excitation spectra (Fig. 31). Their emission spectra were essentially identical (Fig. 32). They postulated that the two conformers resulted from different mechanisms to relieve the steric strain of two cove regions in this

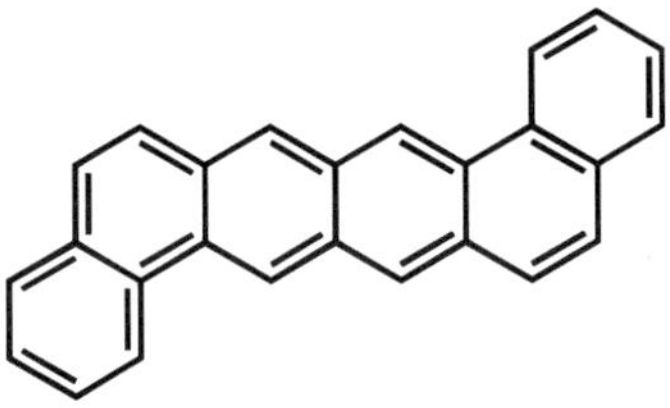

115

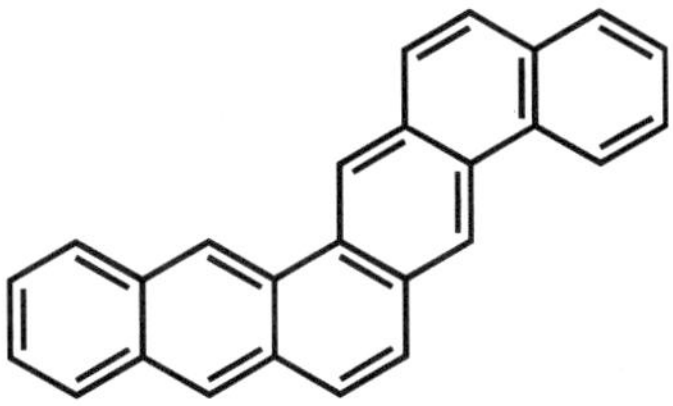

116

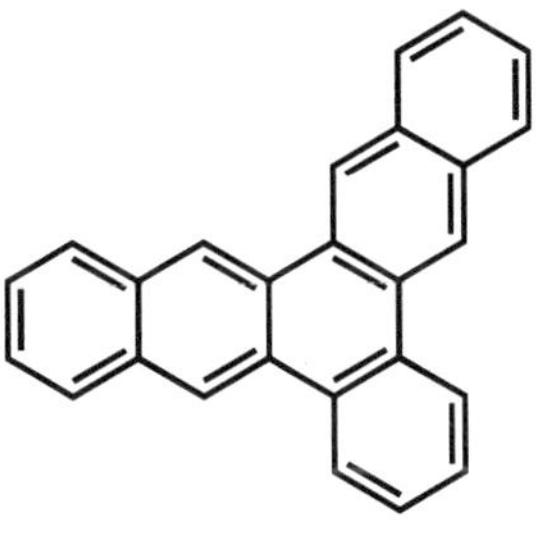

117

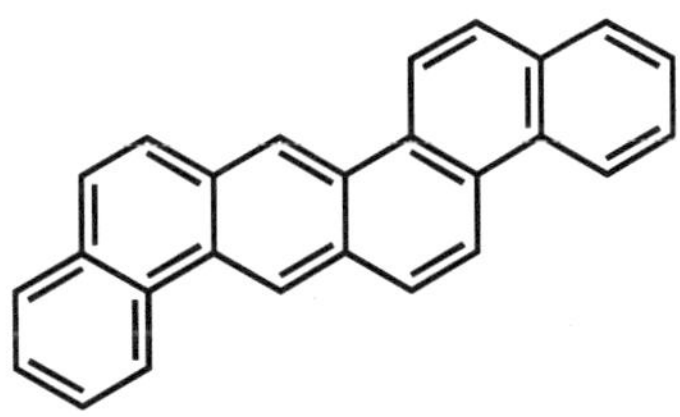

118

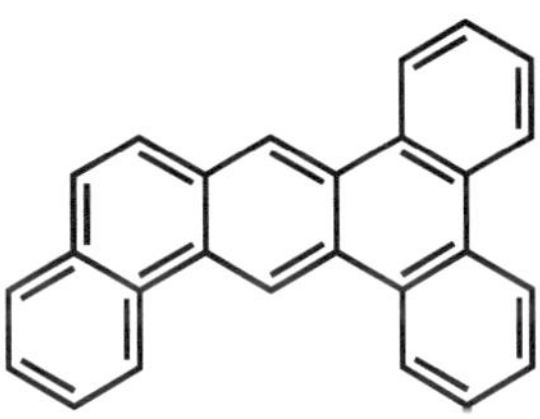

119

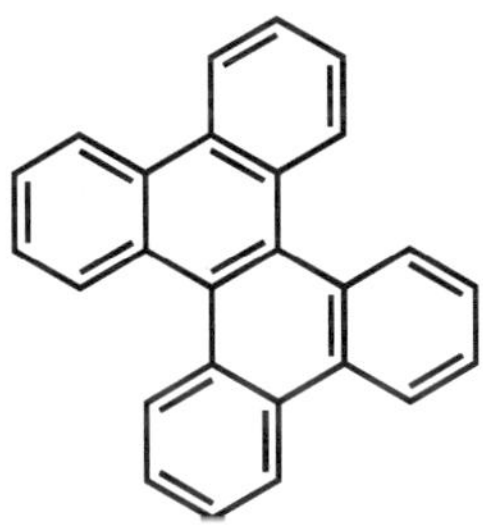

120

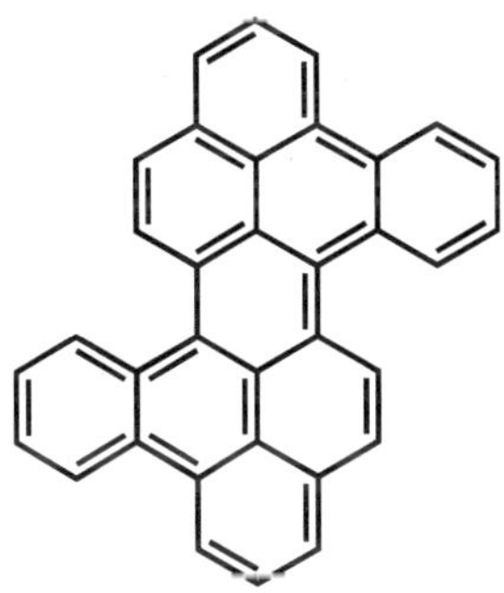

121

LPAH. Molecular models showed that the two conformers were very similar in energy.

Colmsjo and Ostman (33) separated a carbon black extract by a similar combination of normal- and reversed-phase HPLC. The 44 fractions obtained were frozen in *n*-hexane matrices, and Shpol'skii spectra were collected. These showed several LPAHs, including coronene and methylcoronene, but above 24 carbons only dibenzo[*cd,lm*]perylene, naphtho[1,2,3,4-*ghi*]perylene, and naphtho[8,1,2-*abc*]coronene could be unequivocally identified (Figs. 33 and 34).

LPAHs were also isolated in bitumen samples collected near deep-sea hydrothermal vents (34). These samples are produced by the action of the high-temperature waters generated by the hydrothermal vents (areas where magma

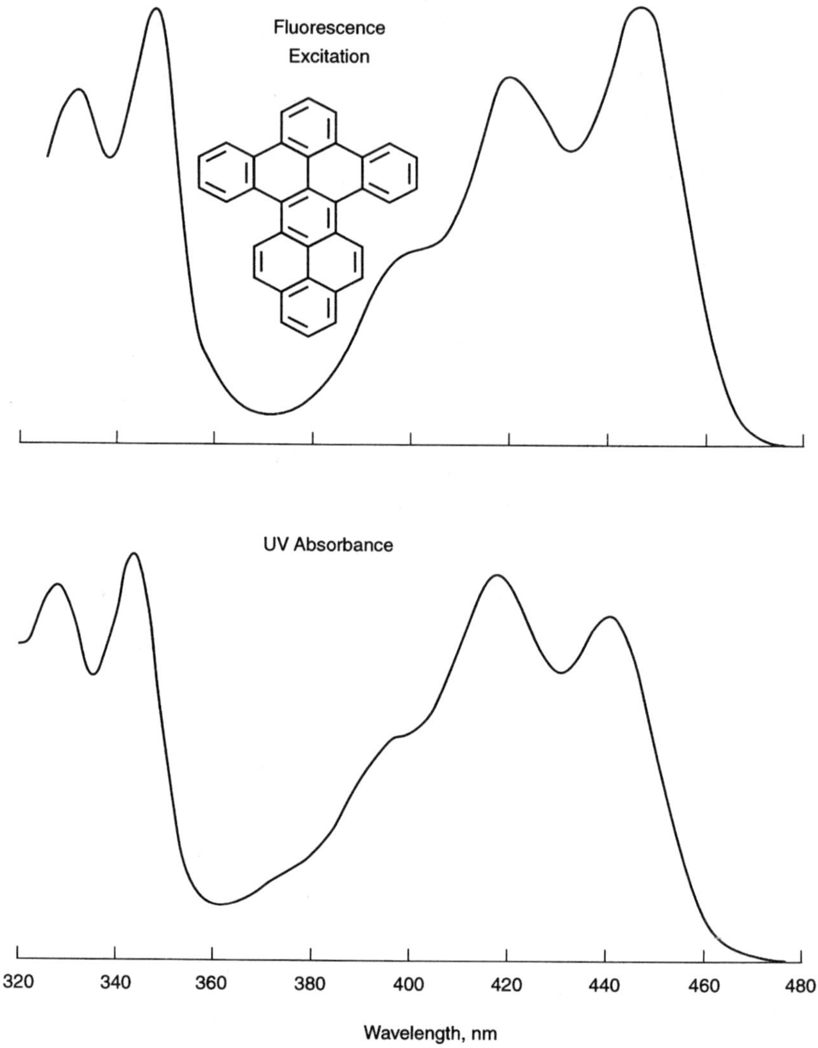

Figure 29. UV absorbance spectrum and fluorescence excitation spectrum of tetrabenzo[*a*,*cd*,*f*,*lm*]-perylene, showing the anomalously low intensity of the highest-wavelength band. From J. C. Fetzer and W. Schmidt, *Spectrochim. Acta A* **45A** 503–505 (1989).

seeps from fault zones in the oceans) on the organic matter on the ocean floor. The samples were first fractionated by adsorbent chromatography, with subsequent reversed-phase HPLC, and individual fractions were examined by Shpol'skii spectroscopy. Several LPAHs were unequivocally found. The presence of these particular LPAHs in bitumen samples was unexpected. It

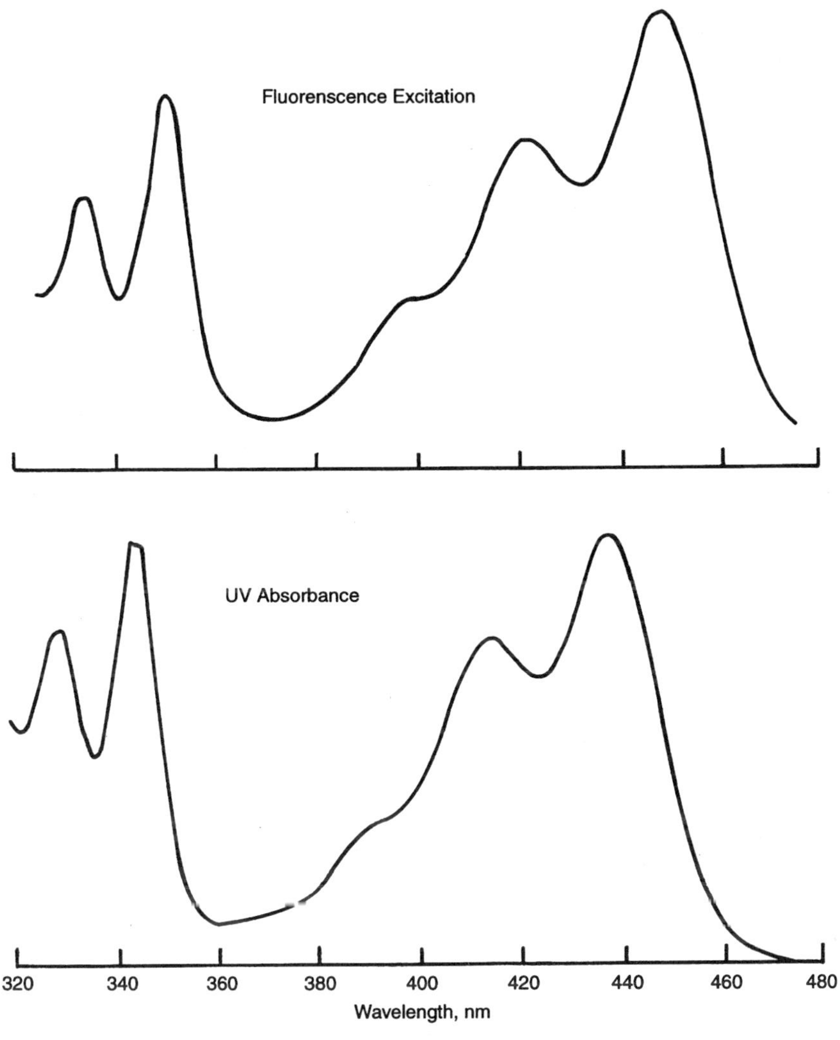

Figure 30. UV absorbance spectrum and fluorescence excitation spectrum of tetrabenzo[*a*,*cd*,*j*,*lm*]-perylene, showing the similarity of the spectra.

had been previously thought, based on analyses for smaller PAHs, that the composition of these hydrothermal bitumens was very similar to petroleum, which does not contain the LPAHs.

Spectra for four of the peaks are shown in Figures 35–38. Note that the spectral differentiating power of Shpol'skii spectra are easily evident in the methylcoronene spectra. The coronene impurity in the "authentic" standard is readily seen in the peaks at 433 and 445 nm. The methylcoronene peak preparatively isolated from the hydrothermal vent sample is much more pure. This group did see the large LPAH dibenzo[*de,kl*]phenanthro[7,8,9,10,1,2-*pqrstuv*]pentaphene (*compound 122*) by conventional fluorescence spectral matching, but did not collect a Shpol'skii spectrum.

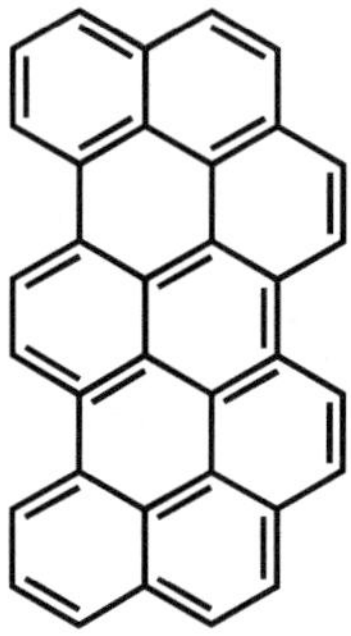

122

Some of the LPAHs examined in this second study are the same as some in the earlier study of Colmsjo and Östman. Comparison of the spectrum obtained for the same LPAH from each study shows great similarities but also some differences due to the conditions used to prepare the matrix. The temperature of the matrices has a great effect because the lower the temperature, the more collapsed the bands will be (with fewer lines). The spectra of dibenzo[*cd.lm*]-perylene, collected at 63 K, in the first study has many more lines than the second, which was collected at 15 K. The spectra of naphtho[8,1,2-*abc*]coronene in each paper cannot be compared because each study looks at only the spectral region of one of the collapsed bands (one showing 406–436 nm and the other showing 430–520 nm).

In both of these studies, the high degree of specificity in the line positioning and patterns in Shpol'skii spectra highlights the demand that either reference spectra or authentic standards be available for comparison. Unfortunately, there are only two reference books that contain any PAH Shpol'skii spectra, and the vast bulk of the spectra are of the smaller PAHs (35, 36). Thus these two books and the papers from these two groups contain all of the useful reference LPAH Shpol'skii spectra that are available. With this limitation, and the even larger limitations of the specialized apparatus and of the few available standard compounds, the great power of Shpol'skii spectra in LPAH identification will in all probability remain unavailable.

Another approach towards narrow-line spectra is the trapping of LPAHs in a matrix formed from large, planar inert molecules. Zander and co-workers (37, 38) have used perhydrocoronene (PHC), the fully hydrogenated form of the PAH.

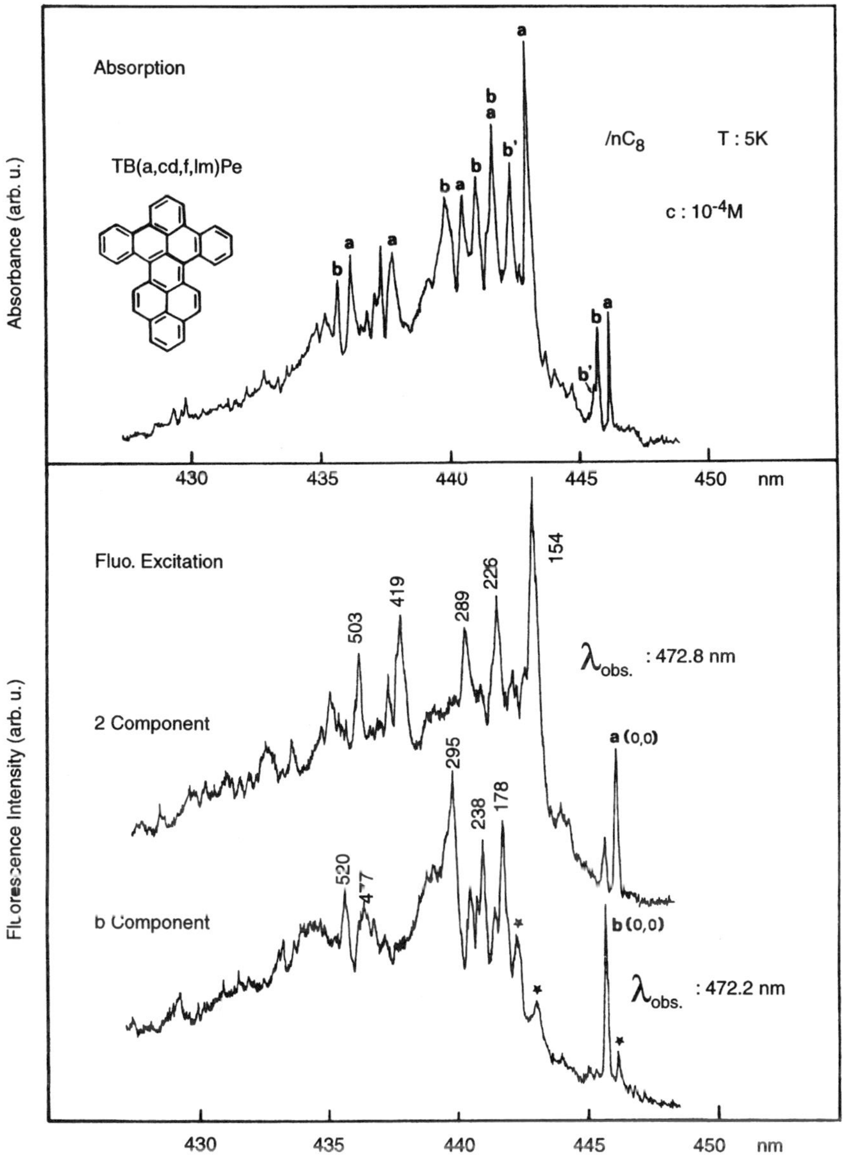

Figure 31. Shpol'skii excitation spectra of tetrabenzo[*a*,*cd*,*f*,*lm*]perylene. From F. J. Morgan, R. Lamotte, P. Garrigues, and J. C. Fetzer, *Polycyclic Aromat. Compds.* **2** 146 (1991).

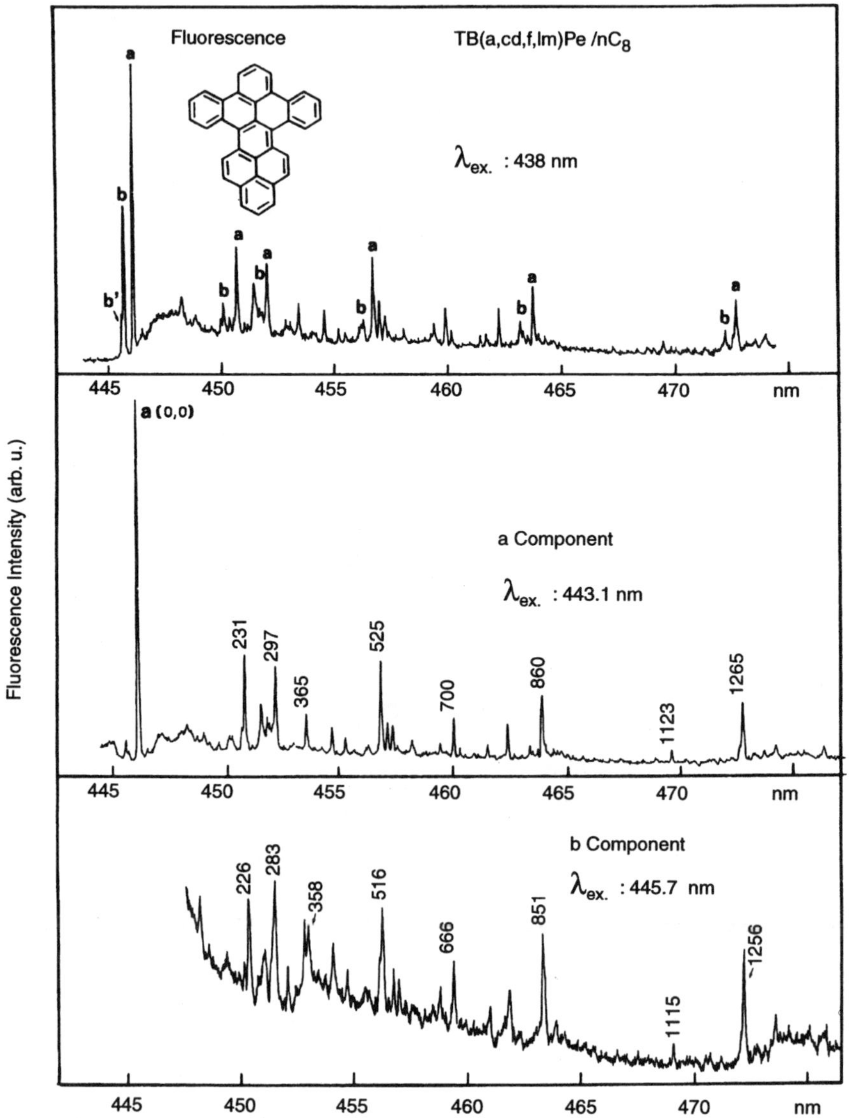

Figure 32. Shpol'skii emission spectra of tetrabenzo[*a*,*cd*,*f*,*lm*]perylene. From F. J. Morgan, R. Lamotte, P. Garrigues, and J. C. Fetzer, *Polycyclic Aromat. Compds.* **2** 147 (1991).

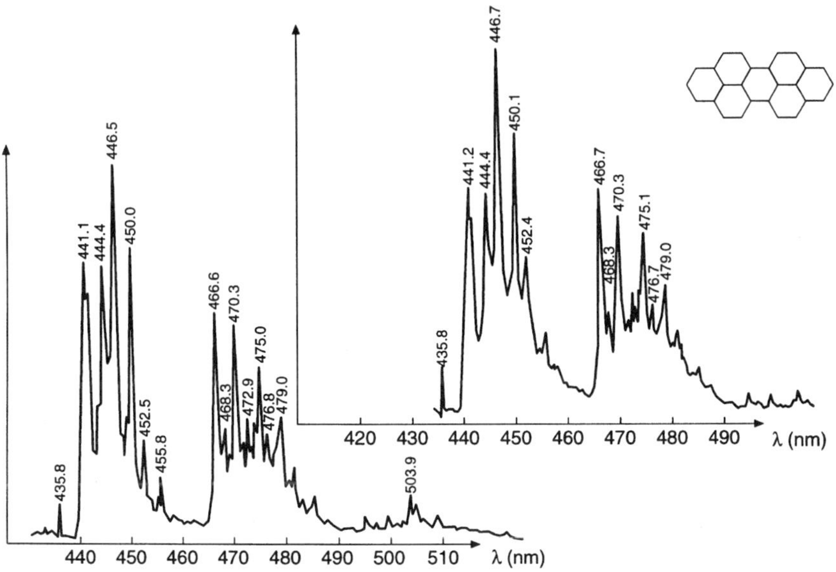

Figure 33. Colmsjo Shpol'skii spectra of fraction isolated from carbon black and standard of dibenzo[*cd,lm*]perylene (offset) From A. Colmsjo and C. Ostman, *Anal. Chim. Acta.* **208** 191, 192 (1988), reproduced with permission from Elsevier Science Publishers BV.

It is a flat, disk-shaped nonpolar molecule. Samples can be prepared by either making solutions of the analyte LPAHs and PHC in a low-boiling solvent and rapidly volatilizing the solvent away or making a molten mixture of PHC and the analyte that is cooled. The resulting mixture is a solid solution of the LPAH in a matrix of orderly arrayed PHC. The spectra of coronene, hexabenzo-[*bc,cf,hi,kl,no,qr*]coronene (*compound 123*) and benzo[*rst*]dinaphtho[2,1,8,7-*defg*:2′,1′,8′,7′-*ijkl*]pentaphene (*compound 124*) were collected in PHC. Their

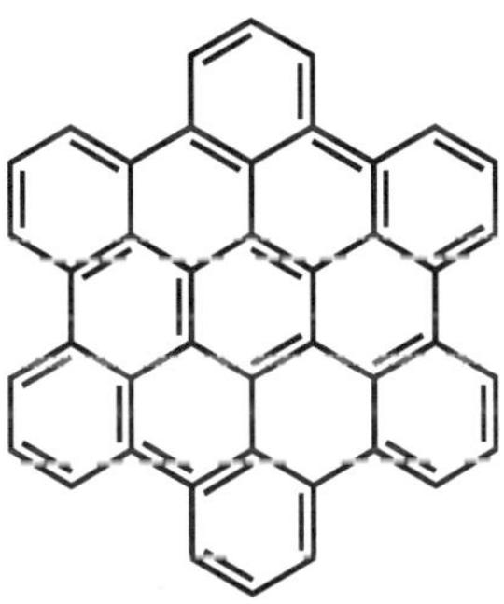

123

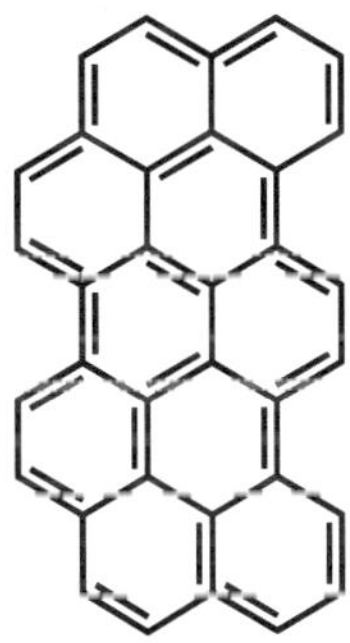

124

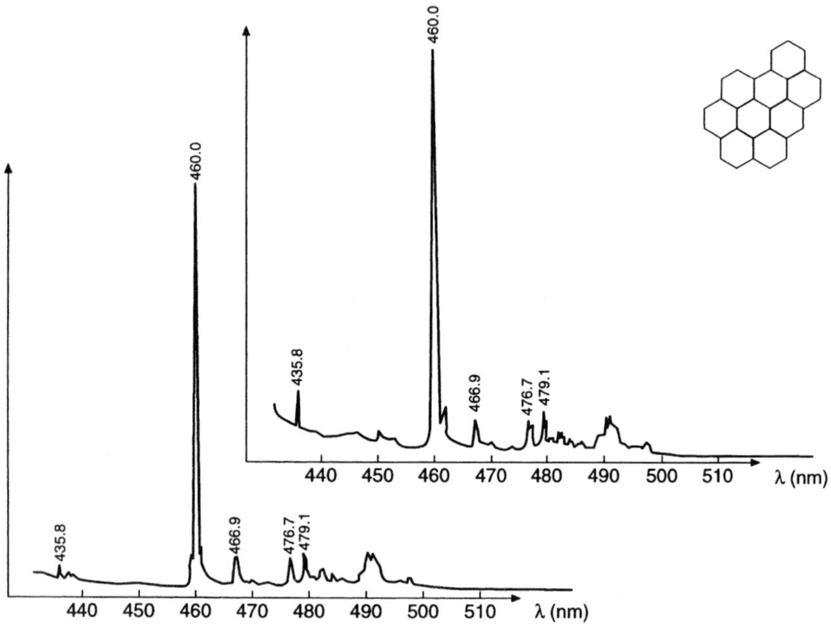

Figure 34. Colmsjo Shpol'skii spectra of carbon black fraction and standard of naphtho[8,1,2-*abc*]coronene (offset). From A. Colmsjo and C. Ostman, *Anal. Chim. Acta.* **208** 191, 192 (1988), reproduced with permission from Elsevier Science Publishers BV.

fluorescence and phosphorescence spectra could both be seen. The coronene sample had a very long phosphorescence lifetime. Using PHC for the matrix, Zander (39) was able to collect the spectra of a series of related very large PAHs with 15, 16, and 17 rings. The first was benzo[1,2,3-*bc*:4,5,6-$b'c'$]dicoronene, and the other two were the products of either one or two additions of a ring at the bridging ring (*compounds 53, 31,* and *32*). The spectra of *compound 32*, benzo[13,14] pentapheno[3,4,5,6,7,8,9,10-$uvwxyza_1b_1c_1d_1$]dinaphtho[2,1,8,7-$defg$:$2',1',8',7'$-$mnop$]heptaphene, are shown in Figure 39.

LPAHs of this size can be very difficult to study because of the low solubility even in "good" LPAH solvents. 1-Methylnaphthalene and 1,2,4-trichlorobenzene (TCB) are two of the best solvents for LPAHs because they are both very strongly interacting solvents and have boiling points well above 200 °C that allow heating of sample solutions to aid in dissolving, which allows heating of the samples. Yet, neither of these solvents could dissolve appreciable amounts of

---

Figure 35. Shpol'skii of hydrothermal vent fraction and coronene. From J. C. Fetzer, B. R. T. Simoneit, P. Garrigues, and H. Budzinski, *Polycyclic Arom. Cmpds.* **9** 114 (1996), reproduced with permission of Gordon and Breach Science Publishers.

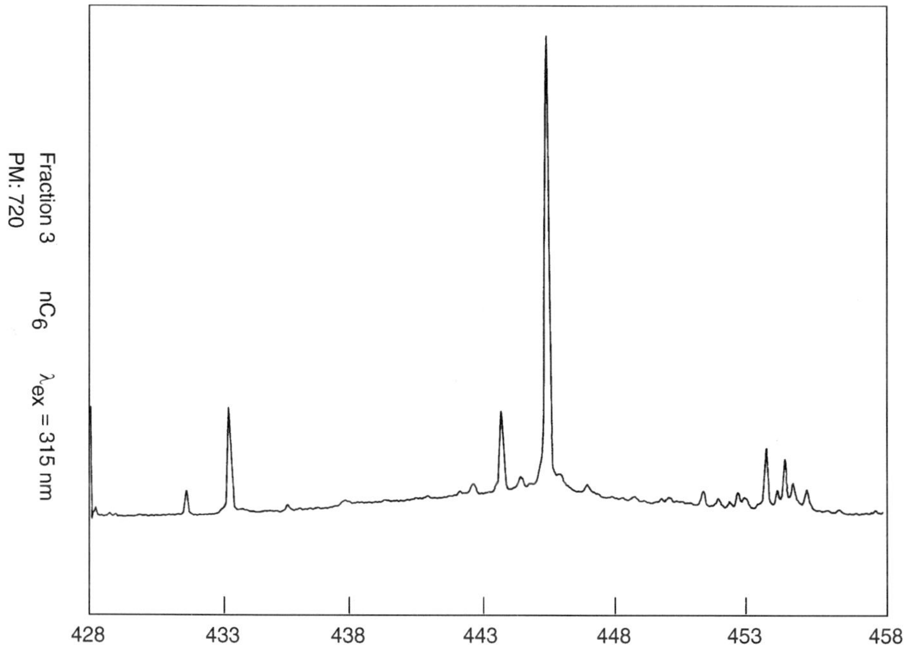
Fraction 3 nC6 λex = 315 nm
PM: 720
428
433
438
443
448
453
458

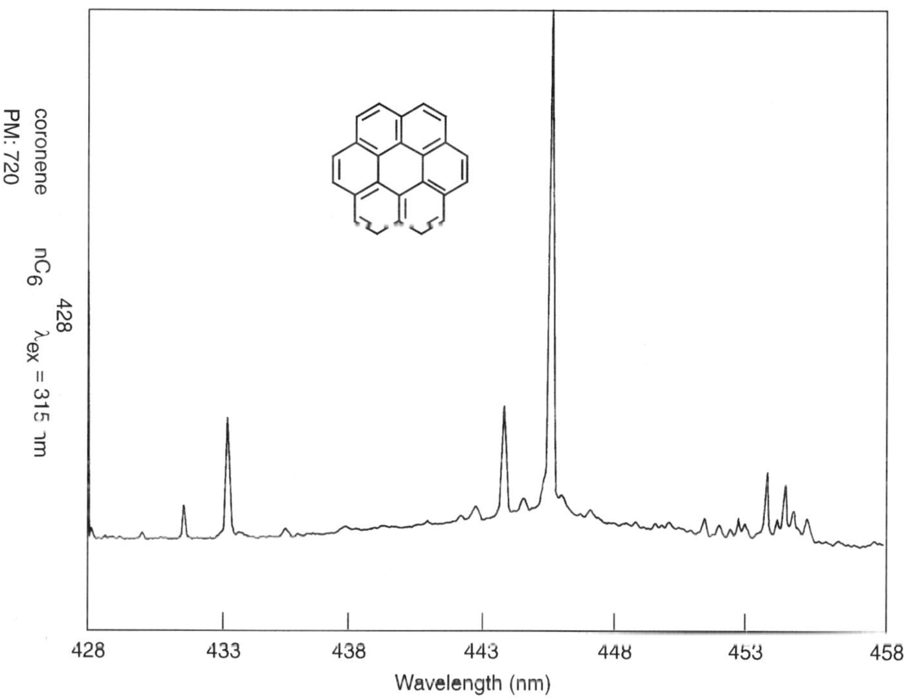
coronene nC6 λex = 315 nm
PM: 720
428
428
433
438
443
448
453
458
Wavelength (nm)

two newly synthesized LPAHs of 16 and 17 rings that Zander studied. On the other hand, spectra of the two in the "solid solutions" of the PHC matrices could be collected.

The spectra of some other highly-condensed LPAHs, including naphtho-[8,1,2-*abc*]coronene and ovalene, were collected for samples in the PHC matrix (40). Three isomers of $C_{28}H_{14}$ were compared. Interestingly, it was found that the spectra in PHC and *n*-hexadecane differ in relative band heights. The use of LPAHs to define so-called solvent polarity scales would predict very similar spectra for these two very nonpolar media. The spectra of four LPAHs, benzo[*pqr*]naphtho[8,1,2-*bcd*]perylene, naphtho[8,1,2-*abc*]coronene, ovalene, and the very large tetrabenzo[*bc,ef,o,uv*]phenanthro[2,1,10,9-*ijkl*]ovalene are shown in Figures 40–43. The corresponding spectra in TCB are also shown to highlight the much richer spectra obtained in PHC matrices.

It is also interesting to note for this and other similarly sized LPAHs that the 0-0 transition is much more prominent in PHC than in solution. This is evidenced by both a low-wavelength band in the emission spectrum and by a larger α band in the excitation spectrum. This band in the excitation spectrum of benzo[*pqr*]naphtho[8,1,2-*bcd*]perylene is over five times more intense (relative to the most prominent band) in PHC compared to TCB. In general, the higher-energy bands in the emission spectra were more prominent in the nonpolar *n*-hexadecane and PHC than in the spectra collected in TCB.

Fluorescence spectra were run in 1-methylnaphthalene, TCB, and PHC for some of the very large "fully benzenoid" LPAHs that were synthesized by Muellen and co-workers through the condensation of polyphenyl polymeric material (41–46). The high degree of delocalization in the π electrons is seen in these spectra. There is an intense band at low wavelength and much smaller bands at higher wavelengths, qualitatively similar to those of coronene or ovalene. The spectra of one of these, tetrabenzo[*bc,de,o,uv*]phenanthro[2,1,10,9-*hijk*]ovalene, in TCB and PHC are shown. The band height differences between the two solvents indicate that this LPAH must have a weak S_0 to S_1 transition (similar to pyrene, coronene, and ovalene), which results in the emission spectral bands being irregular and varying in solvents of different affinities. The spectrum in PHC also has more fine structure owing to the trapping of the molecule and the resulting less collisional broadening than that in the fluid solution spectrum in TCB. This work showed that only a miniscule amount of a $C_{78}H_{26}$ LPAH was soluble in TCB or 1-methylnaphthalene, but much more intense spectra with good signal-to-noise ratios were obtained in PHC.

Figure 36. Shpol'skii of hydrothermal vent fraction and methylcoronene. Note the presence of coronene in the standard material. From J. C. Fetzer, B. R. T. Simoneit, P. Garrigues, and H. Budzinski, *Polycyclic Arom. Cmpds.* **9** 115 (1996), reproduced with permission of Gordon and Breach Science Publishers.

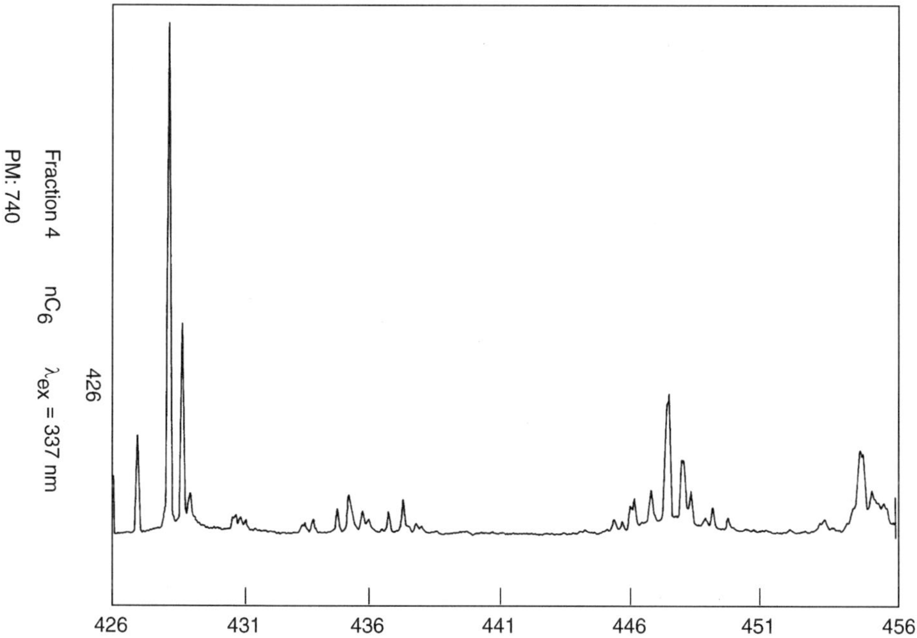
Fraction 4
nC_6
λ_{ex} = 337 nm
PM: 740
426
426
431
436
441
446
451
456

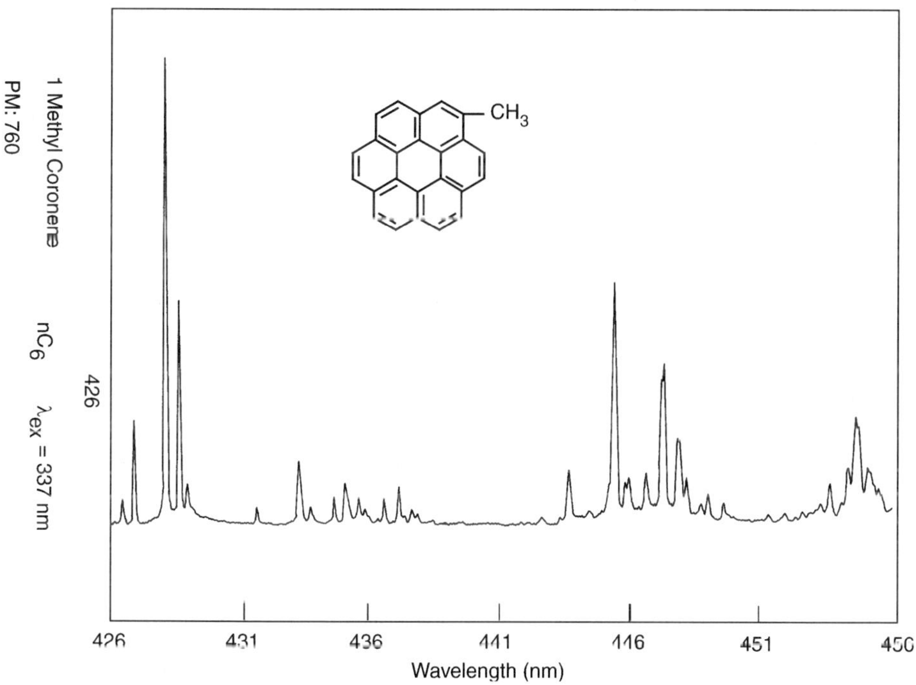
1 Methyl Coronene
nC_6
λ_{ex} = 337 nm
PM: 760
CH_3
426
426
431
436
441
446
451
456
Wavelength (nm)

This approach is probably the only current way to obtain spectra of the very large and extremely insoluble LPAHs. It, however, suffers from one major drawback. PHC is not commercially available, so these few studies, which have shown the usefulness of PHC-matrix spectroscopy, have been done with material that was produced and isolated by those researchers.

4.4 SYNCHRONOUS SCANNING METHODS

For LPAHs with symmetrical fluorescence spectra, another fluorescence scanning technique can reduce the complexity of the spectra of mixtures. For a single LPAH, if both monochromators are scanned simultaneously, with the emission monochromator offset by the Stokes shift, there will only be one peak in the scan. This is located where the LPAH is excited and has emission at the offset wavelength. All the lower-wavelength excitation bands do not have any emission at the offset wavelengths. Concurrently, all the higher-wavelength emission bands are not excited at wavelengths other than the lower offset wavelength. This type of spectrometry is known as synchronous-scanning fluorescence (SSF). Vo-Dinh has written a comprehensive review of the approach (47). He gives examples of the high selectivity seen in SSF spectra.

In the processing of petroleum through catalytic hydrocracking, LPAH products are produced that cause a variety of process problems (48–50). A number of LPAHs that are produced in the process are in the 7- to 10-ring range. These were initially determined by HPLC analysis. The structures of these LPAHs gave an indication of the reactions responsible for their formation. The major processing problem being studied was the formation of a process deposit that clogged reactor lines. The LPAHs found by HPLC only accounted for a small fraction of the deposit. After Soxhlet extraction with dichloromethane of these LPAHs, a reddish residue remained. This residue could be dissolved in TCB. Normal excitation and emission spectra showed this to be the 15-ring PAH benzo[1,2,3-*cd*:4,5,6-*c′d′*]dicoronene (*compound 53*), trivially named dicoronylene. For process samples, the smaller LPAHs interfere with dicoronylene determination by normal fluorescence spectra. SSF could, however, be used to selectively determine dicoronylene and three other similar LPAHs, even in the presence of the smaller PAHs at 1000-times higher concentrations. The Scholl condensation of two pyrene molecules, or two coronene molecules, or a coronene molecule with an ovalene molecule results in formation of very large PAHs very rapidly and readily in this process. These result in dinaphtho[2,1,8,7-*defg*:2′,1′,8′,7′-*ijkl*]pentaphene (*compound 77*) and dinaphtho[2,1,8,7-*defg*:2′,1′,8′,7′-*opqr*]pentacene (*compound 78*), or benzo[1,2,3-*bc*:4,5,6-*b′c′*]dicoronene (the major component of the precipitate), or benzo[10,11]phenanthro[2′,3′,4′,5′,6′:4,5,6,7]chryseno[1,2,3-*hi*]ovalene, trivially named coronylovalene (*compound 125*).

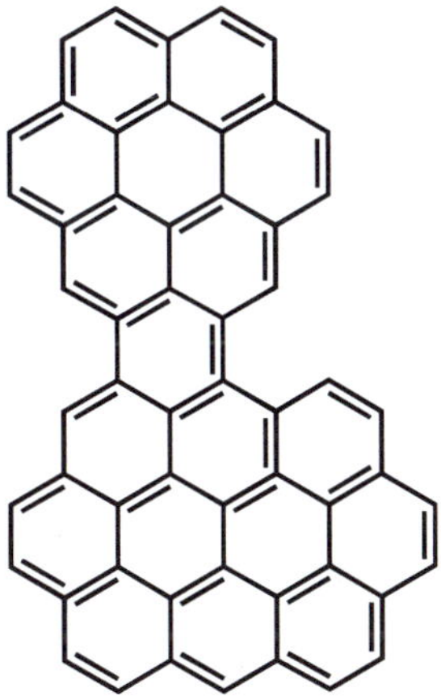

125

Dicoronylene has a Stokes wavelength shift of ~8 nm in TCB. The wavelengths of the maxima of the smaller LPAHs, however, do not correspond to those of dicoronylene. In general, the smaller PAHs, including pyrene, benzo[*ghi*]perylene, coronene, naphtho[8,1,2-*abc*]coronene, and ovalene, all have weak 0-0 transitions and do not even give a peak in the SSF spectrum. In fact, a clear spectral window from 500 to 520 nm was found, if SSF spectra were collected at an 8-nm offset (Fig. 44). This is the wavelength region of dicoronylene's highest wavelength maximum. The first two molecules formed in the Scholl condensations (the pyrene dimers) fluorescence in the ranges of 490–495 nm and 530–535 nm. The last LPAH fluoresces in the range of 545–550 nm. These are the nearest interfering peaks to the main peal of interest and are only bothersome if the pyrene dimerization is very prevalent, where the 490-nm peak is greater than 100 times more intense than a small dicoronylene peak.

In process oil samples, dicoronylene can be detected in SSF spectra run at this offset wavelength down to sub-parts-per-billion levels for standard solutions. In the determination of dicoronylene, the detection limits were typically a few parts per billion by weight in the processed oils that were analyzed. This was in part due to the high wavelengths of the analyte, 505 nm excitation and 513 nm emission, where there were no interferences from the other LPAHs in the samples. Typically, detection limits are in the 100 pg/g to 1 ng/g range for LPAHs such as coronene, dibenzo[*cd,lm*]perylene, naphtho[8,1,2-*abc*]coronene, and ovalene.

This method was subsequently used to determine dicoronylene solubilities in process oils. Its use also allows for monitoring of both dicoronylene formation and the potential saturation of the process oil (and the presumed onset of

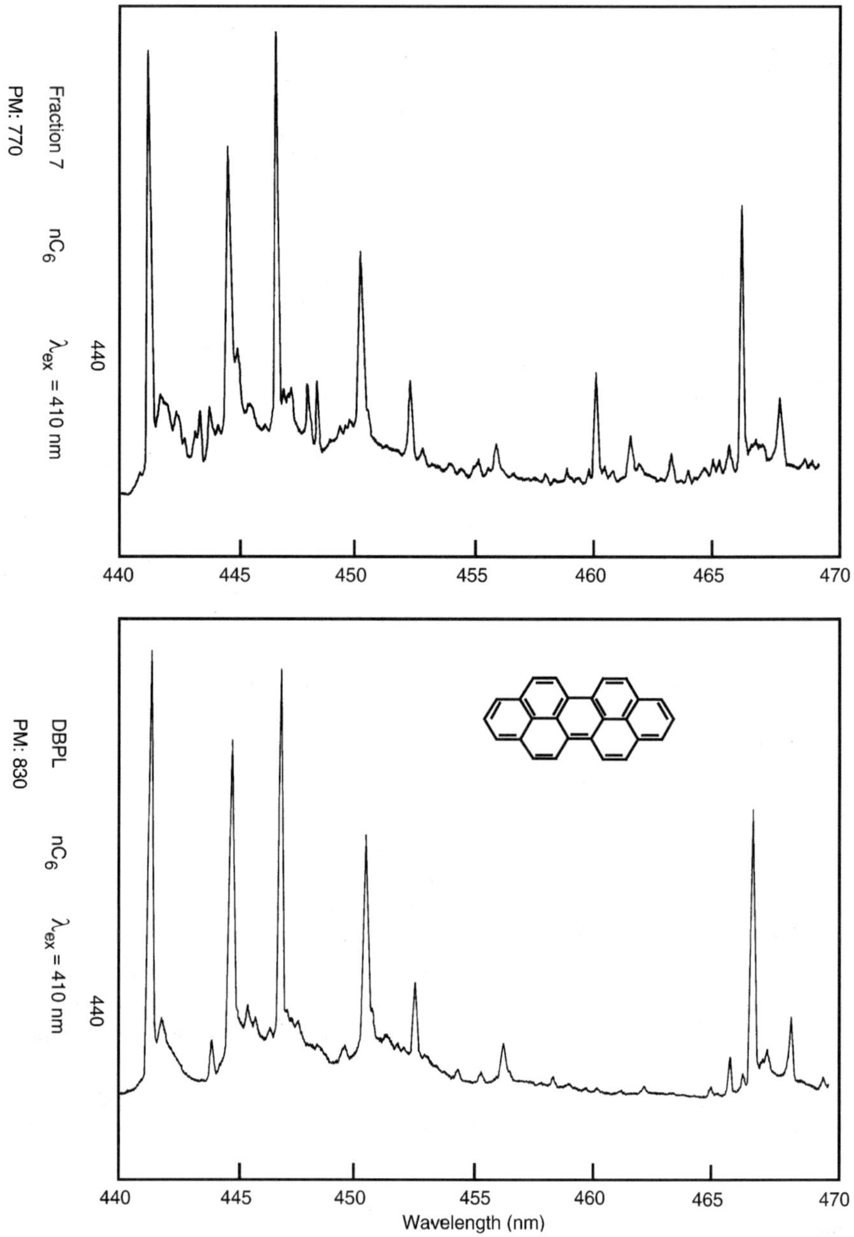

Figure 37. Shpol'skii of hydrothermal vent fraction and dibenzo[*cd,lm*]perylene. Compare to Figure 32 for the same compound. From J. C. Fetzer, B. R. T. Simoneit, P. Garrigues, and H. Budzinski, *Polycyclic Arom. Cmpds.* **9** 116 (1996), reproduced with permission of Gordon and Breach Science Publishers.

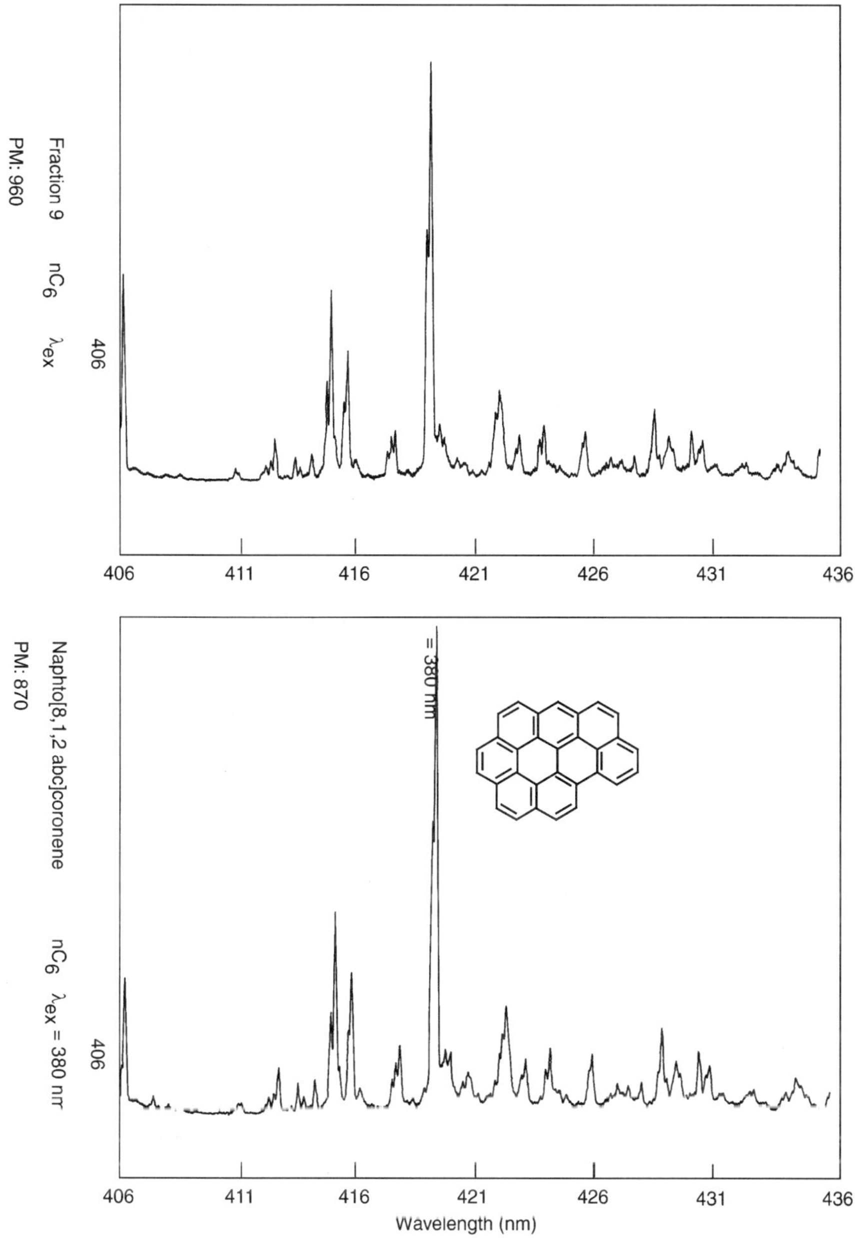

Figure 38. Shpol'skii of hydrothermal vent fraction and naptho[8,1,2-*abc*]coronene. Compare to Figure 33 for the same compound. From J. C. Fetzer, B. R. T. Simoneit, P. Garrigues, and H. Budzinski, *Polycyclic Arom. Cmpds.* **9** 117 (1996), reproduced with permission of Gordon and Breach Science Publishers.

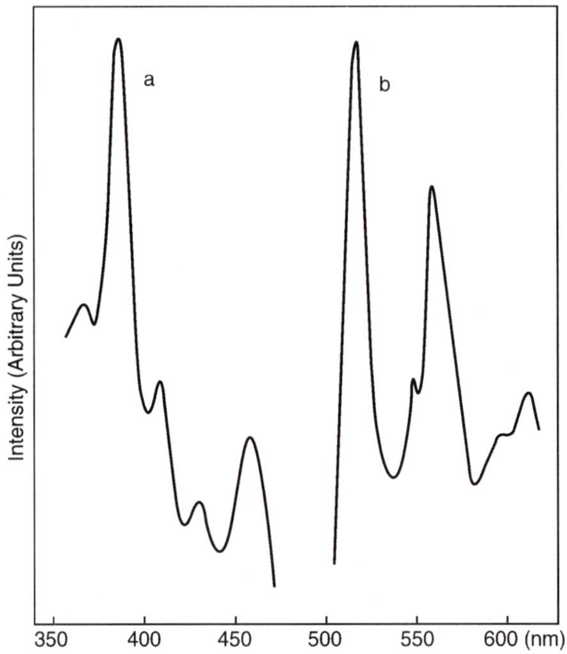

Figure 39. Zander spectrum of 16-ring LPAH. From M. Zander, in *Polycyclic Aromatic Compounds: Synthesis, Properties, Analytical Measurements, Occurrence, and Biological Effects*, P. Garrigues and M. Lamotte, Eds., Gordon and Breach Science Publishers, Yverdon, Switzerland, 1993, p. 326, reproduced with permission of Gordon and Breach Science Publishers.

dicoronylene precipitation). This can then be used to determine the need for and timing of processing steps to prevent the precipitation.

Kershaw and Fetzer (51) presented a review of the use of fluorescence spectrometry in the analysis of petroleum and related materials. In it they describe several applications of SSF, including characterization of coal tar pitches and petroleum fractions.

4.5 3-DIMENSIONAL FLUORESCENCE

One aspect of LPAH fluorescence spectra is that no matter what excitation wavelength is used, the resulting emission spectra only differ in intensity. When normalized, the emission spectra for a single pure LPAH taken at different wavelengths will be perfectly superimposable. The same is true of any excitation spectra collected at different emission wavelengths. If all these separate excitation (or emission) spectra are arrayed in sequential order, then a new type

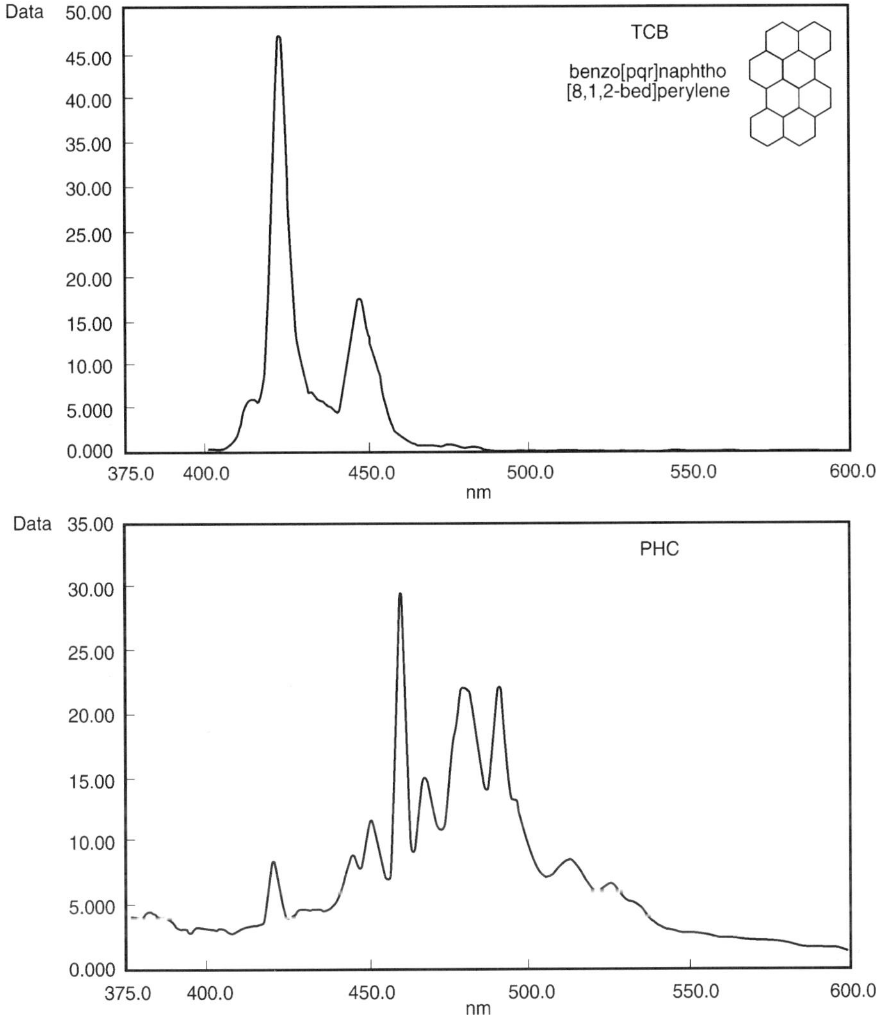

Figure 40. Emission spectra of benzo[*pqr*]naphtho[8,1,2-*bcd*]perylene in 1,2,4-trichlorobenzene and perhydrocoronene.

of spectrum is obtained. This format is known as a 3-dimensional (3D) spectrum (52–54).

The 3D spectra are commonly displayed in one of two forms. They can either viewed at an angle so that the band intensities appear like hills or mountains rising from a plain or viewed from directly above as a contour plot (similar in appearance to a topological terrain map). The latter is usually more diagnostic.

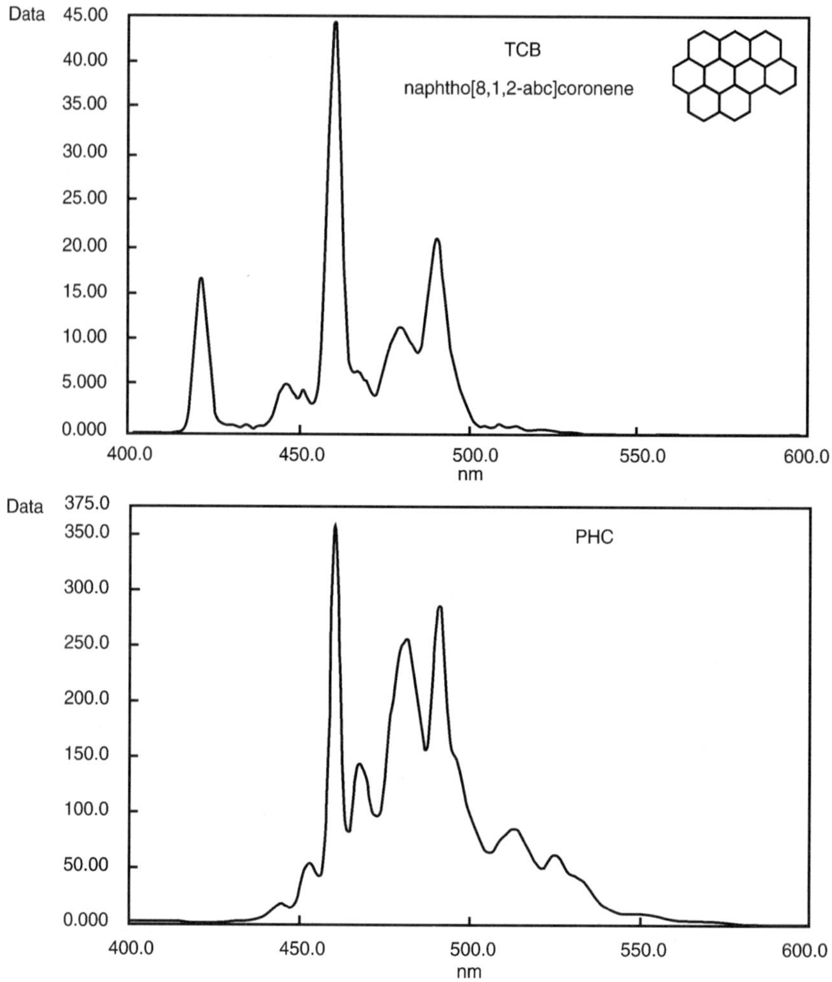

Figure 41. Emission spectra of naphtho[8,1,2-*abc*]coronene in 1,2,4-trichlorobenzene and perhydrocoronene.

Each peak appears as a set of concentric circles or ovals. In the contour plot, a pure compound will give responses so that the peaks only are arrayed perpendicularly in the *x*- and *y*-axes. An example is shown of a mixture of two LPAHs. Each component does this independently of the others so that information from mixtures can be pulled out of a 3D spectrum if each component has a spectral region where the other does not fluoresce.

A series of reaction mixtures were analyzed by 3D fluorescence (55). Smaller PAHs were condensed through Friedel-Crafts condensation to form LPAHs. The

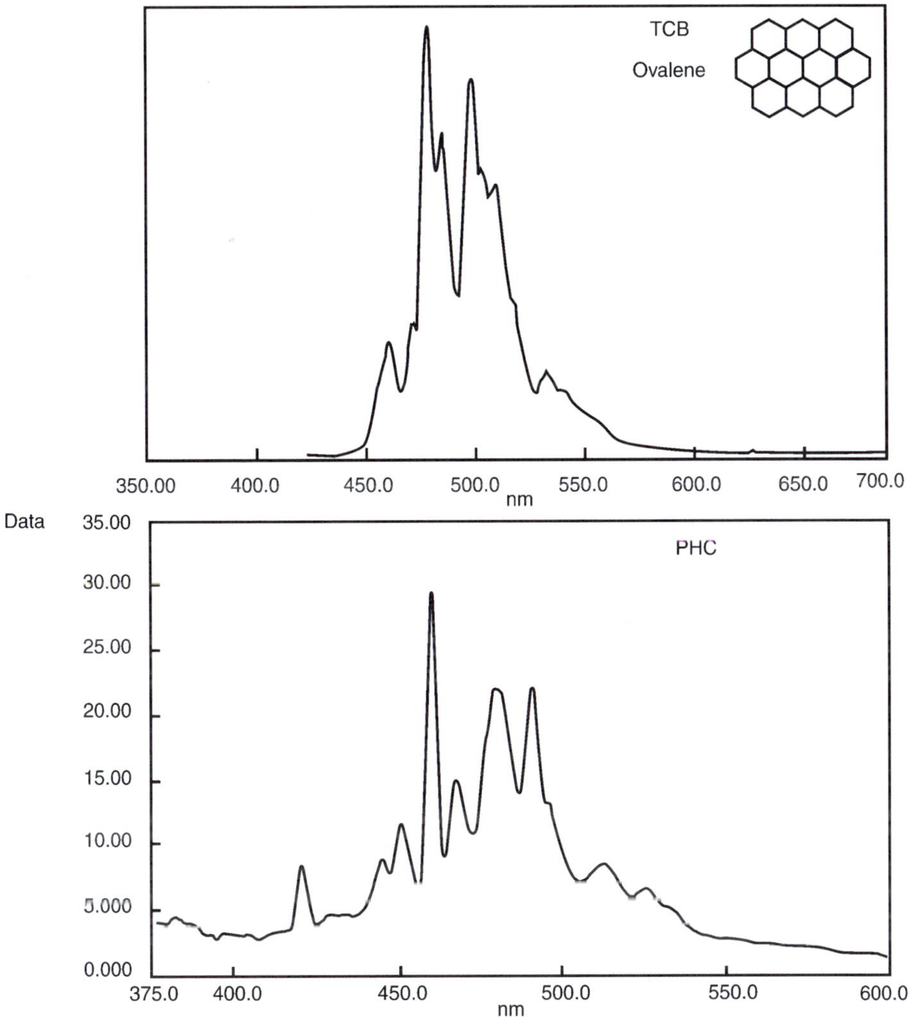

Figure 42. Emission spectra of ovalene in 1,2,4-trichlorobenzene and perhydrocoronene.

product mixtures were dissolved in TCB. Several new LPAHs of 12–21 rings were identified through the use of Clar's annellation theory (described in the chapter on structural determination methods). An example portion is shown in Figure 45. This spectrum is of the reaction product of naphthanthrone by the Clar reaction (zinc dust fusion). Two isomeric 38-carbon LPAHs are seen (*compounds 126* and *127*), although one is in much less abundance (maximum at ~570-nm excitation) and is partially obscured by the more prevalent isomer (maximum at ~540-nm excitation).

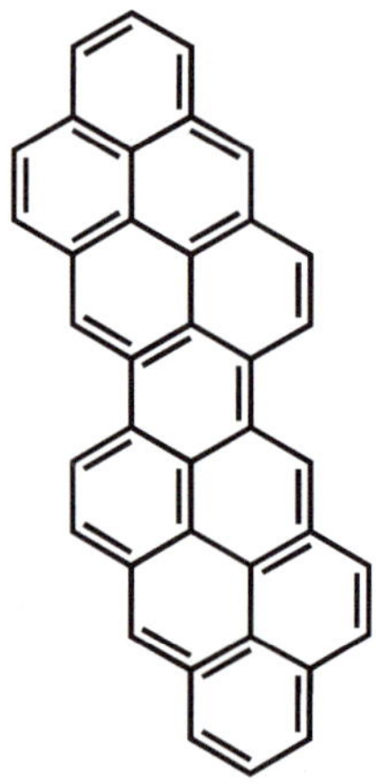

126

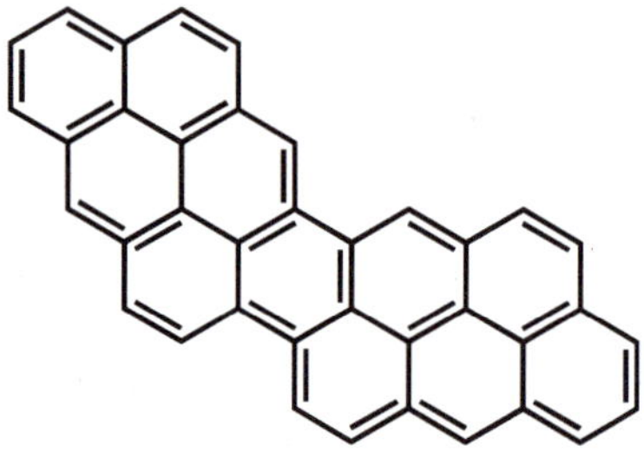

127

4.6 LIFETIME (TIME-RESOLVED) METHODS

The excited state of a PAH has a decay lifetime (56, 57). The emission of a photon by the excited state is governed by probability. The rate of decay in the fluorescence intensity after excitation will decrease exponentially if the excitation is stopped. When pulsed-flash lamps or lasers are used as the excitation source, the emission decay rate can be determined. Structural factors such as the number of π electrons, the number and arrangement of rings, and the solvent used will determine the fluorescence half-life. The lifetimes for emission are usually in the range of picoseconds to a few tens of nanoseconds.

Excitation and emission can be measured at different wavelengths that are chosen for each specific PAH. These can be coordinated with the high-speed optics necessary to measure fluorescence lifetimes. This combination allows determination of several PAHs in mixtures by switching the wavelengths of excitation and emission and measuring the decay of each set. Because each decay is described by an exponential curve, deconvolution of the signal due to more than one component is possible. Unfortunately, there have been no published reports in which this powerful approach has been applied to mixtures of LPAH.

Siemiarchuk, in an unpublished study (56), examined two fractions from the solid deposits found in petroleum hydrocracking processing units. These contained different proportions of the two LPAHs benzo[1,2,3-*bc*:4,5,6-b′*c*′]-dicoronene and benzo[10,11]phenanthro[2′,3′,4′,5′,6′:4,5,6,7]chryseno[1,2,3-*hi*]ovalene (*compounds 53 and 125*) that are produced by the Scholl condensation of coronene and ovalene. In TCB solution the lifetimes were

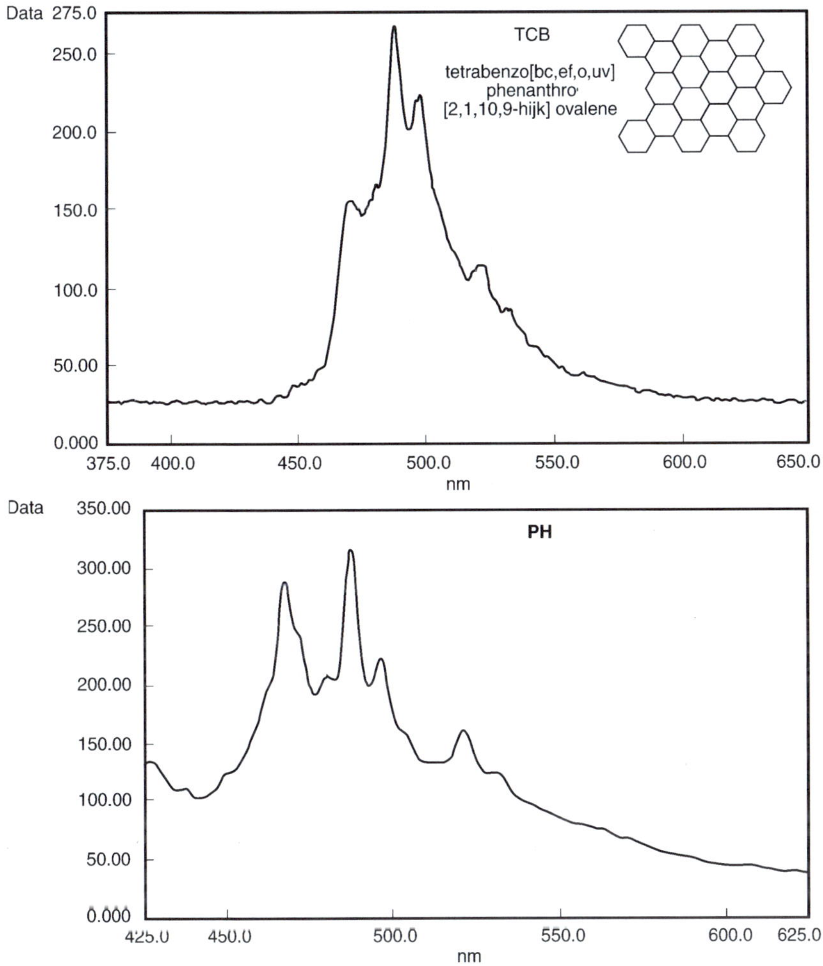

Figure 43. Emission spectra of tetrabenzo[*bc*,*ef*,*o*,*uv*]phenanth[2,1,10,9-*hijk*]ovalene in 1,2,4 trichlorobenzene and perhydrocoronene.

readily seen as different for each LPAH at the three emission wavelengths examined; 510, 535, and 560 nm were used because each of the two fractions contributed different amounts of light at each wavelength. The first wavelength is due solely to the first LPAH, and the other two are at maxima for the second LPAH, but the higher wavelength bands from the smaller LPAH interfere. Figures 46 and 47 show the lifetime decay curves for each wavelength. Excitation was at 360 nm. The deconvoluted lifetimes were found to be 2.61

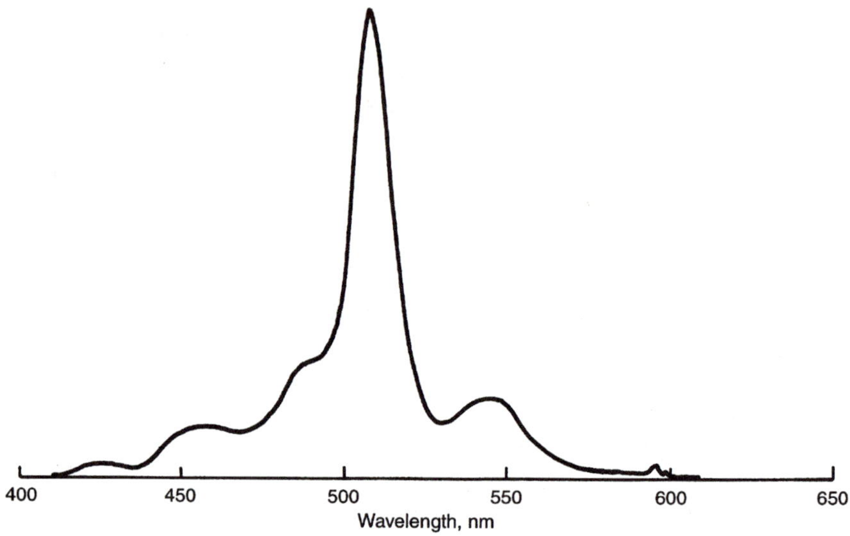

Figure 44. Synchronous-scanning fluorescence analysis for dicoronylene in a petroleum hydrocracking deposit sample. The large peak at 545 nm is from *compound 125*, which is also produced in this process. From J. C. Fetzer, in *Catalysts in Petroleum and Petrochemical Processing 1995*, M. Absi-Halabi, J. Besharah, H. Qabazard, and A. Stanislaus, Eds., Elsevier, 1996, p. 263–271.

($\pm$0.01) ns and 2.95 ($\pm$0.04) ns (Fig. 48). Although these are both very short lived and close to each other in value, they were easily differentiated.

Actually, there is a fourth-dimensional approach available. If the 3D spectrum is collected rapidly and the intensities measured as a function of time on the scale of fluorescence emission, then 3D fluorescence decay spectra can be collected. The research group of McGown (57) has studied this technique for mixtures of isomeric PAHs, calling the technique phase-resolved total luminescence spectroscopy. With fast-scanning optics and multi-channel acquisition devices that can rapidly collect an emission spectrum, the 3D spectrum can be collected through a series of excitation at different wavelengths. Nithipatikom and McGown (58) have reported this approach of adding lifetime-dependent information to both emission and synchronous-scanning spectra. The increased selectivity of detection by adding lifetime and variations on wavelength selection in excitation and emission makes feasible multi-component analysis without the requirement for prior separation. The determination of 9-phenyl-anthracene, 9.10-diphenylanthracene, benzo[*k*]fluoranthene, and benzo-[*a*]pyrene was shown for each mode. In later work, two additional PAHs, 1,3,6,8-tetraphenylpyrene and benzo[*ghi*]perylene, were added to this matrix and similar successful results were obtained. Additionally, the sensitivity of

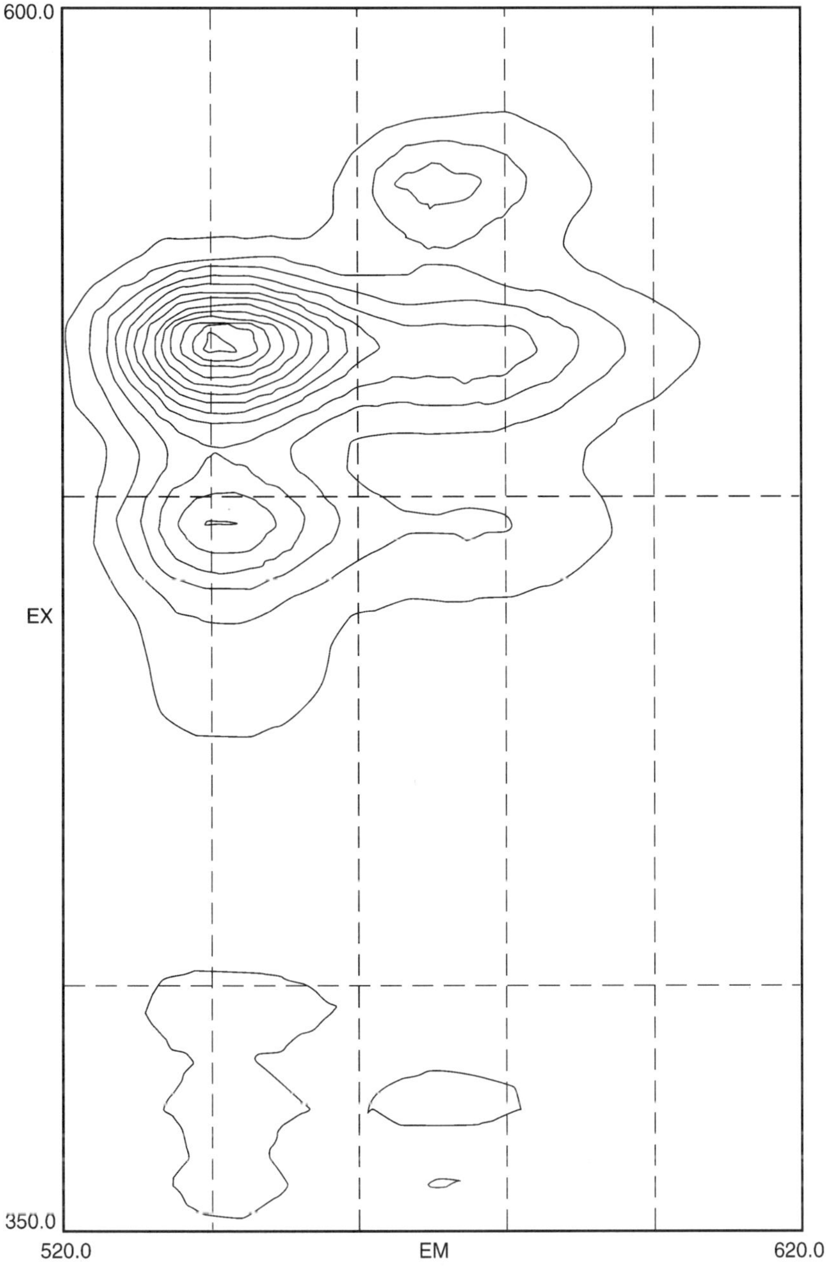

Figure 45. Three-dimensional fluorescence of Clar reaction naphthanthrone product, showing two of the 38-carbon isomers.

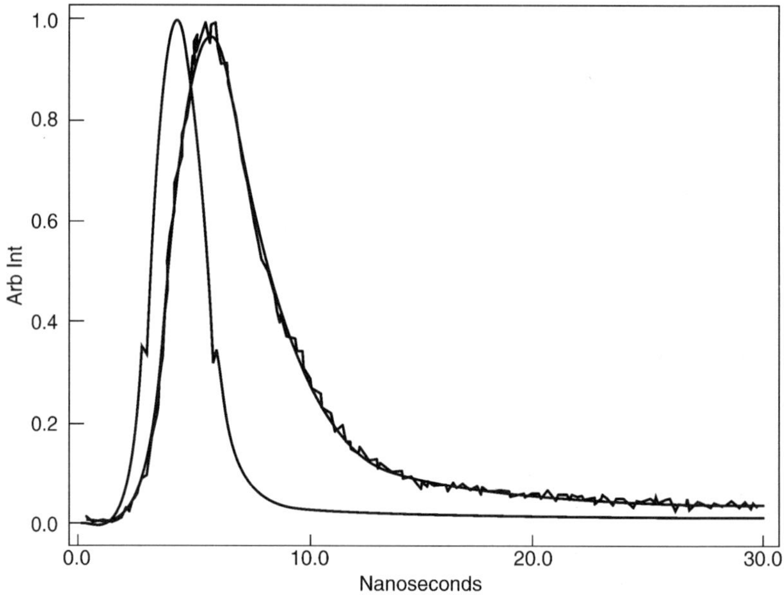

Figure 46. Fluorescence decay spectrum for 360-nm excitation and 510-nm emission of a solution containing a mixture of *compounds 53* and *125*. From A. Seimiarczuk, Photon Technologies International, Inc., personal communication.

the analyses can be lowered because the phase-resolution allows much better elimination of background-scattered light and dark current baseline noise. Unfortunately, there have been no similar reports of these various techniques applied to the LPAHs.

4.7 APPENDIX

The appendix contains fluorescence excitation and emission spectra for several of the LPAHs that have been commonly found. To avoid the solvent-induced intensity changes described above, TCB was used for all of the spectra (because it is the only solvent that can appreciably dissolve all of them).

4.8 PHOSPHORESCENCE SPECTROSCOPY

The PAHs, and particularly the LPAHs, also exhibit phosphorescence (59–61). Several recent books have been written describing phosphorescence. Briefly, phosphorescence is the loss of excited-state energy by emission from a triplet state, rather than from a singlet state as in fluorescence. This is usually

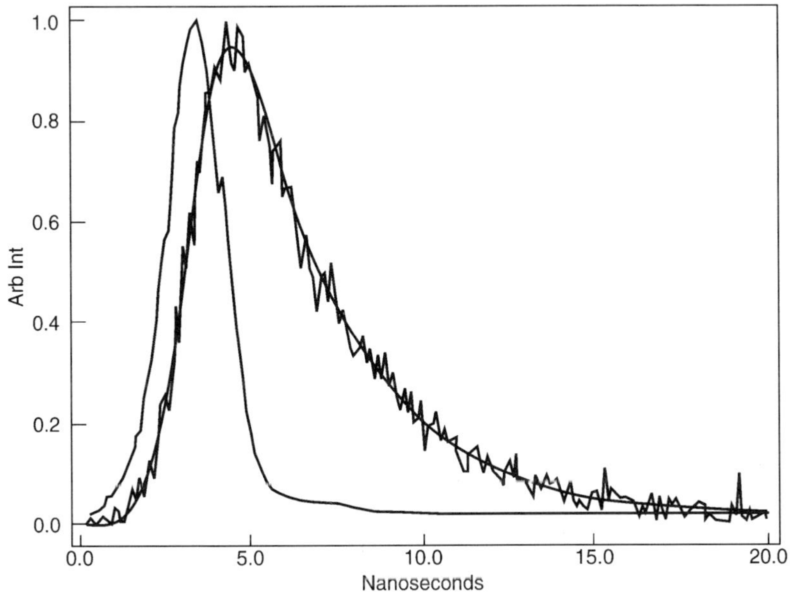

Figure 47. Fluorescence decay spectrum for 360-nm excitation and 535-nm emission of a solution containing a mixture of *compounds 53* and *125*. From A. Seimiarczuk, Photon Technologies International, Inc., personal communication.

accomplished electronically by the LPAH first being excited from ground singlet to excited singlet states.

Phosphorescence spectroscopy is seldom performed in solution because collisional deactivation of the triplet state species is very likely. The LPAHs are trapped in solid matrices, such as frozen solutions of *n*-alkane solvents like *n*-hexane; on adsorbents like cellulose or silica; in a solid solution of a higher-temperature matrix like PHC (perhydrocoronene); or as the LPAH crystals or powders themselves. The latter three approaches can all be done at ambient temperature, and the second has become commonly known as room-temperature phosphorescence (RTP). No RTP results for LPAHs have been reported. The PHC experiments of Zander et al. do include measurement of phosphorescence spectra and lifetimes.

The fluorescence lifetime of the LPAH is sufficiently long that some of the excited-state molecules undergo conversion to an excited triplet state. This results because there is coupling between the molecular orbitals of the excited singlet and triplet states. This is not a highly favored transition because it does not conserve the spin angular momentum states of the molecules (a singlet state has a spin angular momentum of 0 and a triplet state has a value of 1, and so there is a difference from one state to the other).

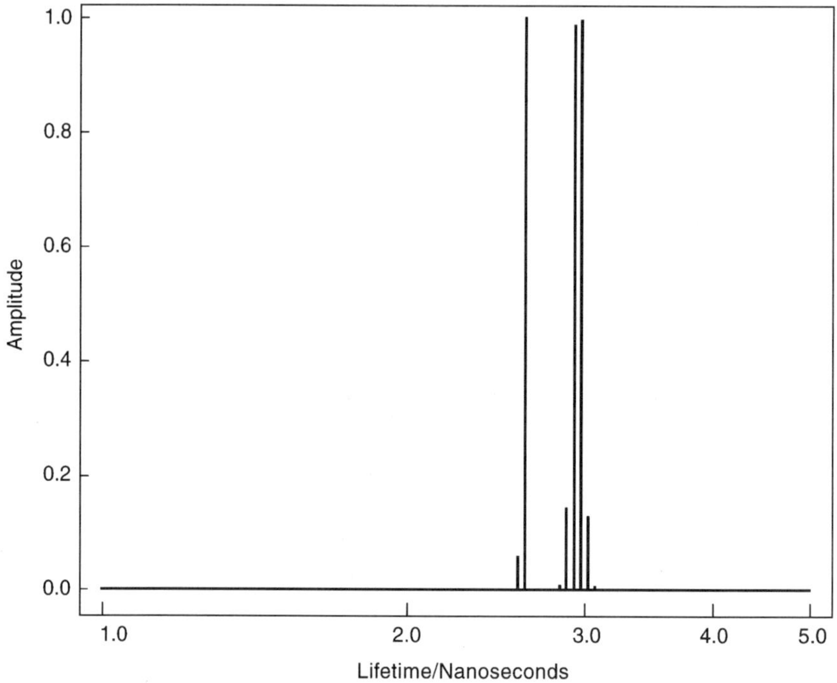

Figure 48. Lifetimes generated from the spectra in Figures 46 and 47.

This conversion occurs through inter-system crossing. Inner-system crossing is the phenomenon in which an excited molecule converts from one spin angular momentum state to another (in the case of excited LPAHs, from singlet to triplet or vice versa). This can occur with or without emission of light.

Once inner-system crossing has occurred, the triplet-state molecules cannot readily lose energy through photon emission to the ground state because this transition is also spin angular momentum forbidden. This results in a long phosphorescence lifetime. For example, coronene in a perhydrocoronene matrix is reported to have a half-life of well over a minute.

There are two major spectrally observed differences between phosphorescence and fluorescence for the same molecule. First, the lifetime is much longer, usually one or more orders of magnitude. Second, it occurs at much higher wavelengths than fluorescence (because of the energy losses during the transitions from singlet to triplet states and vice versa).

Obenland and Schmidt (62) report the low-temperature phosphorescence spectrum of the fully-benzenoid tribenzo[*a*,*g*,*m*]coronene (tribenzo[*fgh*,*pqr*,-za_1b_1]trinaphthylene; *compound 52*) in 1,4-dioxane.

As described earlier, Zander and co-workers used perhydrocoronene as an inert, trapping matrix for the low- and room-temperature fluorescence and phosphorescence spectrometry of several LPAHs.

Ackerman and Hurtubise (63) did room-temperature solid-matrix phosphorimetry of mixtures of PAHs, including coronene. Filter paper impregnated with silicone oil and a tin-containing compound was used to extract and concentrate the PAHs from aqueous samples. A mixture of coronene and benz[*a*]anthracene, at 50 ng/mL each, and phenanthrene and anthracene, at twice that concentration, were extracted from water. The phosphorescence spectrum easily showed the coronene spectrum. They then measured coronene, down to 0.50 ng/mL, in similar PAH mixtures with a 10-fold excess of the other three PAHs.

Zhelezko and co-workers studied the low-temperature fluorescence and phosphorescence of a "dibenzanthranthrene" trapped in a naphthalene matrix (64). Unfortunately, they did not include a structure to define which $C_{30}H_{16}$ isomer they used. The fluorescence emission spectrum showed numerous bands in the range above 620 nm.

REFERENCES

1. J. Michl and E. Thulstrup, in *Spectroscopy With Polarized Light*, VCH, New York, 1986, p. 354–366.
2. E. W. Thulstrup and J. Michl, *Elementary Polarized Spectroscopy*, VCH, New York, 1989.
3. M. Zander and W. Friedrichsen, *Z. Naturforsc. A Phys. Sci.* **47** 1314–1318 (1992).
4. A. Bohnen, K.–H. Koch, W. Luettke, and K. Muellen, *Angew. Chem. Int. Ed. Engl.* **29** 525–527 (1990).
5. S. A. Tucker, H. Darmodjo, W. E. Acree, Jr., J. C. Fetzer, and M. Zander, *Appl. Spectrosc.* **46** 1260–1265 (1992).
6. S. A. Tucker, W. E. Acree, Jr., J. C. Fetzer, and J. Jacob, *Polycyclic Aromat. Compd.* **3** 1–10 (1992).
7. S. A. Tucker, W. E. Acree, Jr., J. C. Fetzer, R. G. Harvey, M. J. Tanga, L. T. Scott, and P. C. Cheng, *Appl. Spectrosc.* **47** 715–722 (1993).
8. S. A. Tucker, H. C. Bates, V. L. Amszi, W. E. Acree, Jr., R. G. Harvey, J. C. Fetzer, and G. Dyker, *Anal. Chimica Acta* **278** 269–274 (1993).
9. S. A. Tucker, H. C. Bates, W. E. Acree, Jr., and J. C. Fetzer, *Appl. Spectrosc.* **47** 1775–1779 (1993).
10. S. A. Tucker, J. M. Griffin, W. E. Acree, Jr., J. C. Fetzer, M. Zander, A. de Meijere, and I. Murata, *Polycyclic Aromat. Compd.* **4** 141–160 (1994).
11. E. Boriessevitch, *J. Luminesc.* **81** 219–224 (1999).
12. H. Matsushita, Y. Esmi, A. Suzuki, and T. Handa, *Bunseki Kagaku* **21** 1471 (1972).

13. J. F. McKay and D. R. Latham, *Anal. Chem.* **45** 1050–1055 (1973).
14. P. A. Peaden, M. L. Lee, Y. Hirata, and M. Novotny, *Anal. Chem.* **52** 2268–2271 (1980).
15. H. Dreeskamp, E. Koch, and M. Zander, *Z. Naturforsch. A* **30** 1311–1214 (1975).
16. M. Zander, U. Breymann, H. Dreeskamp, and E. Koch, *Z. Naturforsch. A* **32** 1561–1563 (1977).
17. U. Breymann, H. Dreeskamp, E. Koch, and M. Zander, Fresen. *Z. Anal. Chem.* **293** 208–210 (1978).
18. U. Breymann, H. Dreeskamp, E. Koch, and M. Zander, *Chem. Phys. Lett.* **59** 68–70 (1978).
19. S. Pandey, W. E. Acree, Jr., and J. C. Fetzer, *Anal. Chim. Acta* **324** 175–181 (1996).
20. S. Pandey, K. A. Fletcher, J. R. Powell, M. E. R. McHale, A.-S. Kauppila, W. E. Acree, Jr., J. C. Fetzer, and R. G. Harvey, *Spectrochima Acta A.* **53** 165–172 (1997).
21. S. Pandey, W. E. Acree, Jr., and J. C. Fetzer, *J. Luminesc.* **71** 189–197 (1997).
22. S. Pandey, W. E. Acree, Jr., B. P. Cho, and J. C. Fetzer, *Talanta* **44** 413–421 (1997).
23. W. E. Acree, Jr., S. Pandey, S. A. Tucker, and J. C. Fetzer, *Polycyclic Aromat. Compds.* **12** 71–123 (1997).
24. S. Pandey, W. E. Acree, Jr., and J. C. Fetzer, *Talanta* **45** 39–45 (1997).
25. J. C. Fetzer, S. Pandey, K. A. Fletcher, and W. E. Acree, Jr., Fresen. *Z. Anal. Chem.* **360** 669–674 (1998).
26. S. Pandey, W. E. Acree, Jr., and J. C. Fetzer, *Mikrochim. Acta* **129** 41–45 (1998).
27. S. Pandey, W. E. Acree, Jr., and J. C. Fetzer, *Talanta* **47** 769–778 (1998).
28. U. Wenzel and H.-G. Loehmannsroeben, *J. Photochem. Photophys. A* **96** 13–18 (1996).
29. A. Colmsjo, *Anal. Chim. Acta* **197** 65–70 (1987).
30. A. Colmsjo and C. Ostman, in *Polynuclear Aromatic Hydrocarbons: Physical* and *Biological Chemistry*, M. Cooke, A. J. Dennis, and G. L. Fisher, Eds., Batelle, Columbus, OH, 1982, p. 201–210.
31. S. A. Wise, B. A. Benner, H. Liu, G. D. Byrd, and A. Colmsjo, *Anal. Chem.* **60** 630–637 (1988).
32. F. J. Morgan, R. Lamotte, P. Garrigues, and J. C. Fetzer, *Polycyclic Aromat. Cmpds.* **2** 141–153 (1991).
33. A. Colmsjo and C. Ostman, *Anal. Chim. Acta.* **208** 183–193 (1988).
34. J. C. Fetzer, B. R. T. Simoneit, P. Garrigues, and H. Budzinski, *Polycyclic Aromat. Cmpds.* **9** 109–120 (1996).
35. L. Nakhimovsky, M. Lamotte, and J. Joussot-Dubien, *Handbook of Low Temperature Electronic Spectra of Polycyclic Aromatic Hydrocarbons*, Elsevier, Amsterdam, 1989.
36. *Spectral Atlas of Polycyclic Aromatic Compounds*, W. Karcher et al., Eds., Reider and Kluwer, Dordrecht, The Netherlands, 1983, 1988, 1991, Vols. 1–3.
37. J. C. Fetzer and M. Zander, *Z. Naturforsch.* **45a** 727 (1990).

38. J. C. Fetzer and M. Zander, *Z. Naturforsch.* **45a** 814 (1990).
39. M. Zander, in *Polycyclic Aromatic Compounds: Synthesis, Properties, Analytical Measurements, Occurrence, and Biological Effects*, P. Garrigues and M. Lamotte, Eds., Gordon and Breach Science, Yverdon, Switzerland, 1993, p. 321–330.
40. J. C. Fetzer, The Seventeenth International Symposium on Polycyclic Aromatic Compounds, Bordeaux, France, October 25–29, 1999.
41. M. Mueller, H. Mauermann-Duell, M. Wagner, V. Endelmann, and K. Muellen, *Chem.-Eur. J.* **1** 1583–1586 (1995).
42. F. Morgenroth, C. Kuebel, and K. Muellen, *J. Mater. Chem.* **1** 1207–1211 (1997).
43. V. S. Iyer, M. Wehmeier, J. D. Brand, M. A. Keegstra, and K. Muellen, *Angew. Chem. Int. Ed. Engl.* **36** 1604–1607 (1997).
44. M. Mueller, V. S. Iyer, C. Kuebel, V. Enkelmann, and K. Muellen, *Angew. Chem. Int. Ed. Engl.* **36** 1607–1610 (1997).
45. M. Mueller, C. Kuebel, F. Morgenrith, V. S. Iyer, and K. Muellen, *Carbon* **36** 827–831 (1998).
46. A. Bohnen, K.-H. Koch, W. Luettke, and K. Muellen, *Angew. Chem. Int. Ed. Engl.* **29** 525–527 (1990).
47. T. Vo-Dinh, in *Modern Fluorescence Spectroscopy*, E. L. Wehry, Ed., Plenum, New York, 1981, Vol. 4, p. 167–192.
48. U. S. patent 5,190,633, March 3, 1993, J. C. Fetzer and D. G. Lammel (to Chevron Research and Technology).
49. J. C. Fetzer, *Polycyclic Aromat. Cmpds.* **7** 269–274 (1995).
50. J. C. Fetzer, in *Catalysts in Petroleum and Petrochemical Processing 1995*, M. Absi-Halabi, J. Besharah, H. Qabazard, and A. Stanislaus, Eds., Elsevier, 1996, p. 263–271.
51. J. R. Kershaw and J. C. Fetzer, *Polycyclic Aromat. Cmpds.* **7** 253–268 (1995).
52. G. Patonay and I. M. Warner, in *Advances in Multidimensional Luminescence*, I. M. Warner and L. B. McGown, Eds., JAI, Greenwich, CT, 1991, p. 1–32.
53. K. Nithipatikom and L. B. McGown, *Anal. Chem.* **58** 2469–2473 (1986).
54. L. B. McGown and K. Nithipatikom, in *Chemical Analysis of Polycyclic Aromatic Compounds*, T. Vo-Dinh, Ed., John Wiley and Sons, New York, 1989, p. 201–218.
55. J. C. Fetzer, *Polycyclic Aromat. Cmpds.* **11** 317–324 (1996).
56. A. Siemiarczuk, Photon Technologies International, personal communication, 1991.
57. L. B. McGown and K. Nithipatikom, in *Advances in Multidimensional Luminescence*, I. M. Warner and L. B. McGown, Eds., JAI, Greenwich CT, 1991, p. 97–109.
58. K. Nithipatikom and L. B. McGown, *Appl. Spectrosc.* **41** 1080–1082 (1987).
59. S. P. McGlynn, T. Azumi, and M. Kinoshita, *Molecular Spectroscopy of the Triplet State*, Prentice-Hall, Englewood Cliffs, NJ, 1969.
60. R. J. Hurtubise, *Phosphorimetry: Theory, Instrumentation, and Application*, VCH, New York, 1990.
61. T. Vo-Dinh, *Room-Temperature Phosphorimetry for Chemical Analysis*, John Wiley and Sons, New York, 1984.

62. S. Obenland and W. Schmidt, in *Polycyclic Aromatic Hydrocarbons and Astrophysics*, A. Leger, L. d'Hendecourt, and B. Boccara, Eds. Reidel, Dordrecht, The Netherlands, 1987, p. 165–172.

63. A. H. Ackerman and R. J. Hurtubise, *Appl. Spectrosc.* **53** 770–775 (1999).

64. F. B. Zhelezko, I. M. Gulis, B. Lounis, and M. Orrit, *J. Appl. Spectrosc.* **66** 344–352 (1999).

CHAPTER

5

INFRARED SPECTROSCOPY OF LPAHs

Infrared (IR) spectra arise from the movements of a molecule's structure. The structures of LPAHs are relatively simple, having only the core aromatic carbon rings and the associated peripheral hydrogen atoms. Thus much of the information that scientists typically find in IR spectra will not be seen. This, however, only means that certain specific structural features can be elucidated from the IR spectrum of a LPAH, not that this type of spectrum has no utility.

Additionally, IR transitions arise from changes in the dipole moments of a molecule or its substructures. The LPAHs are very simple structures, consisting of only carbon rings and peripheral hydrogens. Gross changes in the polarization of the LPAH structure or parts of it are not as likely as for other classes of organic compounds. This leads either to weak transitions or to those with no change in polarization (all the atomic motions are equivalent relative to each other and cancel). The latter occurrences are not seen in IR spectra but rather as features in Raman ones.

The simplest IR spectral feature are the ones due to the peripheral hydrogen atoms. These are all due to one hydrogen atom bonded to an aromatic carbon atom. The complexity found is the saturated hydrocarbon bands is not seen because the only movements are the C-H stretching and wagging involving these two atoms. These bands tend to dominate the spectrum because they are the easiest molecular bond movement. The IR spectra of several PAHs, including coronene and hexabenzo[*bc*,*ef*,*hi*,*kl*,*no*,*qr*]coronene are shown in Figure 49.

There are a variety of substructures in the carbon skeletons of LPAHs that can be found in LPAHs. Differences in ortho- and peri-fused structures result in different molecular shapes. A review of some of the information in Chapter 1 will highlight these differences and how these affect the IR spectrum.

There can be four types of ring structures in a strictly ortho-fused LPAH. These are: the end rings, which contain four unfused carbons, each with a hydrogen atom; a linear-bridged fusion with two carbons on opposite sides of the ring and each having a hydrogen atom; a "bent" fusion with two adjacent, unfused carbon atoms, each with a hydrogen atom; and a "central" fusion in which there are no unfused carbon atoms and no peripheral hydrogen atoms.

Large (C≥24) Polycyclic Aromatic Hydrocarbons: Chemistry and Analysis, by John C. Fetzer.
ISBN 0-471-36354-5

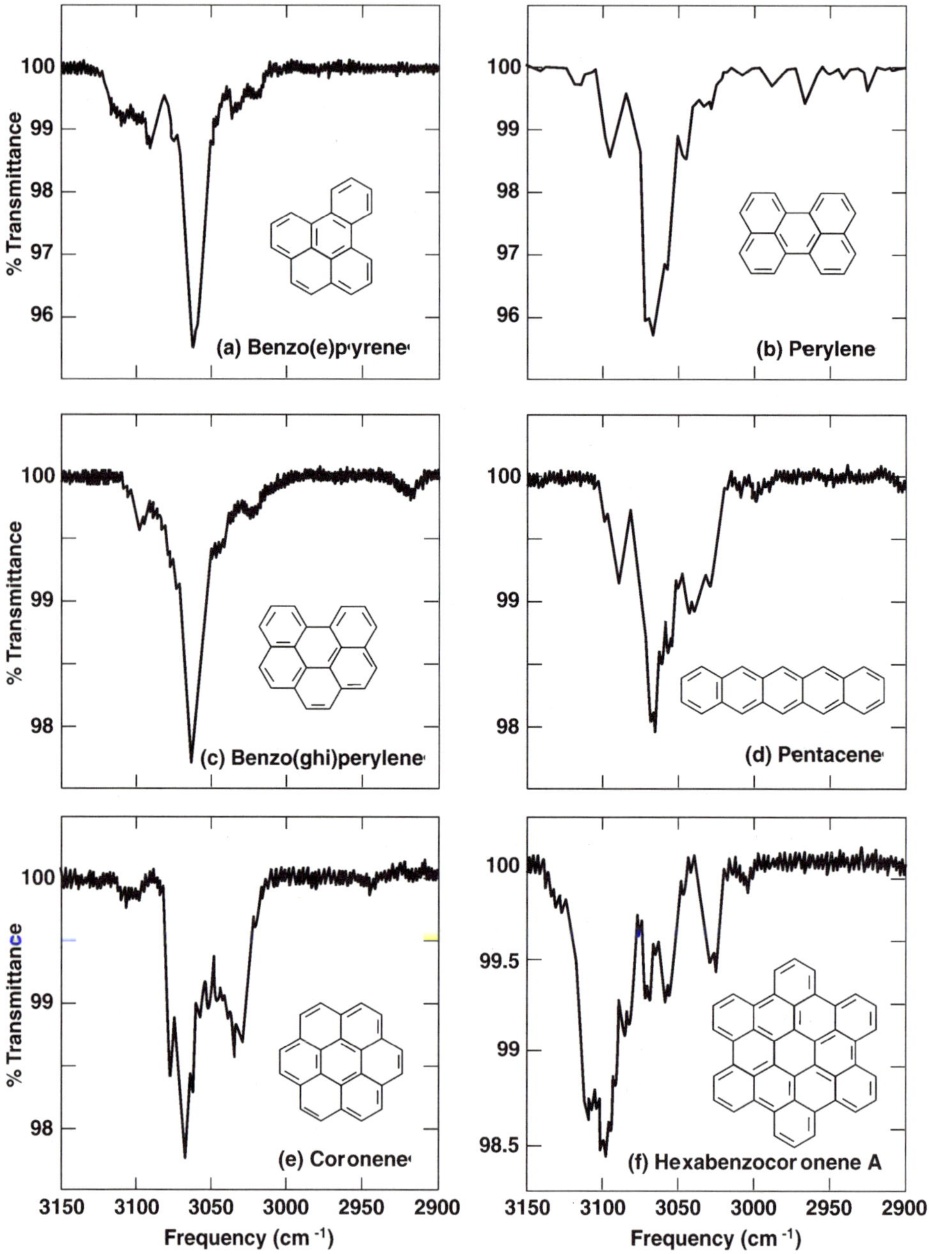

Figure 49. Infrared spectra of several LPAHs. From D. M. Hudgins and S. A. Sandford, *J. Phys. Chem. A*, **102** 345 (1998), reproduced with permission from the American Chemical Society.

Examples of each of these ring types are shown in any of the large ortho-fused structures.

Peri-fused LPAHs can have these structures and also internal fused carbons that are different than those in ortho-fused structures. These internal carbons are conceptually similar to the fourth category above, but the additional condensation in the peri-fused structures makes these more rigid (less prone to molecular movements). This leads to lower-frequency infrared transitions with higher energies and lower intensities because they involve the movement of the more rigid structures. Peri-fused LPAHs can also have a fifth type of peripheral substructure composed of three adjacent carbons, each with a hydrogen atom (as in hexabenzo[*bc*,*ef*,*hi*,*kl*,no,*qr*]coronene).

There are four types of peripheral features that contain carbon-hydrogen bonds. These are substructures containing one to four carbons, each with an attached hydrogen. Because of the normal planarity of the PAH, or at a minimum the planarity in the substructure to which these hydrogen atoms are connected, the most common mode of movement for these four substructures is an out-of-plane wagging. This involves the bending of the two carbon-carbon bonds that join these structures to the rest of the PAH. Also, carbon-hydrogen out-of-plane bending generates bands in the 600–900 cm^{-1} region. Zander lists IR ranges for each type of peripheral structure (1). These are (in cm^{-1}) quatro, 730–750; 750–770; trio, 750–770, 770–800, 800–810; duo, 800–810, 810–860; and solo, 860–910. For example, Obenland and Schmidt (2) report the IR spectrum of tribenzo[*a*,*g*,*m*]coronene. Only the duo band at 811 cm^{-1} and quatro band at 748 cm^{-1} are seen.

IR spectra have found use in limited areas of LPAH research. Zander (3) studied the formation of coal tar pitches and the carbon mesophase by changes in the IR spectra. Several groups of astrophysicists have studied interstellar IR spectra in an attempt to elucidate the type of LPAHs that may be in nebulae and gas clouds. Zander pointed out the universal applicability of IR techniques for LPAH samples because they do not need a solvent (1). Solid mixtures with potassium bromide can be easily made into pellets suitable for spectroscopy.

One technique that might be promising in the analysis of LPAHs, but which has as of yet not been applied to them, is matrix-isolation IR. It has been extensively used for PAHs. In this approach, the sample is trapped in an inert matrix such as frozen nitrogen, argon, or neon. This is usually accomplished by vaporizing the PAH into the inert gas. This mixture is then exposed to a cold surface, and it forms a frozen layer. This is then moved into the spectrometer viewing area. The IR bands of the PAHs become very narrow, on the order of 1 cm^{-1}. This high resolution yields characteristic band patterns for each PAH. For example, the individual methylchrysene isomers were identified and quantified in a mixture. The inherently weak IR absorption in these dilute samples requires the use of Fourier transform IR (FTIR) instruments. The FTIR spectrum also has the added advantage of being more quantitative.

Semmler, Yang, and Crawford (4) have reported on the use of FTIR as a detector for gas chromatography. The gas-phase spectra of many PAHs are reported. They report the 600–900 cm^{-1} to be very spectrally rich with C-H out-of-plane and C-C in-plane bending modes. These form "unique fingerprints" for each of the 33 model compounds. This paper did not report any LPAHs, with indeno[1,2,3-*cd*]pyrene and benzo[*ghi*]perylene at 22 carbons being the largest ones studied. With the development of high-temperature GC and heated transfer lines, this method should also be applicable to the LPAHs of 24 and 26 carbons, if not even larger ones. The uniqueness of gas-phase spectra of the larger number of isomers at these higher carbon numbers cannot be estimated.

Much of the recent research activity in the IR spectrometry of LPAHs has been directed toward characterizing this class of compounds for comparison to interstellar spectral features. That work is described separately in the section on the spectroscopy of LPAHs and its relationship to astrophysics.

There has only been one report of Raman spectra of LPAHs, that of the very symmetrical coronene. Both theoretical and experimental spectra were compared to those of benzene, which belongs to the same point group (5).

REFERENCES

1. M. Zander, *Polycycliche Aromaten*, Teubner, Stuttgart, Germany, 1995, p. 186.
2. S. Obenland and W. Schmidt, in *Polycyclic Aromatic Hydrocarbons and Astrophysics*, A. Leger, L. d'Hendecourt, and B. Boccara, Eds., Reidel, Dordrecht, The Netherlands, 1987, p. 165–1723.
3. M. Zander, *Polycyclic Arom. Cmpds.* **7** 209–221 (1995).
4. J. Semmler, P. W. Yang, and G. E. Crawford, *Vibrational Spectrosc.* **2** 189–202 (1991).
5. U. Fleischer and P. Pulay, *J. Raman Spectrosc.* **29** 473–481 (1998).

CHAPTER

6

MASS SPECTROMETRY

The characteristic properties of the LPAHs determines what types of mass spectrometric (MS) techniques must be used. Normal electron-impact (EI) ionization methods are limited in their utility. This is due to two factors. First, the LPAHs do not readily fragment and create distinct fragment patterns for isomers (as with most other classes of compounds). Second, the empirical formulae for a carbon number only differ by increments of 2 daltons with increasing condensation. This is too close under EI conditions for differentiation from the normal hydrogen losses or $M + 1$ and $M + 2$ isotope peaks that are common for high-carbon-number hydrocarbons.

The most common ion loss is a retro-Diels-Alder loss of an acetylene (M-26). If the structure contains an ortho-fused ring, an M-52 ion is found from loss of a C_4H_4 fragment. The most prominent feature of LPAH spectra are the molecular ion peaks, including those due to multiply ionized molecular ions. These multiply charged ions are characteristic of LPAHs because of their extremely high stability. These are seen at masses corresponding to M divided by the charge on the ion. This, for coronene the ions are at 300, 150, and 100 for the singly, doubly, and triply charged ions. The presence of bay, cove, and fjord regions in the structure influences the relative intensities of the multiply charged ions and the M-15 and M-26 ions (1, 2).

6.1 PROBE AND OTHER SAMPLE INTRODUCTION TECHNIQUES

The LPAHs have extremely high sublimation temperatures; therefore, normal sample introduction methods cannot generally be used. High-temperature sample probes are used. These devices are capable of heating the sample to several hundred degrees by resistive heating of a metallic filament or ribbon. The sample is placed on the cool heating element before introduction into the mass spectrometer vacuum system. The temperature is then increased, volatilizing the sample. The high vacuum in MS systems helps to lower the sublimation

Large (C>=24) Polycyclic Aromatic Hydrocarbons: Chemistry and Analysis, by John C. Fetzer.
ISBN 0-471-36354-5

temperature. The fifteen-ring LPAH benzo[1,2,3-*bc*:4,5,6-*b*′*c*′]dicoronene sublimes at 700° at 1 microtorr (3).

Klempler and Binder (4) extracted soots from diesel exhaust and the pyrolysis of benzene. They observed the clusters of ions, caused by different degrees of condensation and isotope effects, up to 38 carbons. Greinke and Lewis (5) directly placed drops of pyridine dispersions of samples from naphthalene pitches on the inlet probe for field-desorption MS. The heating of the probes then volatilized the components of up to 900 daltons. Fetzer and Rechsteiner (6) examined the extracts from coal tar pitches and a carbon black were examined by field-ionization MS, with a heated sample probe being used for sample introduction. Figures 50–55 are of various fractions from a coal tar pitch and a carbon black. Only 15 LPAHs could be identified (if only alternant PAHs are present!), although masses of up to 1,500 daltons were found. The limitations were a lack of authentic standard compounds and the inability of this approach to differentiate isomers. The LPAHs identified were usually those with unique

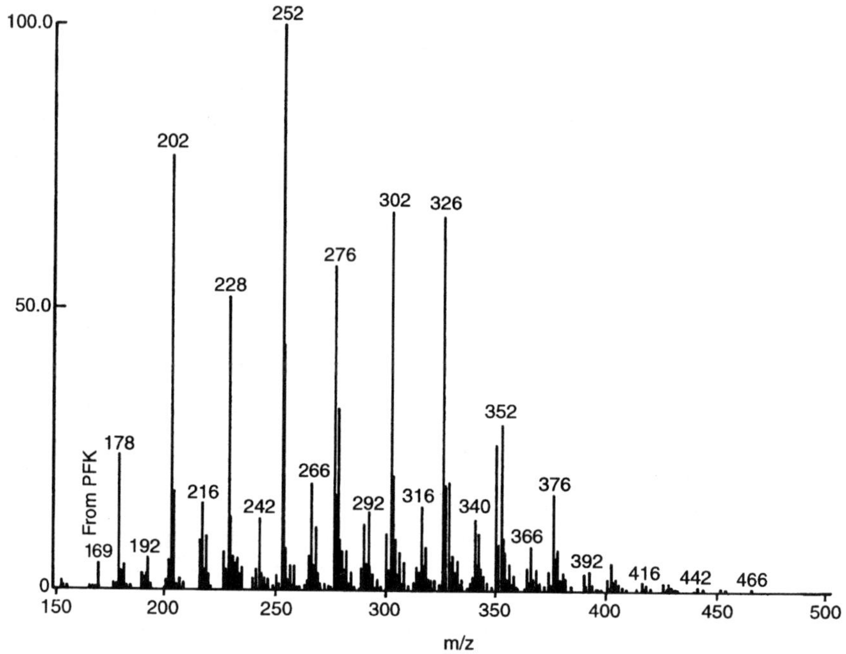

Figure 50. Fetzer and Rechsteiner field-ionization MS with heated probe sample introduction of a coal tar pitch toluene extract. From J. C. Fetzer and C. E. Rechsteiner, *Polynuclear Aromatic Hydrocarbons: Eleventh International Symposium*, M. Cooke, K. Loening, and J. Merritt, Eds., Battelle, Columbus, OH, 1991, p. 259–270.

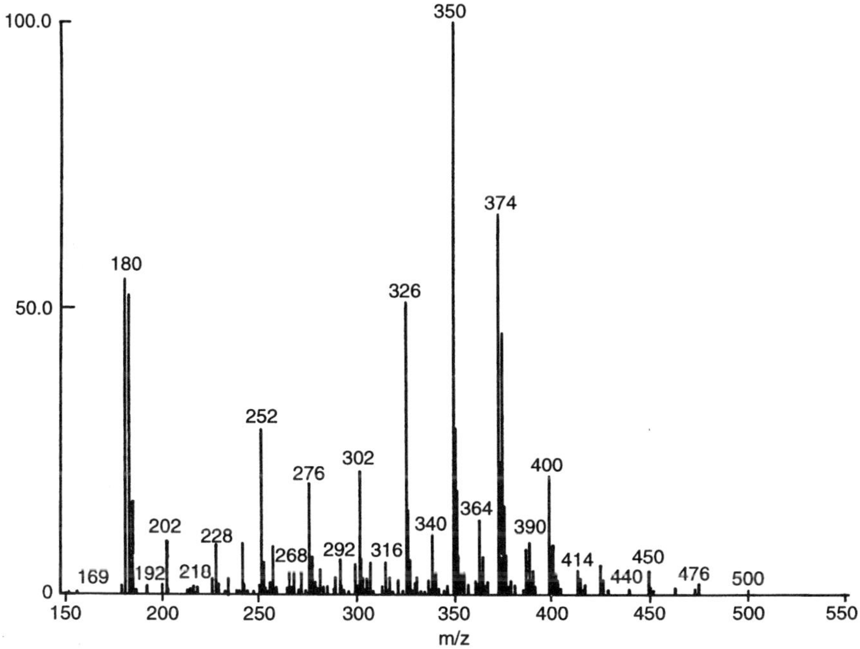

Figure 51. Fetzer and Rechsteiner field-ionization MS with heated probe sample introduction of a coal tar pitch dichloromethane extract. From J. C. Fetzer and C. E. Rechsteiner, *Polynuclear Aromatic Hydrocarbons: Eleventh International Symposium*, M. Cooke, K. Loening, and J. Merritt, Eds., Battelle, Columbus, OH, 1991, p. 259–270.

masses, such as 300 for coronene, 398 for ovalene, and 522 for hexabenzo-[*bc*,*ef*,*hi*,*kl*,*no*,*qr*]coronene. The use of the high-temperature probe, however, showed that solvent extraction—even with very strongly interacting, high-boiling solvents like 1,2,4-trichlorobenzene over several days—did not completely remove even the smaller PAHs. Thermal desorption using this probe showed that solvent extraction does not quantitatively remove the adsorbed PAHs.

More recently, Beck, Keki, Szabo, and Zsuga (7) created similar samples from naphthalene pyrolysis and solvent electrolysis, which were analyzed by probe EI-MS. They observed a pattern of peaks that they interpreted as series. The complex mixtures showed peaks that were separated by either 24 or 26 daltons ($C2$ or C_2H_2 units). Figure 56 shows one of their spectra. It bears a very strong similarity to those seen in Figures 50–55 although this sample was very different.

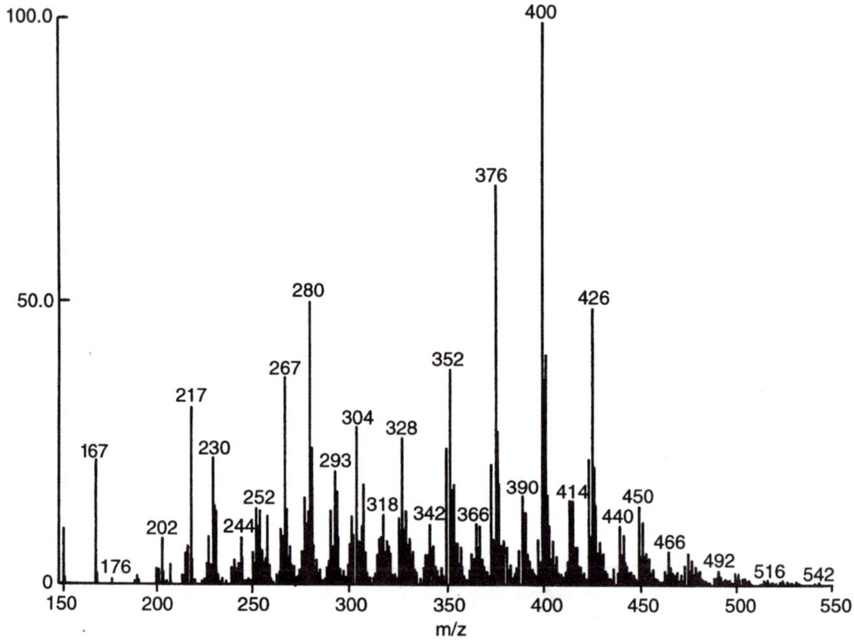

Figure 52. Fetzer and Rechsteiner field-ionization MS with heated probe sample introduction of a coal tar pitch chlorobenzene extract, fractionated by dichloromethane over acidic alumina. From J. C. Fetzer and C. E. Rechsteiner, *Polynuclear Aromatic Hydrocarbons: Eleventh International Symposium*, M. Cooke, K. Loening, and J. Merritt, Eds., Battelle, Columbus, OH, 1991, p. 259–270.

di Lorenzo, Masi, and Guerrini (8) directly introduced soot samples collected from a fuel-oil flame into a MS after a GC analysis did not show LPAHs. They saw masses above 500 daltons. Some of the lower masses were very prominent in the spectrum, such as 374, 398, 400, 424, and 448. The first is naphtho[8,1,2-*abc*]coronene and its isomers. The second is ovalene and its isomers. This contrasted with their results of soots from methane flames. These showed little material above 300 daltons.

LaFleur, Taghizadeh, Howard, Anacleto, and Quilliam (9) directly introduced very high-molecular-weight fractions of PAHs. They looked at samples generated by flame-generated samples where a mixture of ethylene and naphthalene was the fuel. The MS data showed LPAHs well above those found when these same samples were analyzed by gas, liquid, and supercritical fluid chromatographies with the same MS set-up as the detector. The LPAHs seen in this fashion ranged from 32 to 160 carbons.

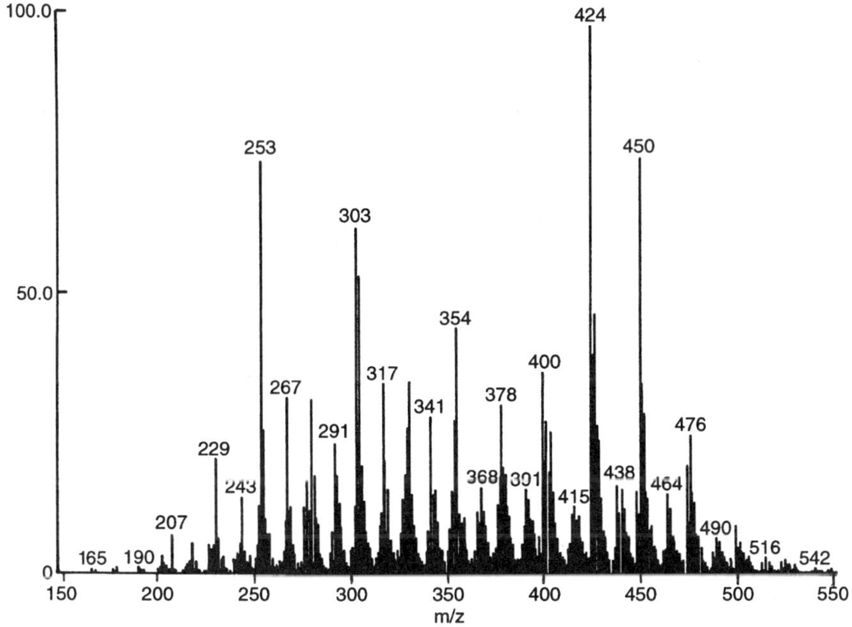

Figure 53. Fetzer and Rechsteiner field-ionization MS with heated probe sample introduction of a coal tar pitch chlorobenzene extract, fractionated by ethyl acetate over acidic alumina. From J. C. Fetzer and C. E. Rechsteiner, *Polynuclear Aromatic Hydrocarbons: Eleventh International Symposium*, M. Cooke, K. Loening, and J. Merritt, Eds., Battelle, Columbus, OH, 1991, p. 259–270.

Pace and Betowski (10) used micro-column HPLC to introduce samples into a particle-beam MS. The separation was with a polymeric octadecyl-silane bonded phase and methanol-tetrahydrofuran gradients. They examined a set of standard LPAHs, from 24 to 36 carbons, and two extracts from soils collected at hazardous-waste sites. Generally, their detection limits were ~ 1 ng. Figures 57–59 are specific ion profiles at three LPAH masses of a contaminated soil extract. Figure 60 shows several individual spectra collected from that run, and Figure 61 shows the spectra under similar conditions for two of the standard compounds.

Ruckmick, Hurtubise, and Silver (11) used a sequence of separation methods to fractionate a coal liquid. Field-ionization MS showed many high-molecular-weight components in the fractions. The even-mass components were assumed to be aromatic hydrocarbons. These authors drew several postulated structures of between 30 and 40 carbons, but they did not identify any specific LPAHs. In similar work, Chen and Qian (11) looked at the composition of solvent-refined coal extracts using field-ionization and field-desorption MS. These materials are

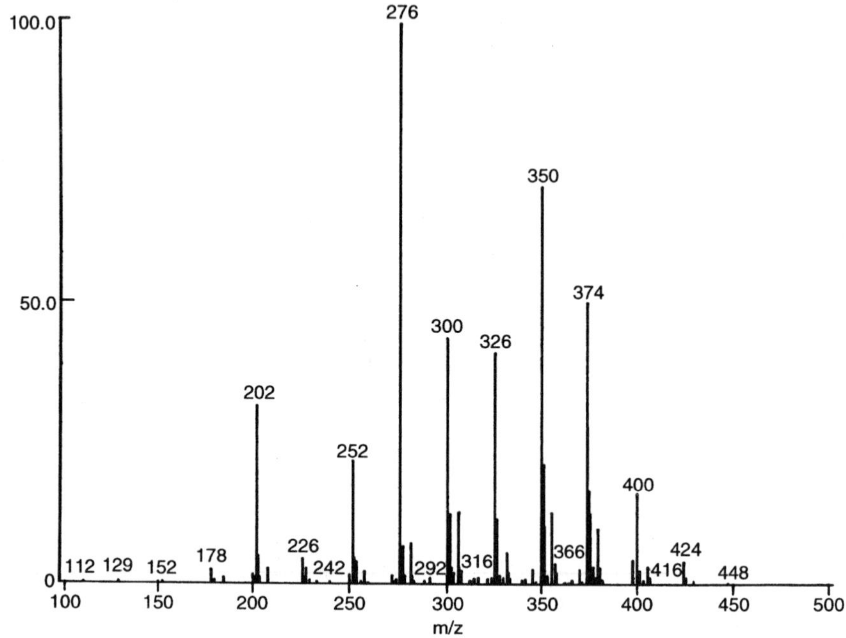

Figure 54. Fetzer and Rechsteiner field-ionization MS with heated probe sample introduction of the dichloromethane extract of a carbon black. From J. C. Fetzer and C. E. Rechsteiner, *Polynuclear Aromatic Hydrocarbons: Eleventh International Symposium*, M. Cooke, K. Loening, and J. Merritt, Eds., Battelle, Columbus, OH, 1991, p. 259–270.

very complex, containing LPAH masses of peri-fused structures of up to 32 carbons.

6.2 LASER DESORPTION

Another approach to introducing LPAHs into the MS is laser desorption (LDMS). A burst from a laser is focused on the sample. The intense heat volatilizes the sample components. LDMS requires little sample because of the small, but intense, focus of the laser. Volatilization is quick and efficient and can be confined to very small areas.

LDMS was the analytical method used by both sides of a recent controversial claim involving PAH analysis (13–17). One research group claimed to have found signs of life in Martian meteorites by evidence of PAHs, whereas another

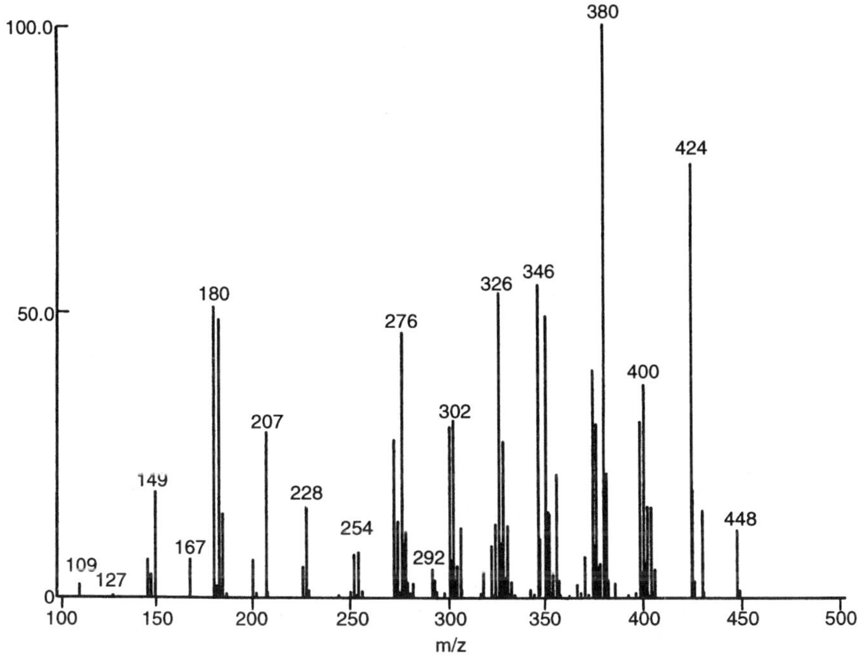

Figure 55. Fetzer and Rechsteiner field-ionization MS with heated probe sample introduction of the chlorobenzene extract of a carbon black Note the 380 ion, indicating desorption of the sulfur-containing PACs. From J. C. Fetzer and C. E. Rechsteiner, *Polynuclear Aromatic Hydrocarbons: Eleventh International Symposium*, M. Cooke, K. Loening, and J. Merritt, Eds., Battelle, Columbus, OH, 1991, p. 259–270.

group interpreted their LDMS data as refuting this claim and they explained the PAHs as terrestrial contamination or the result of abiotic processes.

Extremely large LPAHs were found by LDMS in the products of coronene pyrolysis (18). These included the previously known condensed dimer, of mass 596, but additional species of three, four, and five condensed coronene molecules were found (of molecular masses of 892, 1188, and 1484 daltons). Additionally, LPAHs of 300–700 daltons were seen in this sample. These resulted from condensation of the main component coronene with the trace levels of impurities, such as pyrene, benzo[*ghi*]perylene, ovalene and other PAHs found in the commercial coronene sample used.

Soot particles from shale oil retorting were also examined with LDMS (19). The masses of many LPAHs, such as $C_{26}H_{14}$ and $C_{36}H_{14}$, were seen. Soot particles in the size range of 1–10 μm could be individually examined. LDMS

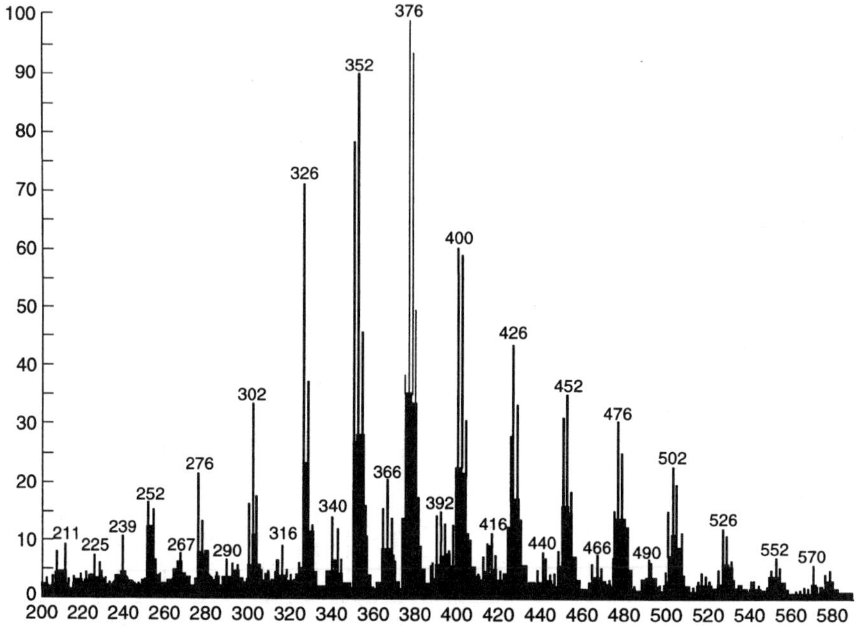

Figure 56. Beck MS results that bear a striking resemblance to the spectra in Figures 50–55, even though they are of a very different carbonaceous material. From M. Beck, S. Keki, P. T. Szabo, and M. Zsuga, *Tetrahedron* **55** 1803 (1999), reproduced with permission from Elsevier Science Publishers BV.

was used to directly determine the LPAHs on cloud-water filtrates (19). The laser could be focused directly onto particles on the filter. A difference in distribution of LPAHs was seen between 193-nm and 248-nm laser beams, indicating excitation wavelength discrimination. Selectivity in ionization could be seen for individual sections of some particles.

The LPAHs in coal tar pitches was analyzed by LDMS (21, 22). Masses of up to 12,000 daltons were found, but the predominant peaks were in the range of 100–400 daltons. A smaller envelop of peaks occurred from 1,000 to 3,000 daltons. Ions at 300 ($C_{24}H_{12}$), 326 ($C_{26}H_{14}$), 350 ($C_{28}H_{14}$), 352 ($C_{28}H_{16}$), 374 ($C_{30}H_{14}$), 376 ($C_{30}H_{16}$), 398 ($C_{32}H_{14}$), 400 ($C_{32}H_{16}$), and 424 ($C_{34}H_{16}$) were some of the ions assigned to LPAHs. Herod et al. later examined many standard LPAHs and a mixture of C-60 and C-70 fullerenes to compare to the earlier results. They saw no evidence of secondary reactions in the ion source. From this they concluded that the high-mass peaks seen in the coal tar pitches were not method artifacts but were due to real components of the samples.

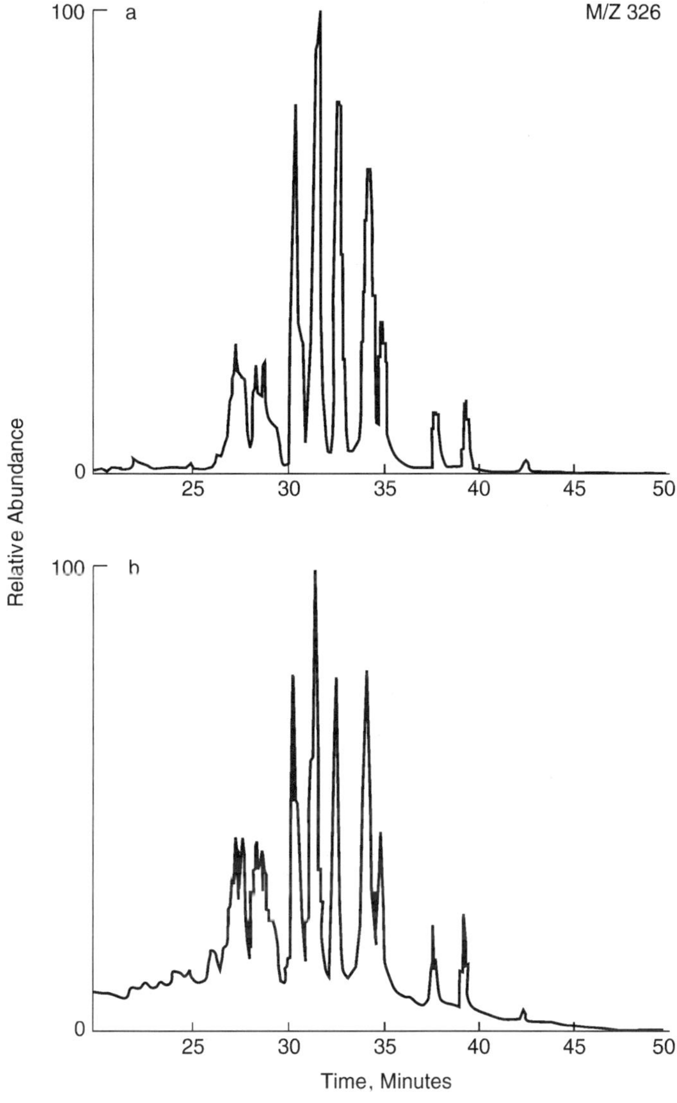

Figures 57–59. Selective ion profiles of the HPLC-MS analysis of a contaminated soil extract. From C. M. Pace and L. D. Betowski, *J. Am. Soc. Mass Spectrom.* **6** 600, 605, and 606 (1995), reproduced with permission from Elsevier Science Publishers BV.

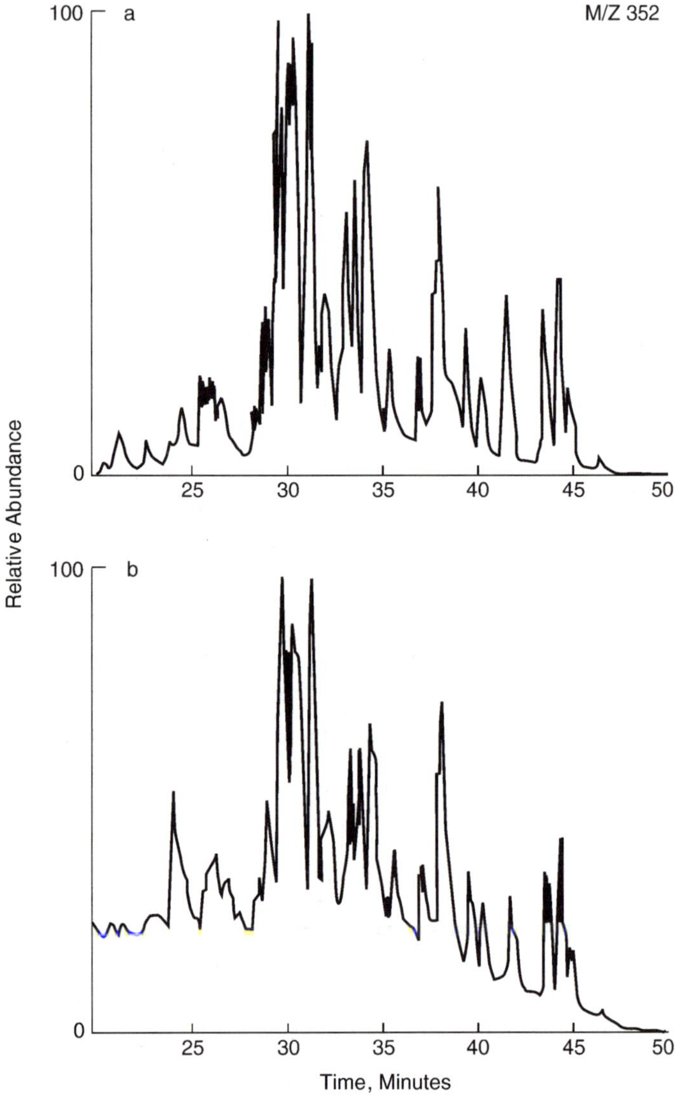
100
a
M/Z 352
0
25
30
35
40
45
50
Relative Abundance
100
b
0
25
30
35
40
45
50
Time, Minutes

Figures 58

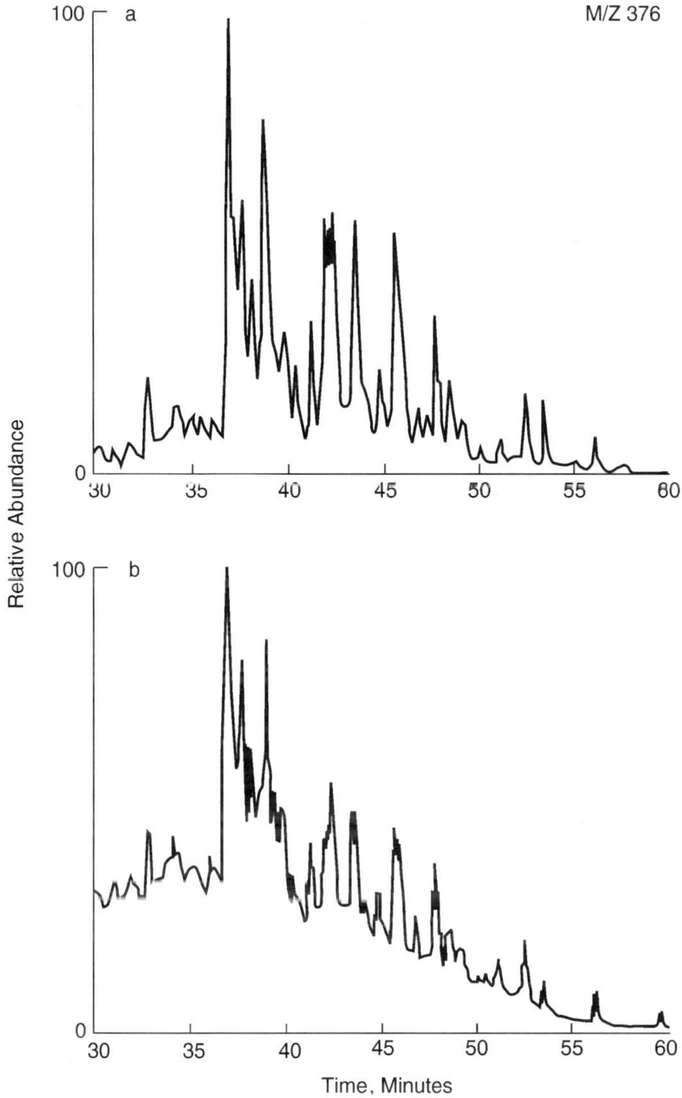
100
a
M/Z 376
0
30
35
40
45
50
55
60
Relative Abundance
100
b
0
30
35
40
45
50
55
60
Time, Minutes

Figures 59

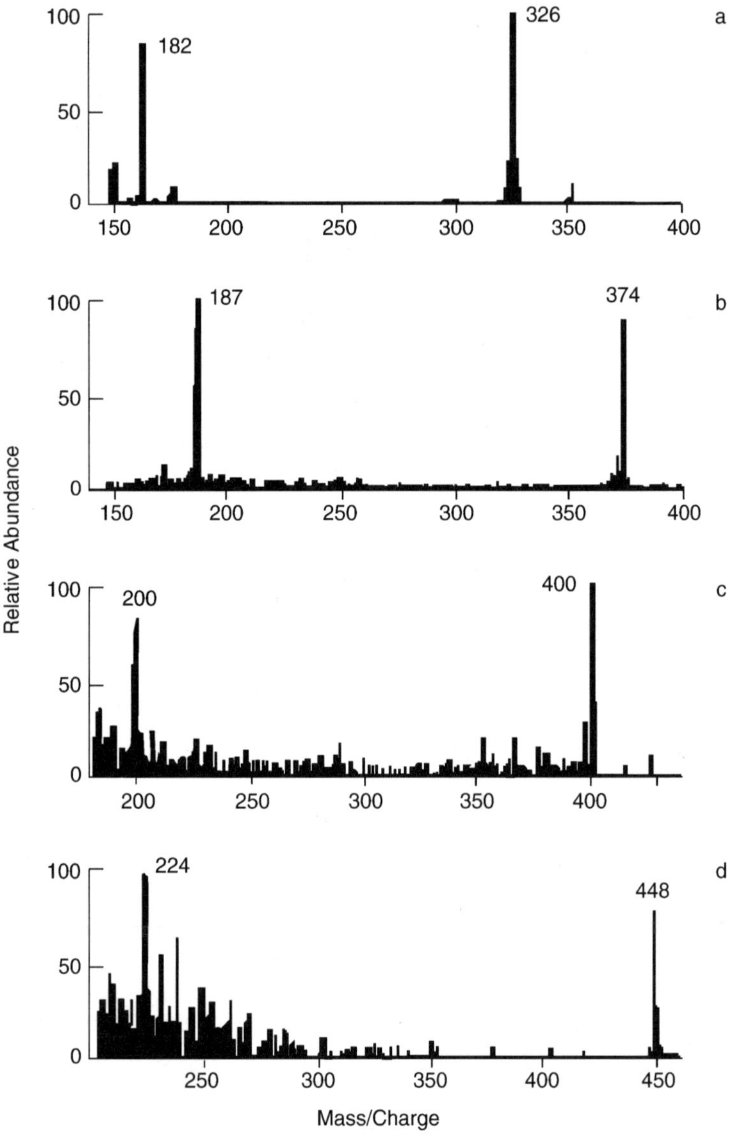

Figure 60. Individual mass spectra collected during the run shown in Figures 57–59. From C. M. Pace and L. D. Betowski, *J. Am. Soc. Mass Spectrom.* **6** 600, 605, and 606 (1995), reproduced with permission from Elsevier Science Publishers BV.

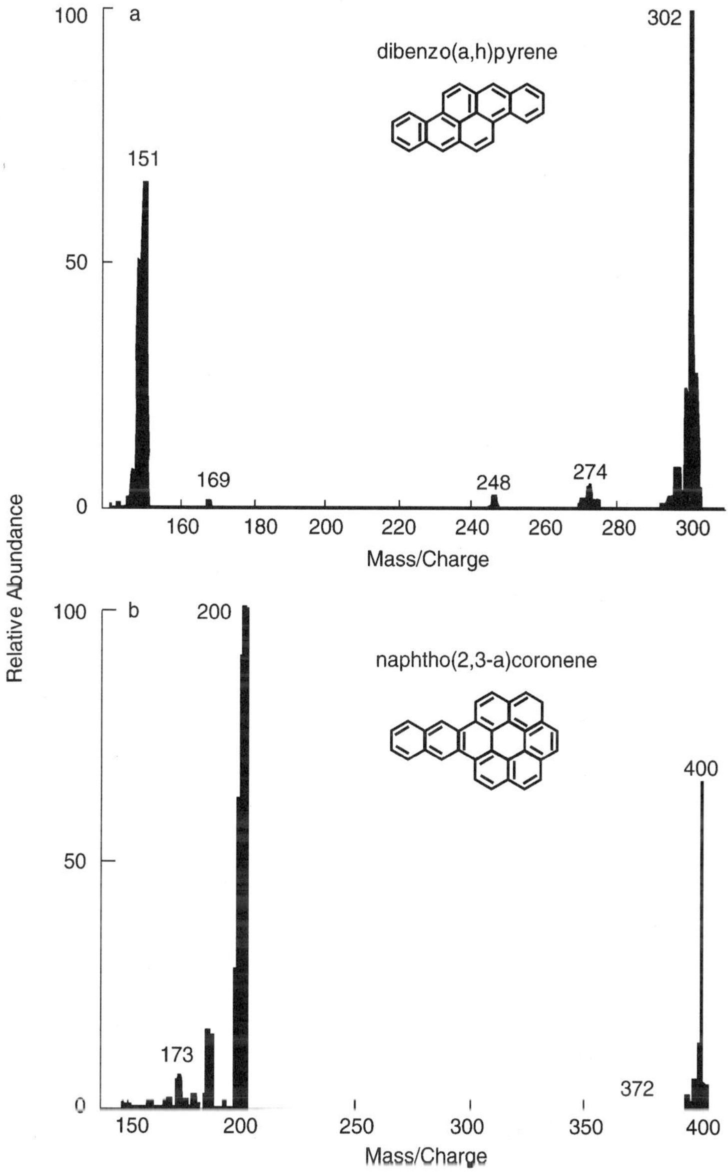

Figure 61. Spectra of two standard LPAHs under conditions identical to those used in Figures 57–60. From C. M. Pace and L. D. Betowski, *J. Am. Soc. Mass Spectrom.* **6** 600, 605, and 606 (1995), reproduced with permission from Elsevier Science Publishers BV.

6.3 SELECTIVE IONIZATION TECHNIQUES FOR ISOMER DIFFERENTIATION

Different LPAH isomers possess different ionization energies and electron affinities. Simonsick and Hites(23) used a mixture of methane and argon as a selective chemical ionization reagent. A total-ion chromatogram is shown in Figure 62. The reagent mixture reacted differently with LPAH isomers. The relative heights of the molecular ion (M), protonated molecular ion (M + 1), and ethylated molecular ion (M + 29) were used to differentiate isomers (A portion of the data correlation scheme is shown in Figure 63). One key area was the use of molecular modeling to calculate LPAH ionization potentials (IPs). A correlation between the IP and the ion ratios was made (Fig. 64), relying on the known values of several smaller PAHs and of a few LPAHs. This allowed some identifications to be made in the absence of authentic standards. Their approach also relied on high-temperature GC columns capable of separating LPAHs of molecular weights 350 and 374.

Suzuki, Kaneko, and Tsuchiya (24) used methanol chemical ionization to differentiate LPAHs from heteroatom-containing polycyclic aromatic com-

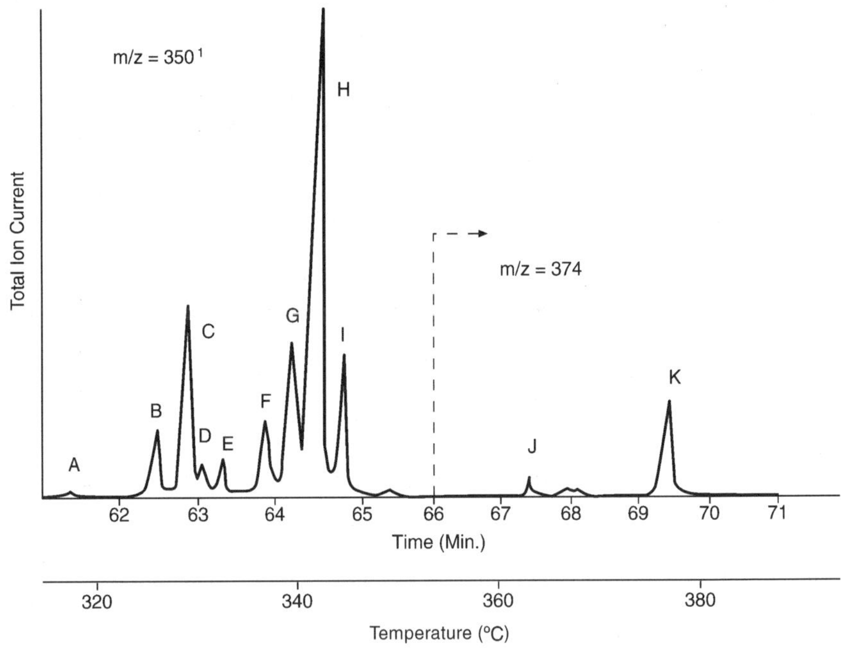

Figure 62. Total ion chromatogram of a carbon black extract. From W. J. Simonsick and R. A. Hites, *Anal. Chem.* **58** 2116 (1986), reproduced with permission from the American Chemical Society.

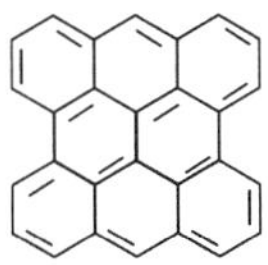

Phenanthr o[1,10,9,8-opqr s]perylene

Compound designation - 350(1)
IP (MNDO) = 7.251 eV
M+1/M (calc) = 0.197

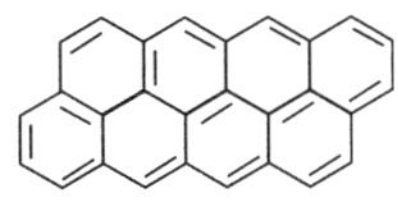

Perinaphthacenonaphthacene

Compound designation - 350(2)
IP (MNDO) = 7.296 eV
M+1/M (calc) = 0.218

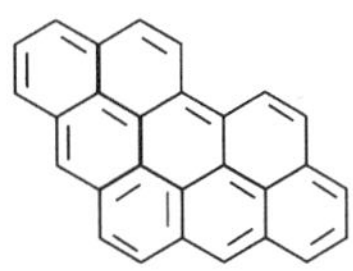

Phenanthr o[1,10,9,8-opqr s]pentaphene

Compound designation - 350(3)
IP (MNDO) = 7.374 eV
M+1/M (calc) = 0.257

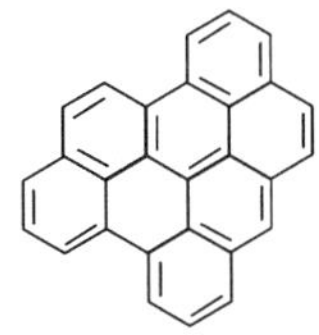

Benzo[cd]naphtho[3,2,1,8-pqr s]perylene

Compound designation - 350(4)
IP (MNDO) = 7.610 eV
M+1/M (calc) = 0.421

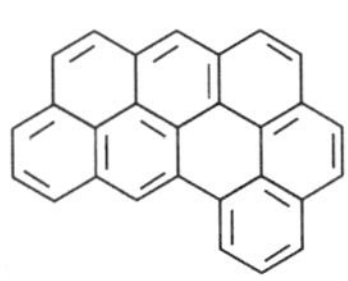

Benzo[lmn]naphtho[22,1,8-qr s]perylene

Compound designation - 350(5)
IP (MNDO) = 7.658 eV
M+1/M (calc) = 0.465

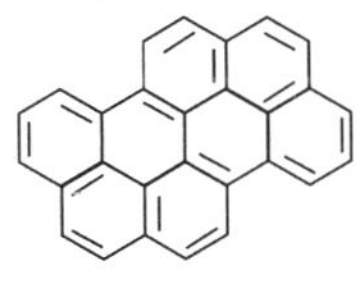

Benzo[pqr]naphtho[8,1,2-bcd]perylene

Compound designation - 350(6)
IP (MNDO) = 7.699 eV
M+1/M (calc) = 0.506
M+1/M (exp) = 0.497

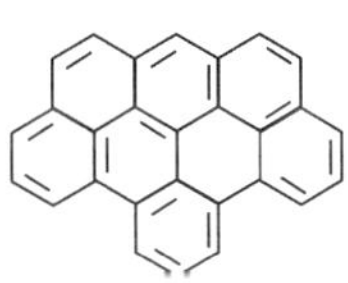

Phenanthr o[5,4,3,2-abcde]perylene

Compound designation - 350(7)
IP (MNDO) = 7.923 eV
M+1/M (calc) = 0.792

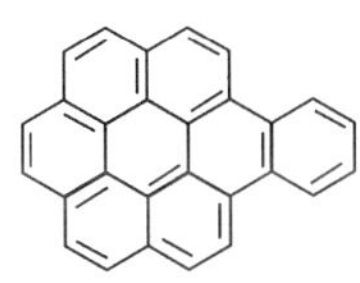

Benzo[a]cor onene

Compound designation - 350(8)
IP (MNDO) = 7.936 eV
M+1/M (calc) = 0.814
M+1/M (exp) = 0.698

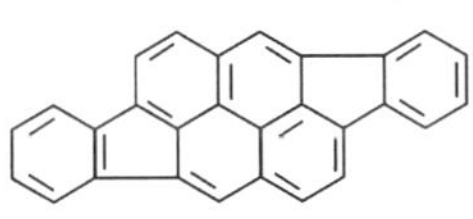

Diindeno[1,2,3-cd:1',2',3'-jk]p yrene

Compound designation - 350(9)
IP (MNDO) = 8.019 eV
M+1/M (calc) = 0.958

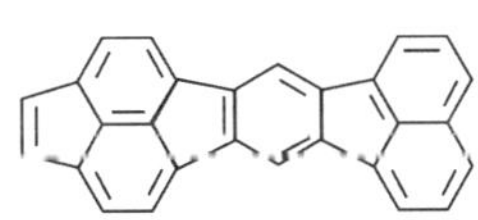

Acenaphtho[1,2-k]cyc lopenta[cd]fluoranthene

Compound designation 350(10)
IP (MNDO) = 8.188 eV
M+1/M (calc) = 1.326

Figure 63. Calculated ionization potentials and mass spectral ion ratios for several $C_{28}H_{14}$ LPAHs. From W. J. Simonsick and R. A. Hites, *Anal. Chem.* **58** 2118 (1986), reproduced with permission from the American Chemical Society.

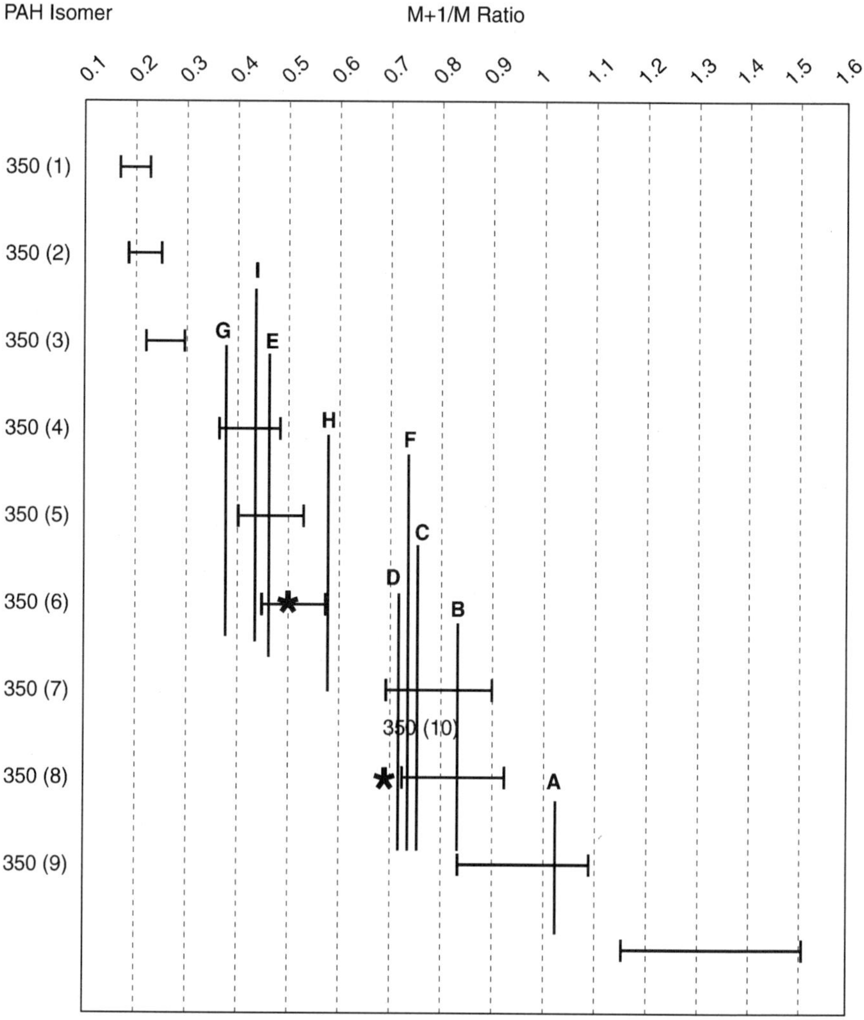

Figure 64. Correlation graph used to assign structures to peaks seen in Figure 62. From W. J. Simonsick and R. A. Hites, *Anal. Chem.* **58** 2118 (1986), reproduced with permission from the American Chemical Society.

pounds. They assigned example structures to the masses observed. Because their methodologies for the LPAHs could not distinguish isomers since they did not use a large reference compound set, these can only be deemed to be possible structures among the huge number of isomers.

Resonance-enhanced multi-photon ionization (REMPI) is a technique using lasers to selectively excite molecules. The laser excites a specific electronic transition, so that the over-effect is very selective ionization in the mass spectrometer source. Ahrens, Keller, Kovacs, and Homann (25) examined the PAHs, LPAHs, fullerenes, and their radicals produced in fuel-rich hydrocarbon flames. They found that a 265-nm laser was better for PAHs of 20 or fewer carbons, but a laser of 208 nm was better for larger ones and fullerenes. Some of the LPAHs seen included coronene, benzo[*a*]coronene, benzo[*pqr*]-naphtho[8,1,2-*bcd*]perylene, ovalene, and decacyclene.

REFERENCES

1. T. Ueda, M. Ohno, H. Harada, S. Iwashima, J. Aoki, and T. Kan, *Org. Mass Spectrosc.* **18** 105–109 (1983).
2. T. Ueda, R. Yano, M. Ohno, S. Iwashima, M. Takekawa, J. Aoki, and T. Kan, *J. Chem. Soc. Perkin Trans.* **2** 1195–1198 (1985).
3. J. C. Fetzer, unpublished results.
4. N. Klempler and H. Binder, *Anal. Chem.* **55** 2104–2106 (1983).
5. R. A. Greinke and I. C. Lewis, *Carbon* **22** 305–314 (1984).
6. J. C. Fetzer and C. E. Rechsteiner, in *Polycyclic Aromatic Hydrocarbons: Eleventh International Symposium*, M. Cooke, K. Loening, and J. Merritt, Eds., Battelle, Columbus, OH, p. 259–270 (1991).
7. M. Beck, S. Keki, P. T. Szabo, and M. Zsuga, *Tetrahedron* **55** 1799–1806 (1999).
8. J. A di Lorenzo, S. Masi, and R. Guerrini, *Le Revista dei Combustibili* **30** 46–51 (1976).
9. A. L. LaFleur, K. Taghizadeh, J. B. Howard, E. Anacleto, and M. A. Quilliam, *J. Am. Soc. Mass Spectrom.* **7** 276–286 (1996).
10. C. M. Pace and L. D. Betowski, *J. Am. Soc. Mass Spectrom.* **6** 597–607 (1995).
11. S. C. Ruckmick, R. J. Hurtubise, and H. F. Silver, *Fuel* **66** 1677–1683 (1986).
12. J. Chen and S. Qian, *Fuel Sci. Technol. Int.* **6** 687–722 (1988).
13. D. S. McKay, E. K. Gibson, K. L. Thomas-Keprta, H. Vali, C. S. Romanek, S. J. Clemett, X. D. F. Chillier, C. R. Maechling, and R. N. Zare, *Science* **273** 924–930 (1996).
14. S. J. Clemett, M. T. Dulay, J. Seb Gillette, X. D. F. Chillier, T. B. Majajan, and R. N. Zare, *Faraday Discuss.* **109** 417–436 (1996).
15. D. S. McKay, E. K. Gibson, K. L. Thomas-Keprta, and C. S. Romanek, *Proc. SPIE Int. Soc. Opt. Eng.* **3111** 2–6 (1997).
16. L. Becker, D. P. Glavin, and J. L. Bada, *Geochim. Cosmochim. Acta* **61** 475–481 (1997).
17. L. Becker, D. P. Glavin, and J. L. Bada, *Proc. SPIE Int. Soc. Opt. Eng.* **3111** 36–50 (1997).
18. C. Joblin and R. N. Zare, unpublished results.

19. T. Mauney, F. Adams, and M. R. Sine, *Sci. Total Environ.* **36** 215–24 (1984).
20. M. J. Dale, O. H. J. Downs, K. F. Costello, S. J. Wright, and P. R. R. Lagridge-Smith, *Environ. Pollut.* **89** 123–129 (1995).
21. J. E. Parker, C. A. F. Johnson, P. John, G. P. Smith, A. A. Herod, B. J. Stokes, and R. Kandiyoti, *Fuel* **72** 1381–1391 (1993).
22. A. A. Herod, B. J. Stokes, P. Hancock, R. Kandiyoti, J. E. Parker, C. A. F. Johnson, P. John, and G. Smith, *J. Chem. Soc. Perkin Trans.* **2** 499–506 (1994).
23. W. J. Simonsick and R. A. Hites, *Anal. Chem.* **58** 2114 (1986).
24. S. Suzuki, T. Kaneko, and M. Tsuchiya, *Kankyo Kagaku* **6** 511–520 (1996).
25. J. Ahrens, A. Keller, R. Kovacs, and K.-H. Homann, *Ber. Bunsen-Gas* **102** 1823–1839 (1998).

CHAPTER

7

ASTROPHYSICAL SPECTROSCOPY

One research area that involves PAHs that has become very active over the past two decades is the spectroscopy of LPAHs and their relevance to astrophysics. A review of the history of the spectroscopy of interstellar material will describe how this linkage has arisen. Leger and Allamanoda were the first to propose that PAHs were very prevalent in interstellar space. Their proposals arose to explain several spectrometric behaviors seen for interstellar bodies (1–3).

The IR and UV spectra of interstellar nebulae and dust clouds can be collected by attaching the appropriate spectrometers to a telescope that focuses on the object of interest. Over the past three decades, these spectra have been collected by ground-based, airborne, and orbiting observatories. Two types of spectra can be obtained, emission spectra, where the light is emitted by the interstellar bodies through fluorescence, phosphorescence, or other photon-emitting processes, or absorbance spectra, where stellar light passes through intermediate interstellar material and is modified by absorption processes. In either case, spectral features that have been named the diffuse interstellar bands, or DIBs, are seen. Many of these bands could be readily assigned to simple molecules, such as hydrogen gas, methane, acetylene, carbon monoxide, and water, that were expected and known to be in the interstellar medium.

In the IR range of 3–15 μm, there were bands seen that have are commonly called the unidentified IR emission bands, or UIRs. The UIRs do not match the spectral features of any simple molecules. In an attempt to match the UIRs, Leger and d'Hendecourt (1) collected the spectra of several planar LPAHs, including coronene, ovalene, naphtho[1′,8′:4,10]anthra[1,9,8-*abc*]coronene, and benzo[1,2,3-*bc*:4,5,6-*b*′*c*′]dicoronene. They compared these to those of astronomical ones such as those in the Orion nebulae M42 and M43. The similarities were striking (Fig. 65). Many of these bands match the spectral locations of those of PAHs, and most particularly of LPAHs. These include bands at 3.3, 6.2, 7.7, and 8.6 μm. These bands arise from carbon-carbon bond motions in the fairly constrained skeleton of the LPAHs.

In addition, bands at 11.3 and 12.7 μm are seen in the interstellar spectra. These two bands coincide with the carbon-hydrogen bond bands in LPAH

Large (C>=24) Polycyclic Aromatic Hydrocarbons: Chemistry and Analysis, by John C. Fetzer.
ISBN 0-471-36354-5

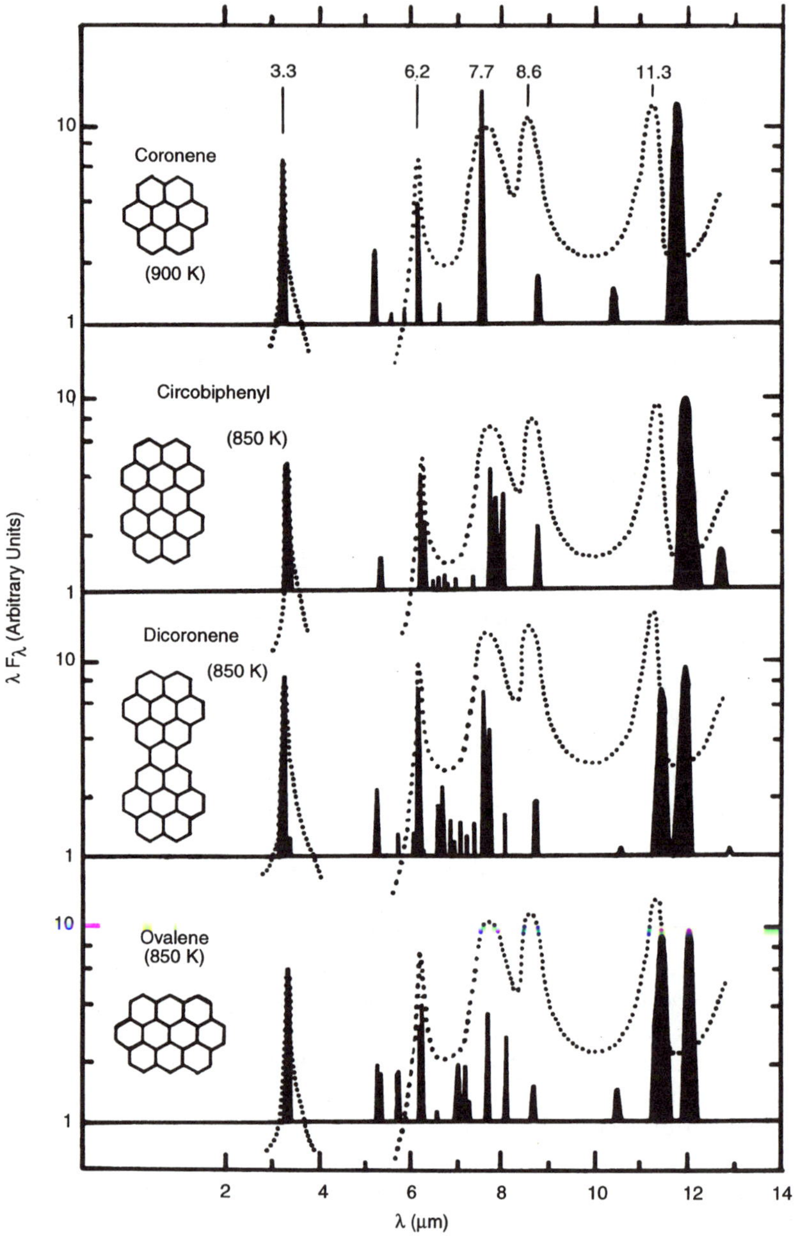

Figure 65. Spectra of several LPAHs superimposed on the spectrum of a region of the Orion Nebula M43. From A. Leger, L. d'Hendecourt, and D. Defourneau, *Astron. Astrophys.* **216** 156 (1989), reproduced with the permission of the European Southern Observatory.

spectra; These are seen somewhere in the range of 11–12.5 μm. Some researchers estimate that the LPAHs in the interstellar medium must have from 50 to 70 total ring carbons because of the high wavelengths involved. Also from these spectral features it is estimated that LPAHs of >200 to 300 ring carbons are not present because there are no very high-wavelength features.

The relative intensity of the IR bands suggests that the size of the ring system is more on the order of 100 carbon atoms if the interstellar species are neutral molecules. This is only very loosely estimated, because useful spectra to determine the carbon-to-hydrogen ratio have only been collected for neutral, much smaller LPAHs. The extrapolation may be very poor. Allamandola (4) and Allamandola, Tielens, and Barker (5) have reviewed the evidence for LPAHs as the sources of these spectral features (4, 5). These reviews list the IR emission bands of many PAHs and describes the structural arrangements, peripheral hydrogen atoms, and LPAH size and how they effect on the emission bands. Leger, d'Hendecourt, and Defourneau (6) discussed the photophysics involved and give examples of comparison of the spectra of nebulae and those of several LPAHs, including coronene, ovalene, and benzo[1,2,3-*cd*:4,5,6-*c*′*d*′]dicoronene.

Many novel spectroscopic methods have been used to obtain PAH and LPAH spectra under conditions simulating those in space. There are two main forms that the PAHs are thought to be found in, adsorbed species on interstellar dust and ices or as free gas-phase molecules. From the variety of observations of different objects and different spectral behavior, both forms are highly probable. Joblin, Boissel, Leger, d'Hendecourt, and Defourneau reviewed many of the aspects of the spectrometry of gaseous LPAHs (7). The IR emission of the LPAHs is temperature dependent, with the bands becoming broader and less intense with elevated temperatures.

Inert frozen neon or argon has been used to collect UV and IR spectra of neutral and cationic PAHs and simulate intersteller ices. Hudgins and Sandford and other co-workers have collected the matrix-isolated IR spectra of a wide variety of PAHs (8–15), including the LPAHs coronene and hexabenzo[*bc*,*ef*,-*hi*,*kl*,*no*,*qr*]coronene. The matrices were created by volatilizing the PAH samples into an argon atmosphere, which was then exposed to a spectrometer window cooled to 10 K. A film rapidly develops through deposition on the cooled surface. The volatilization of the PAH sample served the dual purpose of sample introduction and sample purification. If there were impurities, they were generally either more volatile than the PAH and were removed earlier during the temperature gradient to the sample sublimation point or they were less volatile and were left in the sample holder.

The IR spectra of the PAHs that were collected (Fig. 64 in Chapter 6) show many features and are dominated by the carbon hydrogen bands. There also are bands due to much larger molecular motions. The out-of-plane wagging of the extended ring in benzo[*e*]pyrene was seen as a strong band in the 6–7-μm region. This motion involves four carbons and the hydrogen atoms attached to each, and so the strong intensity at that location is not common. Hudgins et al. found specific bands due to the presence of one, two, three, or four carbon-hydrogen structures. These structural features are highlighted in the figure. They termed the first three features "solo," "duo," and "trio" structures.

These low-temperature, truly inert matrices allow the collection of the spectra of the highly reactive cations. The spectra of neutral PAHs, not surprisingly, resemble those collected in the low-temperature n-alkane matrices of Shpol'skii spectroscopy. Spectra in high-temperature, gas-phase conditions and in a supersonic jet have also been collected to simulate the isolated conditions of the individual molecules (16–17). This removes the effects of the matrices. The gas-phase spectra of PAHs are generally shifted very much more to the blue than solution- or matrix-isolated spectra, because the solvent interactions extend the π orbitals of the PAHs and reduce the energies needed to be absorbed. Naphthalene cation was examined in the gas phase. The spectrum showed two bands that could be identified as ones seen in the matrix-isolation studies. These were at 670.65 and 648.84 nm; they shifted 3.45 and 3.16 nm lower, respectively, than those seen in frozen neon did.

Because some of the spectral features are at higher wavelengths than those found for the neutral LPAHs, it was proposed that the species are actually the cations because these absorb and emit at much higher wavelengths than the neutral species in both the UV and IR spectra. The cation spectra are very different than the spectra of the same neutral molecules. This is especially true for those PAHs of 24 or fewer carbons (Figs. 66 and 67). There is still very-

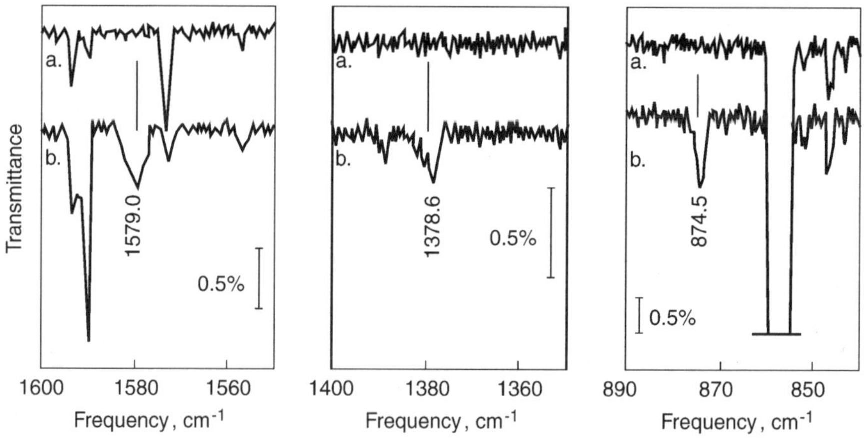

Figure 66. Mid-IR spectrum of coronene cation trapped in a 10K argon matrix. a. before photolysis, b. after 8-minute irradiation. From D. M. Hudgins and L. J. Allamandola, *J. Phys. Chem.* **99** 3041 (1995), reproduced with permission from the American Chemical Society.

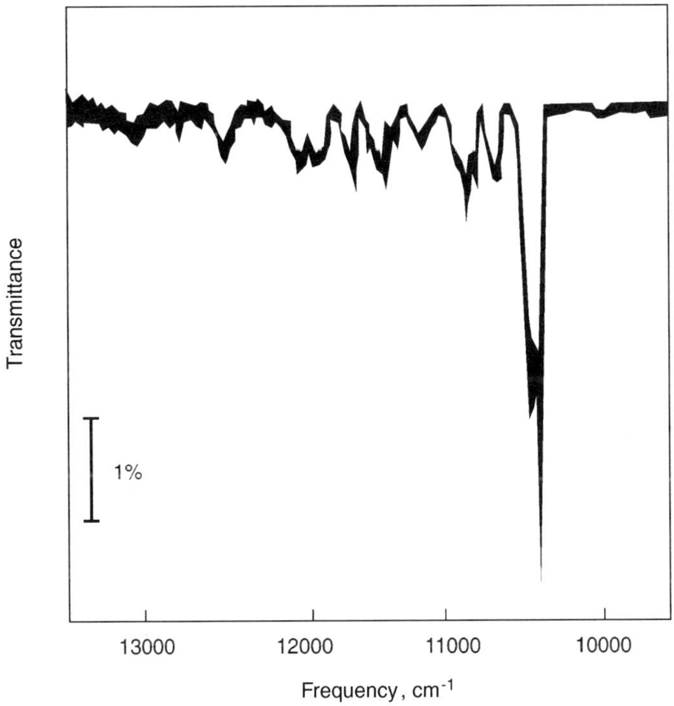

Figure 67. The near-IR spectrum of coronene cation. From D. M. Hudgins and L. J. Allamandola, *J. Phys. Chem.* **99** 3043 (1995), reproduced with permission from the American Chemical Society.

active debate on whether the interstellar conditions would favor the neutral LPAHs or their radical cations. Some experimental data for PAH radical cations have been collected and show the same types of spectral features as some of those observed for certain nebulae and dust clouds.

Allamandola, Hudgins, and Sandford (18) combined the neutral and cation spectra of PAHs of 10–48 ring carbons. These were compared to the emission spectra of two interstellar objects and showed a strong agreement (Figs. 68–70). Hudgins and Allamandola (19, 20) matched certain spectral features to those found in neutral and cation PAH spectra. Bands seen in some interstellar spectra at 10.8 and 11.1 μm are found in the cation spectra and are ascribed to the motion of a carbon-hydrogen in a "solo" structure. Hudgins and Allamandola showed that the small spectral features seen above 11 μm in interstellar spectra, which had previously only been assigned to "solo" structures in neutral PAHs, could also arise from combination bands of "duo" structures in cations.

In another paper, Allamandola and Hudgins, along with several other coworkers, made a very high-spatial-resolution observation (21). They used this to

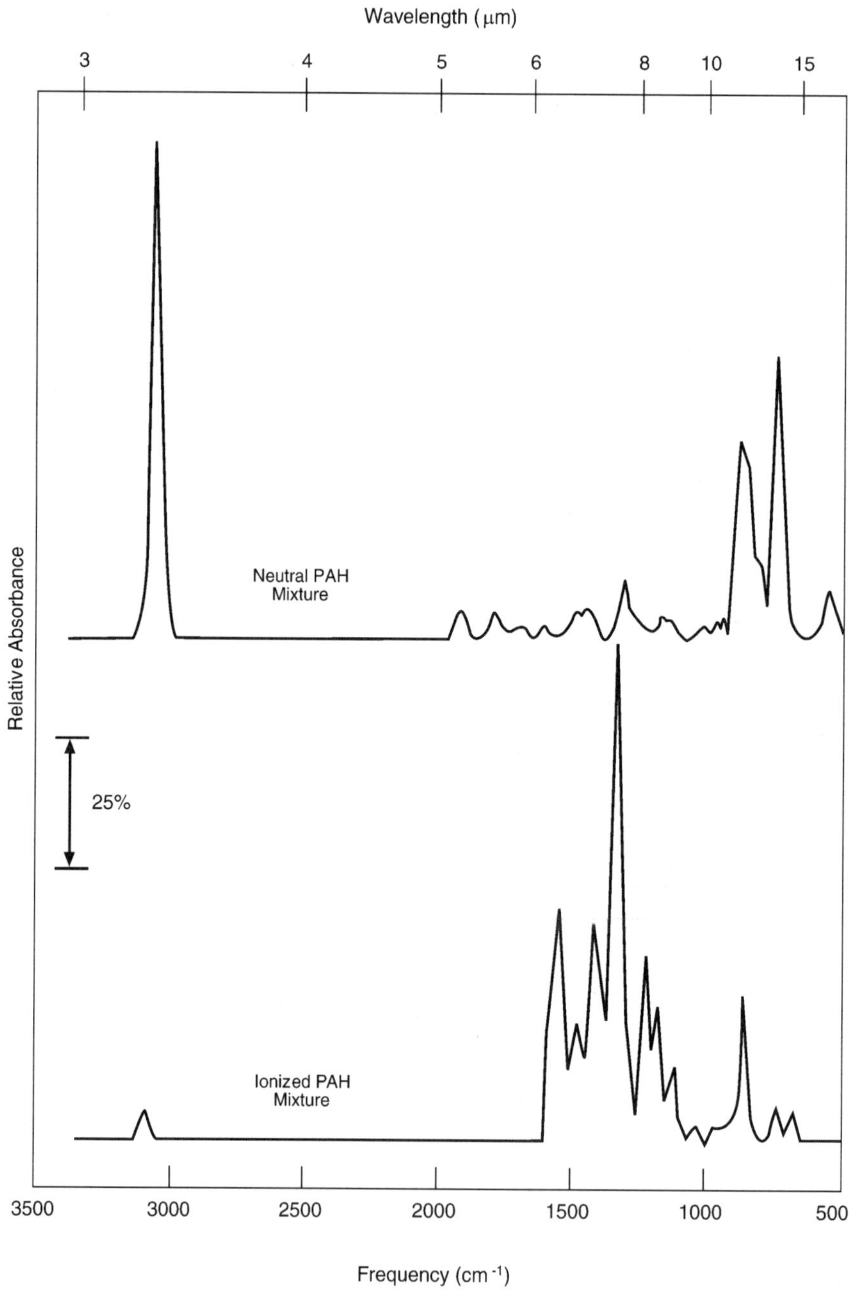

Figure 68. Comparison of the IR spectra of neutral and cationic LPAHs, using a mixture of six compounds. From D. Hudgins, NASA Ames Research Center.

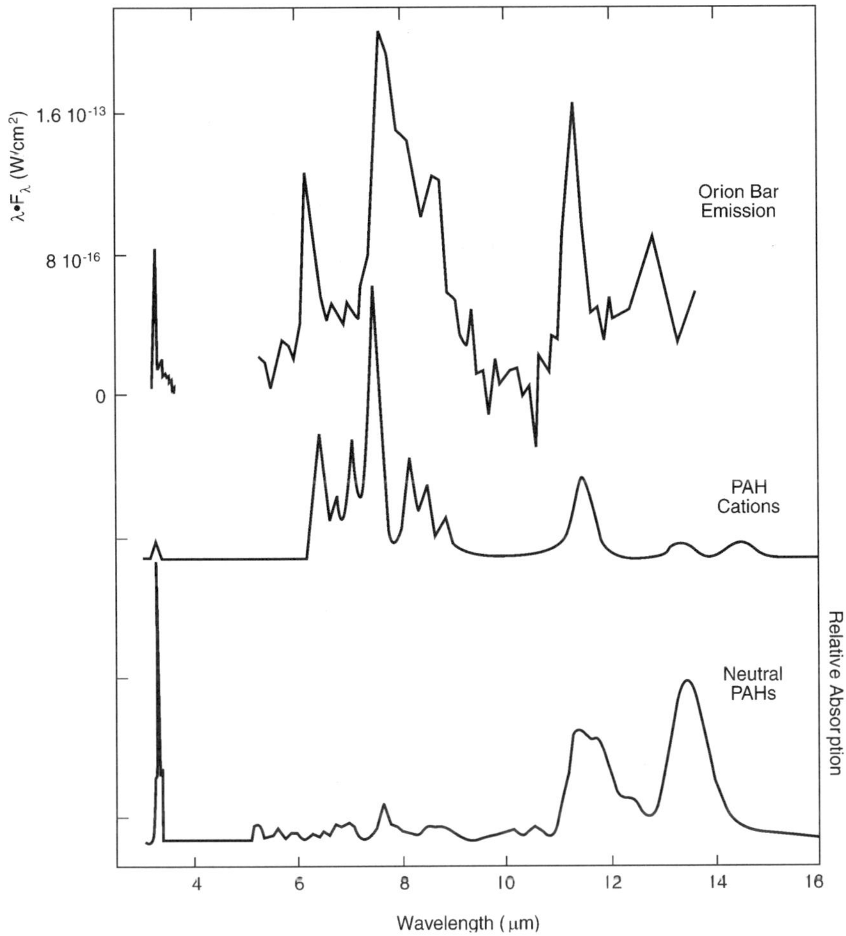

Figure 69. Comparison of the IR spectra of a mixture of LPAHs in the neutral and cationic forms and the Orion Bar Emission. From D. Hudgins, NASA Ames Research Center.

show that the spectral features in one nebula had the spectral features of cationic LPAHs closer in to the energy source and those of neutral LPAHs progressively further away.

Hudgins and Allamandola (22–24) also compared the spacing between two of the emission lines in the cationic spectra of 11 PAHs of up to 48 ring carbons (Fig. 71). These bands, sitting around 6.5 and 7.5 μm, were more separated in the larger-carbon number LPAHs. For pyrene (16 carbons), the separation was <1.0 μm. For ovalene (32 carbons) it was ~ 1.1 μm, and for benzo[1,2,3-*bc*:4,5,6-*b′c′*]dicoronene (48 carbons) it was ~ 1.3 μm. From this correlation,

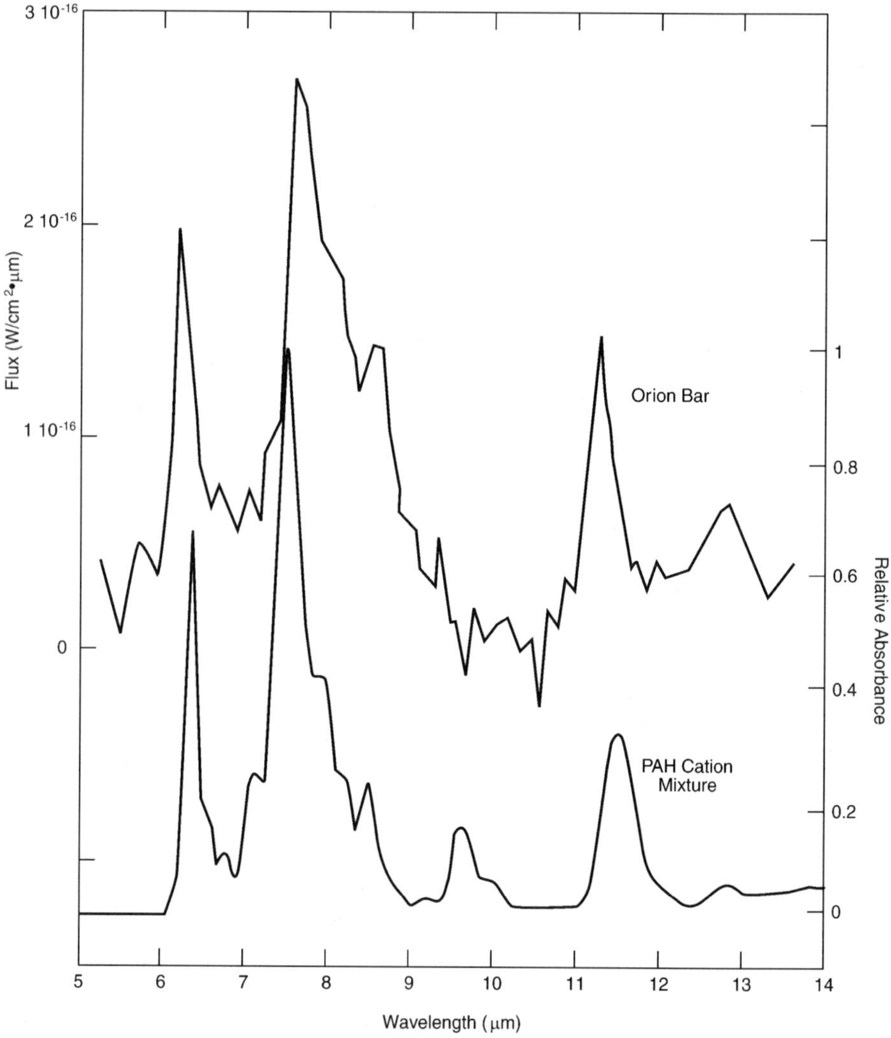

Figure 70. Close-up comparison of the IR spectra of a mixture of LPAHs in the cationic form and the Orion Bar Emission. Note that this LPAH mixture was of slightly different composition than that of Figure 69. From D. Hudgins, NASA Ames Research Center.

they concluded that the LPAHs in interstellar space, with spectral emission bands at 6.2 and 7.6 μm, must include some of a size range larger than their model compounds set. The lower band in the model compounds does not move much with increasing carbon number and matches the position of the lower of those pairs of bands seen in the interstellar spectra. The shift in the upper band

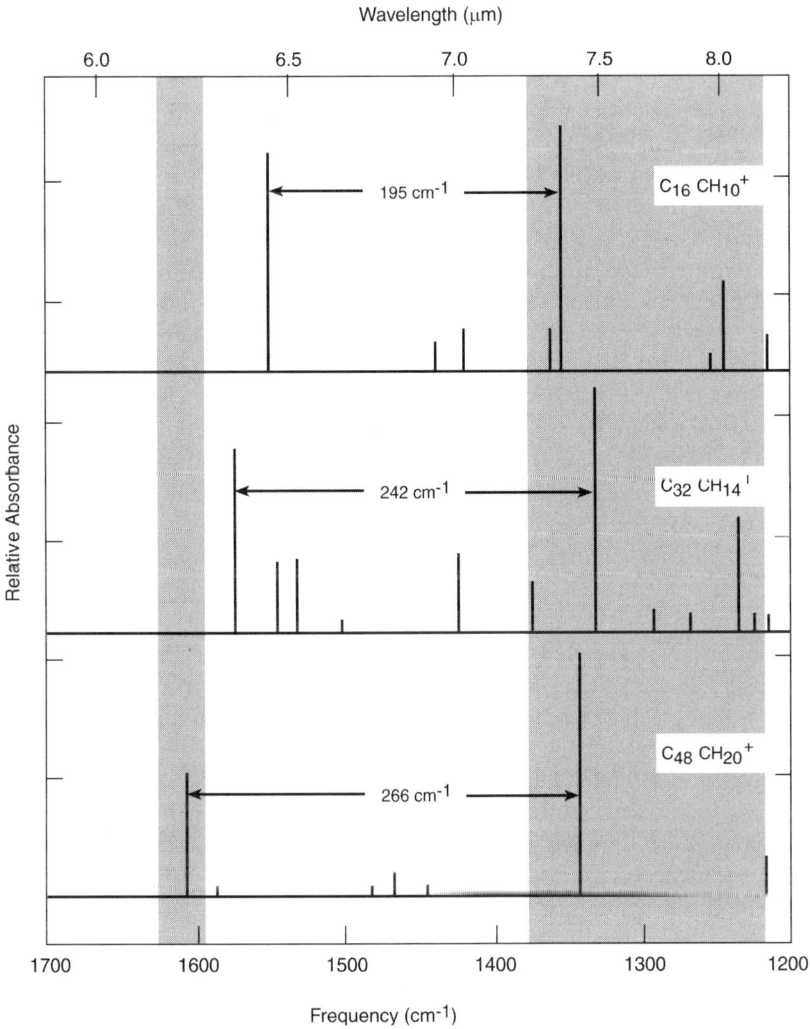

Figure 71. The spectral spacing features found by Hudgins and Allamandola. These were used to correlate width vs. carbon number to estimate the LPAH size in interstellar spectra. From D. M. Hudgins and L. J. Allamandola, *Astrophys. J.* **513** L71 (1999), reproduced with the permission of the authors and the American Astronomical Society.

corresponds to a LPAH size of 50–80 ring carbons for the interstellar cation species. The interstellar spectra also indicate that PAHs of as small as 20 ring carbons could also be present in the interstellar mixtures.

In an effort to predict what IR spectral features may arise from PAHs, Langhoff (25) used molecular modeling to generate the spectra of neutral, cationic, and anionic PAHs. One conclusion from this work was that many of the UIRs found in bodies such as the Orion nebula could be due to the cations of moderately-sized PAHs which have a high degree of symmetry. Coronene and ovalene were cited as examples of such LPAHs. The calculated spectra of neutral and anionic forms did not give good matches to the interstellar spectra.

The LPAHs, however, are not the only likely candidate molecules being proposed for these IR spectra features. Bernstein, Sandford, and Allamandola (26) examined the partially- and totally hydrogenated PAHs (including the totally hydrogenated form of the LPAH coronene). These molecules have spectral bands that match some of the lower-wavelength bands at 3.4 and 3.51 μm (Fig. 72). The polyacetylenes (long carbon chains of alternating single and triple bonds that can exist in the high vacuum of interstellar space), polymantanes (or diamondoids), and fullerenes have also been proposed as candidates and their spectra studied. These other hydrocarbon species are not thought to be the dominant ones in interstellar space, but rather their spectral features match those of some of the minor bands that cannot be ascribed to neutral or cation PAHs. Several of these alternative candidates may cause some of the IR spectral features, but other observed effects cannot be due to them. Among these are the red luminescence observed in certain reflection nebulae. Aihara proposed LPAH luminescence as the source of these unique spectral observations (27). The Red Rectangle is such a nebula. It surrounds the binary star HD44179. Large peri-fused LPAH structures were proposed because they have low-energy emission, which is uncommon in many other classes of simple compounds.

In fact, this search for matches to interstellar spectra drove some of the early work on fullerenes. Huffman and Kratschmer were looking for carbon-containing species to compare with the interstellar spectra. They first stumbled on how to make the fullerenes in macroscopic quantities by the now-famous production route through vaporization of carbon rods in an electrical arc. They were interested, at first, in fullerenes only as the possible answer to the identity of the UIRs.

Salama, in a recent review on the field (28), stated that the current view among many astrophysicists is that the DIBs arise from large neutral PAHs, whereas the emission features arise from very large PAHs or their cations. Some astrophysicists find that these spectral features are so common that LPAHs may comprise up to 20% of the total cosmic carbon (29). Allamandola (4) states, "They are probably more abundant than all of the other known interstellar polyatomic molecules combined."

Although PAHs have been found in cometary nuclei, interstellar dust grains, and meteorites, no LPAHs have yet been seen (30). The distribution of PAHs suggests that this may be due only to the low sensitivity of the analyses. PAHs of 10–20 carbons were found in ever-decreasing amounts as carbon number went

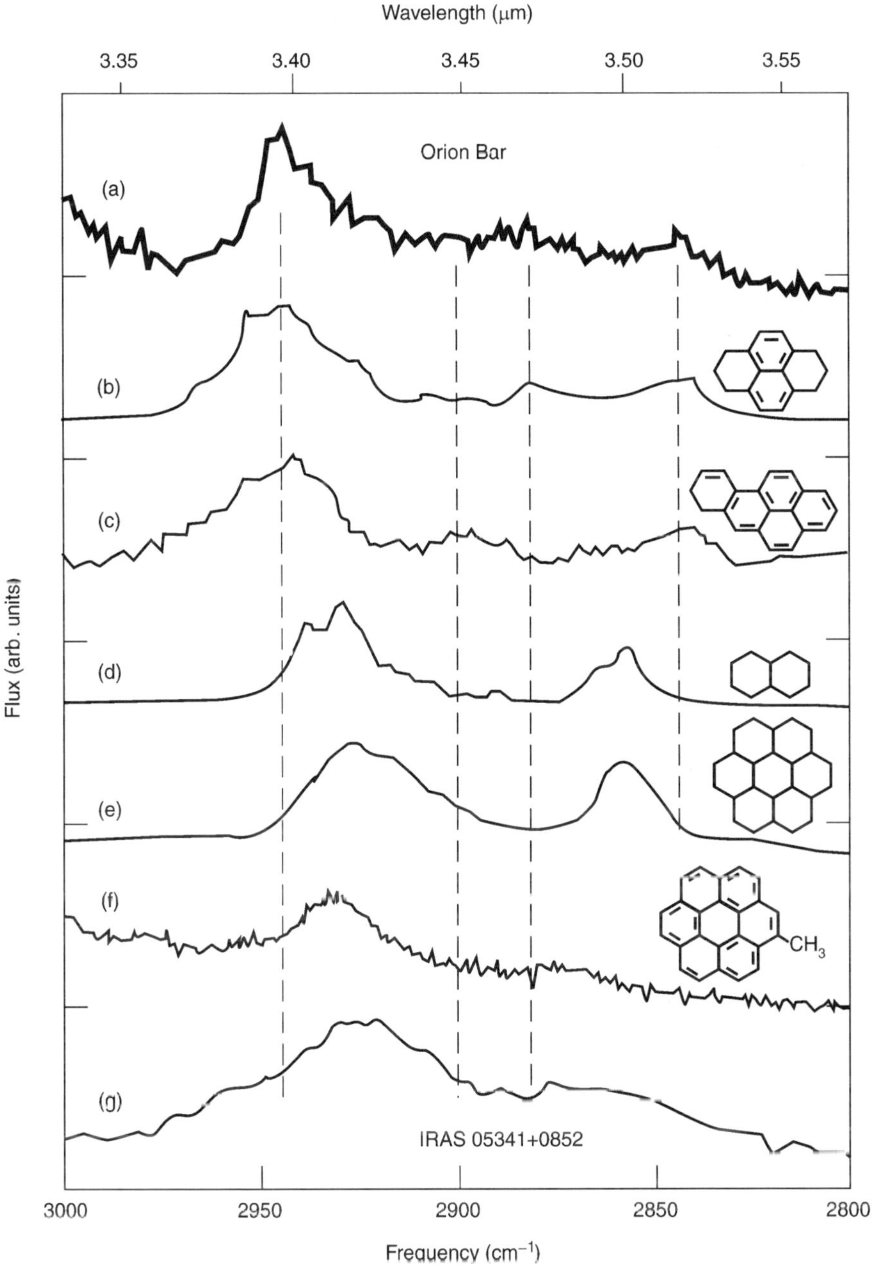

Figure 72. Comparison spectra of some PAHs and some partially and totally hydrogenated forms. From M. P. Bertstein, S. A. Sandford, and L. J. Allamandola, *Astrophys. J.* **472** L129 (1998), reproduced with the permission of the authors and the American Astronomical Society.

up in each of these materials. Bernstein and co-workers (31) recently studied the effects of high-intensity UV radiation on PAHs in matrices simulating ices under astrophysical conditions (such as might be found on interstellar and interplanetary dust grains). They found a wide range of alcohols, quinones, and ethers. These types of oxygenated PAHs have been found in meteorites.

Sagan et al. (32) made the simulated atmospheres of Jupiter and the Saturnian moon Titan (simple mixtures of methane, ammonia, and water) These were exposed to heating, irradiation by light and radioactivity, and electrical discharge. The tholins, the resinous brownish material that results, were found to contain naphthalene, phenanthrene, pyrene, and chrysene.

Messenger et al. examined the carbonaceous-grains in two "ancient" meteorites (32), the Murcheson and Acfer 094. These were thought to be caused by circumstellar carbon. With laser-ablation MS, masses indicative of phenanthrene, pyrene, perylene, or their isomers, were among those seen. Becker and Bunch used similar methods to examine the Allende meteorite (33). They saw evidence for naphthalene, phenanthrene, and several of their simpler alkyl derivatives. The PAHs pyrene, chrysene, triphenylene, perylene, corannulene, and dibenzo[*def,mno*]chrysene were also found. The only LPAHs seen were methyl-, dimethyl-, and trimethylcoronenes. In addition to these PAHs, they saw the C-60 and C-70 fullerenes.

REFERENCES

1. A. Leger and D. d'Hendecourt, *Astron. Astrophys.* **146** 81–85 (1985).
2. L. Allamandola, in *Galactic and Extragalactic Infrared Spectroscopy*, M. F. Kessler and J. P. Phillips, Eds., Reidel, Dordrecht, The Netherlands, 1984, p. 3–35.
3. *Polycyclic Aromatic Hydrocarbons and Astrophysics*, A. Leger, L. d'Hendecourt, and N. Boccara, Eds., Reidel, Dordrecht, The Netherlands, 1987.
4. L. Allamandola, *Topics Curr. Chem.* **153** 1–25 (1990).
5. L. J. Allamandola, A. G. G. M. Tielens, and J. R. Barker, *Astrophy. J. Supp. Ser.* **71** 733–775 (1989).
6. A. Leger, L. d'Hendecourt, and D. Defourneau, *Astron. Astrophys.* **216** 148–164 (1989).
7. C. Joblin, P. Boissel, A. Leger, L. d'Hendecourt, and D. Defourneau, *Astron. Astrophys.* **299** 815–846 (1995).
8. D. M. Hudgins and L. J. Allamandola, *J. Phys. Chem.* **99** 3033–3046 (1995).
9. D. M. Hudgins and L. J. Allamandola, *J. Phys. Chem.* **99** 8978–8986 (1995).
10. C. W. Bauschlicher, S. R. Langhoff, S. A. Sandford, and D. M. Hudgins, *J. Phys. Chem.* **101** 2414–2422 (1997).
11. D. M. Hudgins and L. J. Allamandola, *J. Phys. Chem.* **101** 3472–3477 (1997).
12. D. M. Hudgins and S. A. Sandford, *J. Phys. Chem. A* **102** 329–343 (1998).

13. D. M. Hudgins and S. A. Sandford, *J. Phys. Chem. A* **102** 344–352 (1998).
14. D. M. Hudgins and S. A. Sandford, *J. Phys. Chem. A* **102** 353–360 (1998).
15. S. R. Langhoff, C. W. Bauschlicher, D. M. Hudgins, S. A. Sandford, and L. J. Allamandola, *J. Phys. Chem. A* **102** 1632–1646 (1998).
16. D. Romanini, L. Biennier, F. Salama, A. Kachanov, L. J. Allamandola, and F. Stoeckel, *Fren. Chem. Phys. Lett.* **303** 165–170 (1999).
17. J. Szczepanski and M. Vala, *Nature* **363** 699–701 (1993).
18. L. J. Allamandola, D. M. Hudgins, and S. A. Sandford, *Astrophys. J.* **511** L115–L119 (1999).
19. D. M. Hudgins and L. J. Allamandola, *Astrophys. J.* **513** L69–L73 (1999).
20. D. M. Hudgins and L. J. Allamandola, *Astrophys. J.* **516** L41–L44 (1999).
21. G. C. Sloan, T. L. Hayward, L. J. Allamandola, J. D. Bergman, B. DeVito, and D. M. Hudgins, *Astrophys. J.* **515** L65–L68 (1999).
22. D. M. Hudgins and L. J. Allamandola, *J. Phys. Chem.* **99** 3033–3046 (1995).
23. D. M. Hudgins and L. J. Allamandola, *J. Phys. Chem.* A **101** 3472–3477 (1997).
24. C. Joblin, F. Salama, and L. Allamandola, *J. Chem. Phys.* **110** 7287–7297 (1999).
25. S. R. Langhoff, *J. Phys. Chem.* **100** 2819–2841 (1996).
26. M. P. Bertstein, S. A. Sandford, and L. J. Allamandola, *Astrophys. J.* **472** L127–L130 (1998).
27. J. Aihara, *Bull. Chem. Soc. Jpn.* **60** 3143–3147 (1987).
28. F. Salama, in *Solid Interstellar Matter: The ISO Revolution*, L. d'Hendecourt, C. Joblin, and A. Jones, Eds., EDP Sciences, Les Ullis, France, 1999, p. 1–23.
29. T. Henning and F. Salama, *Science* **282** 2204–2210 (1998).
30. E. Dwek, *Astrophys. J.* **475** 565 (1997).
31. M. F. Bernstein, S. A. Sandford, L. J. Allamandola, J. S. Gillette, S. J. Clemett, and R. N. Zare, *Science* **283** 1135–1138 (1999).
32. C. Sagan, B. N. Khare, W. R. Thompson, G. D. McDonald, W. R. Wing, J. L. Bada, T. Vo-Dinh, and E. T. Arakara, *Astrophys. J.* **414** 399–405 (1993).
33. S. Messenger, S. Amari, X. Gao, R. M. Walker, S. J. Clemett, X. D. F. Chillier, R. N. Zare, and R. S. Lewis, *Astrophys. J.* **502** 284–295 (1998).
34. L. Becker and T. E. Bunch, *Meteor. Planet Sci.* **32** 479–487 (1997).

CHAPTER

8

CHROMATOGRAPHY

Samples containing LPAHs usually have both a wide range in carbon number and a large number of isomers for any particular carbon number. This combination gives a complexity that almost always necessitates the use of a separation or even several sequential separations, with other subsequent analyses being used for identification and quantitation of the individual LPAHs. For this review, these subsequent steps would include the use of in-line detectors.

8.1 CHROMATOGRAPHIC SAMPLE FRACTIONATION

The complexity of LPAH mixtures can overwhelm even the most sophisticated chromatographic techniques currently in use. To make the analysis of a sample possible, sample fractionation is commonly used. The two most-used approaches are preliminary chromatography and vacuum sublimation. For the chromatographic fractionation, adsorption chromatography is often used.

Adsorbents, such as silica or alumina, separate the LPAHs roughly by the number of π electrons. Small PAHs elute with alkanes, aromatic hydrocarbons such as toluene or the xylenes, or halogenated aromatic solvents such as chlorobenzene. The LPAHs can be eluted with dichloromethane, ethers, or ethyl acetate. In some instances these solvents may not be able to fully dissolve the desorbed LPAHs. This occurrence is usually shown by the cloudy appearance of the undissolved LPAHs just behind the solvent front. To solubilize these fractions, some better solvent, like the halogenated aromatics, can be blended with the elution solvent. The more polar solvents, such as the ketones and alcohols, will elute the more polar non-hydrocarbon (heteroatom containing) species.

Heteroatom-containing polycyclic aromatic compounds (PACs) are often found as impurities in samples containing LPAHs. The oxygen- and nitrogen-containing species are readily removed by adsorption chromatography because they are much more polarizable or have greater hydrogen-bonding or acid-base character than the LPAHs. Sulfur-containing PACs (or S-PACs), on the other

Large ($C \geq 24$) Polycyclic Aromatic Hydrocarbons: Chemistry and Analysis, by John C. Fetzer
ISBN 0-471-36354-5

hand, are not very polar and have little or no hydrogen-bonding or acid-base character. S-PACs often remain as contaminants in LPAH fractions after initial attempts at heteroatom removal by adsorption chromatography. Complexation, such as through argentation chromatography—the use of a silver containing salt to selectively absorb sulfur-containing PACs—has been used in petroleum analyses to selectively remove the S-PACs from the aromatic hydrocarbon fractions.

8.2 HIGH-TEMPERATURE GAS CHROMATOGRAPHY

The common polysiloxane-based GC phases (which are usually substituted with methyl and phenyl groups) are usable to 250–300°C without rapid degradation of either the stationary phase or the outer column material itself and short column lives (1–5). Under these conditions, the LPAHs of 24- and 26-ring carbons elute with good resolution of isomers. The latter group, however, has broad peaks because of their elution during the isothermal end-of-run conditions. The larger 28-carbon LPAHs give very broad peaks at long retention times, with very little separation of isomers. The practical limitation of these phases is, therefore, only at the lower end in molecular weight of the LPAHs.

Special columns must be used for the GC analysis of the LPAHs of this and higher carbon numbers. The higher boiling points and lower volatility of those LPAHs requires very high operating temperatures, which are not compatible with the normally used GC phases. This has been accomplished in a variety of ways, including the use of metal jacket columns, bare silica- and alumina-coated columns, and highly cross-linked phases.

Several research groups have shown that the phases with the greatest power to separate PAH isomers, such as the bonded liquid crystal phases, are not thermally stable enough to be used for the higher-carbon-number LPAHs. They do show large resolution of the smaller PAHs. These phases are also still only in limited use because the methyl-phenyl phases are more readily available, more commonly used for most other analyses (and so column switching between analyses is avoided), and cheaper.

Lee and Hites (1) separated the LPAHs from carbon black extracts. They used a methyphenylpolysiloxane column and MS detection and saw structures of up to 34 carbons, which they assigned to a diphenyl-substituted $C_{22}H_{14}$ species. Schmidt, Grimmer, Jacob, and Dettbarn (2, 3) identified many 24-carbon LPAHs in the effluent from burning hard coal by GC-MS with a similar column. Many peaks of 26 carbons eluted toward the end of the chromatograms but were not identified. They later extended their identifications by synthesizing several more isomers. Wise, Benner, Liu, Byrd, and Colmsjo (4) used similar columns (DB-5) to separate the 24-carbon LPAHs in extracts from coal tars. They used GC-MS and were unable to identify many peaks because of a lack of available standards.

Simoneit and co-workers (5–7) used high-temperature GC-MS to examine the PAHs in absorbance-chromatography fractions of bitumens from deep sea hydrothermal vent areas. They found LPAHs of 24, 26, 28, and 30 carbons. Similarly to the other groups using GC, they were not able to identify any specific isomers. The high-molecular-mass region of one of these chromatograms is shown in Figure 73. The chromatograms are a close-up view of the 20-carbon peak cluster and a total-ion chromatogram annotated with observed molecular masses.

Bemgard et al. (8–11) have separated several LPAHs of up to 32-ring carbons (400 daltons) by using a variety of 10 commercial columns and 2 in-lab specially coated columns. They found that four of the column types were stable over

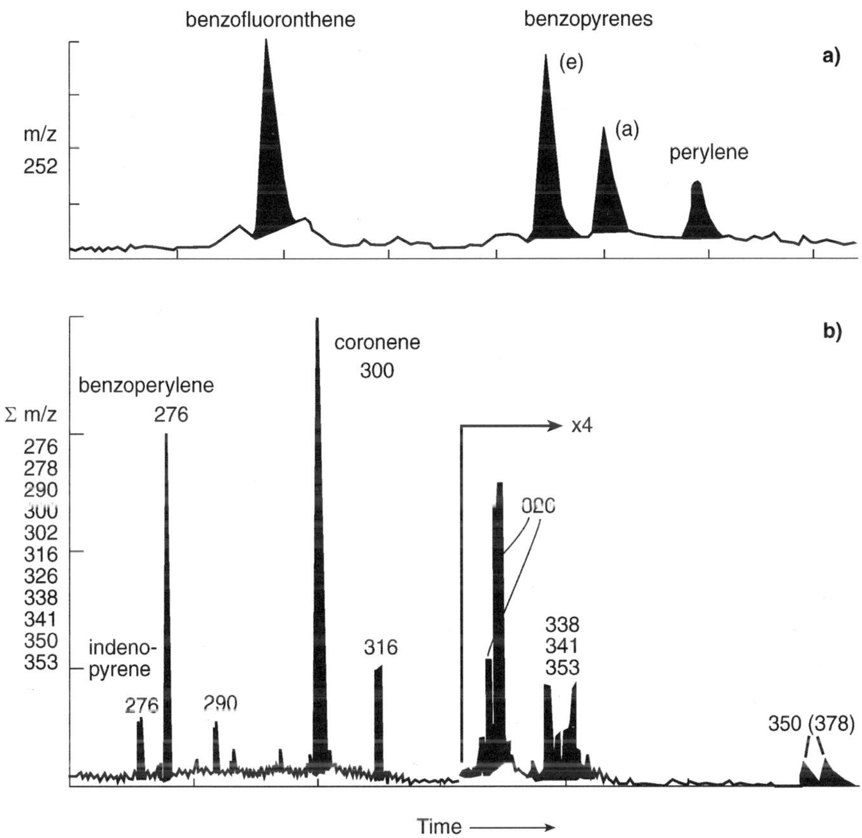

Figure 73 GC-MS chromatogram of a hydrothermal bitumen fraction from a sample collected in the Middle Valley , northern Juan de Fuca Ridge of the northeastern Pacific Ocean. From B. R. T. Simoneit, W. D. Goodfellow, and J. M. Franklin, *Appl. Geochem.* 7 263 (1992), reproduced with permission from Elsevier Science Publishers BV.

extended periods of use with a final column temperature of 370°C. The four column types were a metal-jacketed, thin-walled fused silica capillary column with a methyl-trifluoropropyl polysiloxane phase; a polyimide-coated fused silica column with a 5 % phenyl methyl polysiloxane phase; a metal-jacketed, thin-walled fused silica capillary with 5 % phenyl methyl polysiloxane; and a (noncommercial) polyimide-coated fused silica column with a biphenyl-substituted silarylene/siloxane copolymer phase. They found that sample loading had to be very small because of the low solubilities of the LPAHs in these phases. Their initial studies of the $C_{26}H_{16}$ isomers also showed that nonplanarity greatly effects retention. The more nonplanar isomers are much less retained, implying that there had been less interaction with the stationary phases. Of the four columns, the methyl-trifluoropropyl phase showed the greatest amount of this early retention for the nonplanar isomers.

They observed several clusters of peaks for each carbon number. They found three isomers of $C_{26}H_{14}$, seven of $C_{26}H_{16}$, and nine of $C_{28}H_{14}$ in the coal tar sample. They were later able to extend the elution range to LPAHs of 34 ring carbons (424 and 426 daltons, $C_{34}H_{16}$ and $C_{34}H_{18}$), 36 carbons (448 and 450 daltons, $C_{36}H_{16}$ and $C_{36}H_{18}$), and 38 carbons (472 daltons, $C_{38}H_{16}$). They used the last three types of columns for this work. The first type was not used because it was concluded that its low solubility for the LPAHs greatly reduced the resolution of isomers. They looked at both retention of model compounds and separation of real samples. They ran an HPLC fraction containing the LPAHs in the coal tar used for a standard reference material and the dichloromethane and chlorobenzene extracts (12) of the same carbon black sample that Lee and Hites (1) had used. The in-laboratory manufactured column had much better resolution than either of the two commercially available columns and only moderate bleed. Comparison chromatograms of some of their GC runs are shown in Figures 74–76.

In a related study, this group used molecular descriptors like molecular connectivity index, length, height, ionization potential, and quadrupole moment to predict the retention time for PAHs up to $C_{28}H_{14}$. The retention scale was defined in terms of the retention of the "bent" series of PAHs of phenanthrene, chrysene, picene, and benzo[*c*]picene (*compound 128*), first suggested by Lee and co-workers (13, 14). Because the next member of the series, dibenzo[*c*,*m*]-picene (*compound 129*), was not available, the authors substituted the isomeric and somewhat similarly shaped benzo[*c*]naphtho[2,1-*k*]tetraphene (*compound 130*, but they used the incorrect name of dinaphtho[2,1-*a*:2′,1′-*h*]anthracene). The prediction for the peri-fused PAHs was much better than that for the ortho-fused PAHs.

Ever after the LPAHs have been separated by a GC column, there is a fundamental weakness to be overcome. The commonly-used GC detectors, such as the flame-ionization or thermal conductivity detectors, are not selective

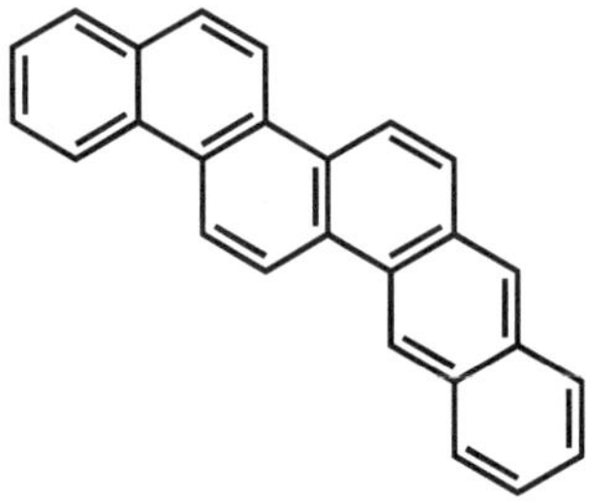

128

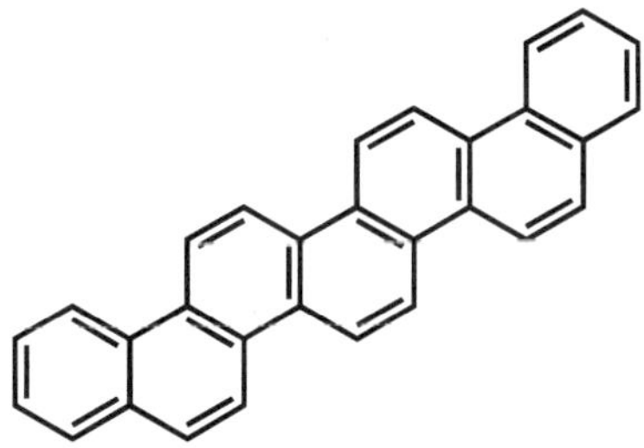

129

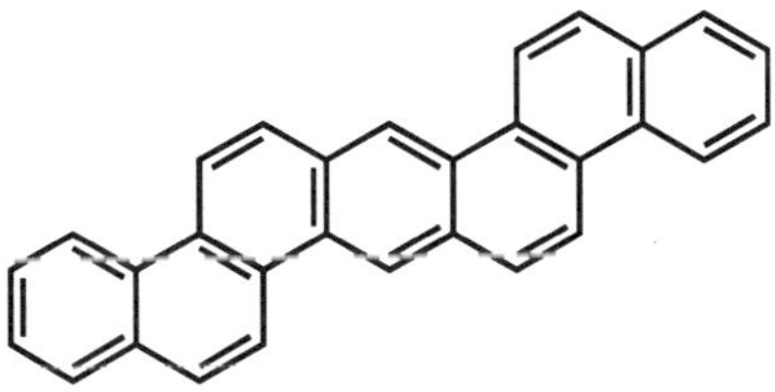

130

enough to differentiate LPAH isomers. The flame-ionization detector is valued for its universality. It sees all isomers with equal responses. A MS detector in the normal electron-impact mode only sees the molecular ion (see the chapter on the mass spectrometry of LPAHs).

Simonsick and Hites (15) showed that special chemical ionization reagent gases can be used, however, so that isomers will appear to have different fragmentation patterns. They used methane as the reagent and then used the M-to-M + 1 ratio to compare to calculated ionization potentials (IPs), to assign structures to several LPAH peaks. They separated the same extract as had been earlier studied by Lee and Hites. They saw eight $C_{28}H_{14}$ isomers and four $C_{30}H_{14}$ isomers. The comparison of calculated IPs to ion ratios and the possession of two isomers from each set (which were used to compare retention times as well as ion ratios) let them assign probable structures to all 12 peaks. An example of this approach of coupling high-temperature GC, chemical-ionization MS, and this specific data analysis is shown in Figures 62–64. A note of caution was given. Even with the high resolution of capillary GC, they observed an ion ratio for one of their standards, benzo[*pqr*]naphtho[8,1,2-*bcd*]perylene (*compound 30*), that did not match the corresponding peak in the carbon black separation. They relied on the GC retention time and a separate preparative HPLC separation and UV analysis of the peaks to confirm this component and assumed that there must be coelution that changed the observed ion ratio. This LPAH, as well as two of its isomers and two of the $C_{30}H_{14}$ LPAHs, were confirmed to be present in these extracts in a later HPLC analysis

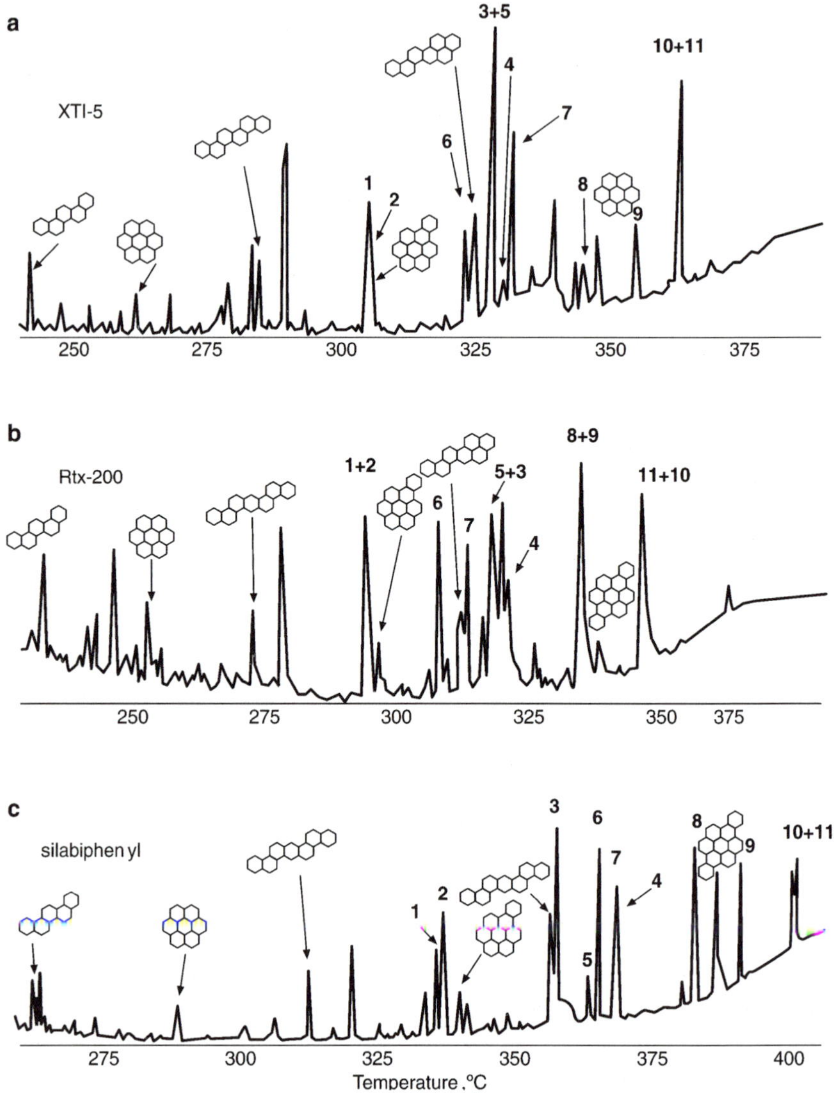

Figure 74. High-temperature gas chromatograms of LPAH standards on three phases, a. the methyl-trifluorylpropyl phase, b. the methyl-diphenyl polysiloxane phase, and c. the biphenyl silarylene phase. From. A. Bemgard, A. Colmsjo, and B.-O. Lundmark, *J. Chromatogr.* **630** 290 and 294 (1993), reproduced with permission from Elsevier Science Publishers BV.

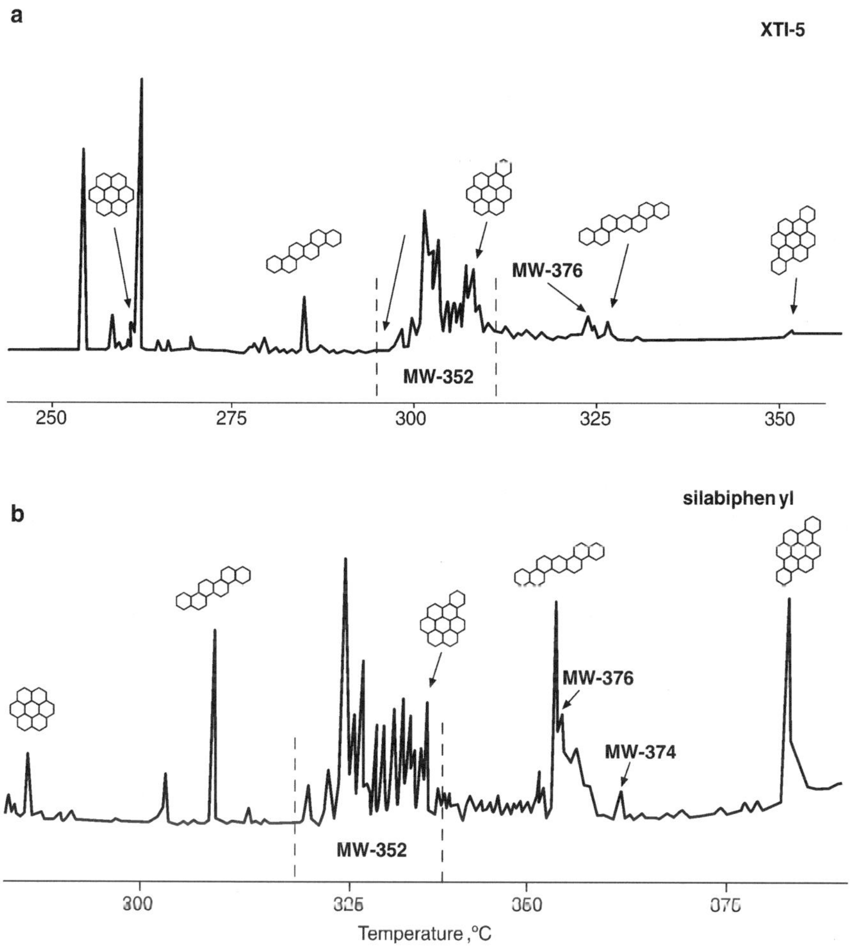

Figure 75. Comparison chromatograms of the high-molecular-weight HPLC fraction of SRM 1597, coal tar pitch. The top chromatogram is the methyl-diphenyl polysiloxane column and the lower chromatogram is for the biphenyl silarylene phase. From A. Bemgard, A. Colmsjo, and B.-O. Lundmark, *J. Chromatogr.* **630** 290 and 294 (1993), reproduced with permission from Elsevier Science Publishers BV.

by Fetzer and Rechsteiner, for which many more LPAH standards were available, including phenanthro[5,4,3,2-*efghi*]perylene (*compound 48*) and diindeno[1,2,3-*cd*:1′,2′,3′-*jk*]pyrene (*compound 131*), specifically cited by Simonsick and Hites as possibilities. In some cases the retention times of these authentic standards did not match the retention of the MS work, although similar HPLC columns were used. This again highlights the importance of standard compounds.

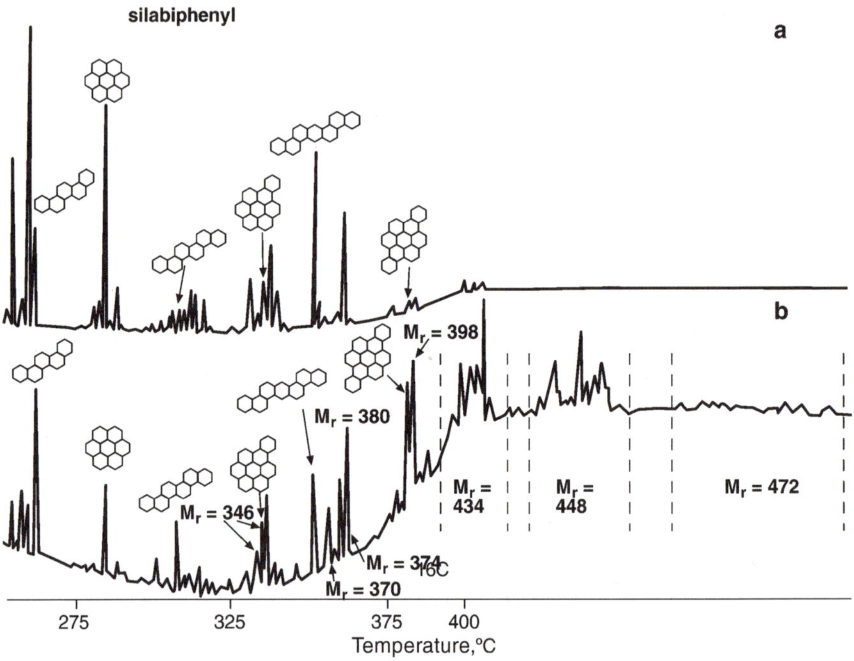

Figure 76. Chromatograms of the dichloromethane and chlorobenzene extracts of a carbon black under identical conditions on the biphenyl silarylene column. From A. Bemgard, A. Colmsjo, and B.-O. Lundmark, *J. Chromatogr.* **630** 290 and 294 (1993), reproduced with permission from Elsevier Science Publishers BV.

Brown and Searl (16) report on a GC method used to fractionate processed petroleum materials. The fractions are then analyzed by UV absorption spectroscopy. They report the results for isolating 42 PAHs, including coronene and naphtho[2,3-*h*]pentaphene (*compound 132*).

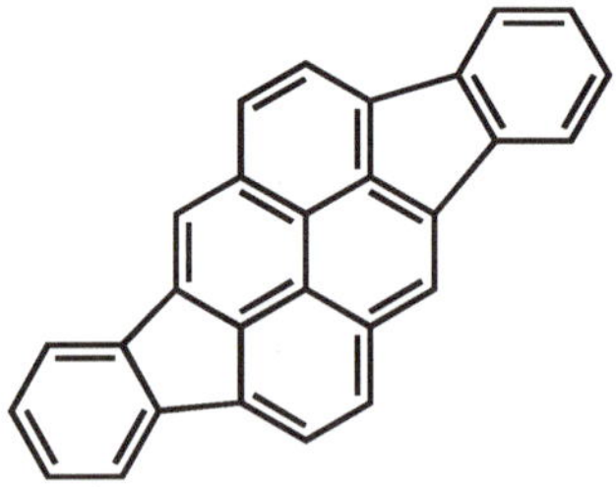

131

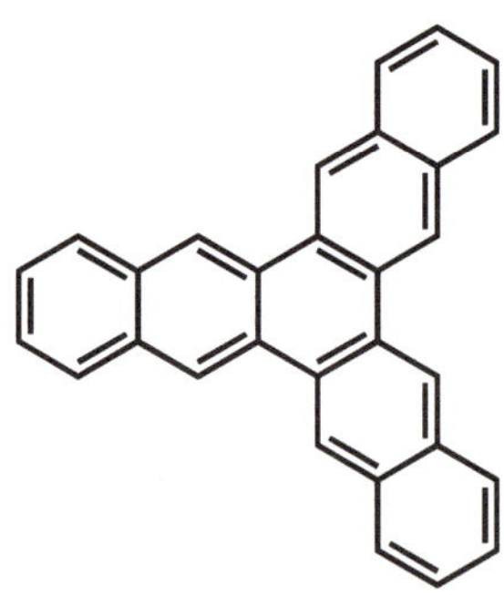

132

8.3 THIN-LAYER CHROMATOGRAPHY

There has only been limited application of thin-layer chromatography (TLC) to LPAH analysis because it is inherently only a moderate-resolution separation technique. This limits its capability to separate the many isomers of a particular empirical formula, let alone totally separate the many species of different carbon numbers and degrees of condensation from each other. In early work, Matsushita, Esmi, Suzuki, and Handa (17) used two types of TLC plates to separate a coal tar pitch. The sample was first fractionated by sequential liquid-liquid extractions. The benzene extract was developed on alumina plates. Individual spots (seen under a UV lamp) were scraped off, extracted, dried, redissolved, and spotted on acetylated cellulose plates. Once again the plates were placed under a UV lamp, and individual spots were scraped off and extracted. Fluorescence spectra were used for identification. Over 90 spots were separated, but only five LPAHs [coronene (*compound 13*), dibenzo[*b,def*]chrysene (*compound 133*), benzo[*rst*]pentaphene (*compound 134*), dibenzo-[*cd,lm*]perylene (*compound 38*), and dibenzo[*h,rst*]pentaphene (*compound 44*)] could be identified because of a paucity of available standards or reference spectra.

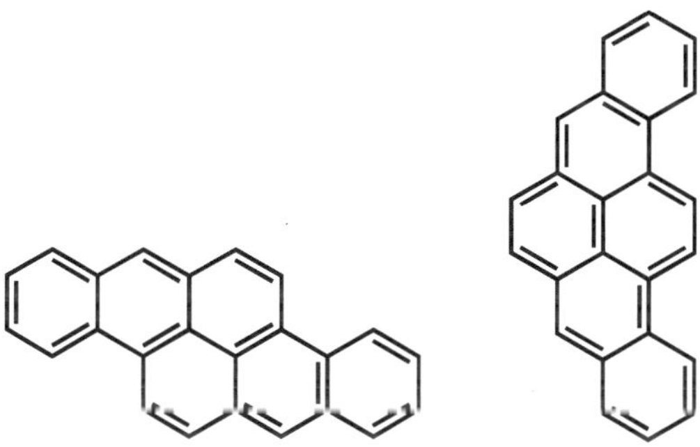

133 **134**

Pierce and Katz used a similar approach to identify a dozen LPAHs extracted from an air particulate sample (18, 19). They also used benzene as the extraction solvent, alumina plates for the initial fractionation, and acetylated cellulose to give individual spots. Acetylated cellulose plates of 20, 30, and 40% were compared, with the highest showing the best resolution of standard $C_{24}H_{14}$ isomers. After their initial success with this scheme, they then made plates coated with a mixed phase of two parts alumina and one part 40% acetylated cellulose. This allowed the use of two-dimensional sequential development. Four LPAHs were found.

McKay and Latham (20) used TLC to isolate the LPAHs in a crude oil. The oil was first distilled. One of the heavier cuts, from 335 to 500°, was fractionated by column chromatography on an ion-exchange resin (used mainly to differentiate heteroatom molecules from hydrocarbons). They found one of the spots to be coronene by comparison of the fluorescence spectra to standards.

8.4 HIGH-PERFORMANCE LIQUID CHROMATOGRAPHY

8.4.1 Normal-Phase HPLC

For the LPAHs, normal-phase HPLC is less versatile than the reversed-phase mode (21). This is because the separation mechanism is less selective toward separating isomers. The retention of LPAHs in normal-phase HPLC is controlled by the induced dipole in the LPAHs created by the polar bonded-phase moieties and the aromatic π electrons. Because isomers contain the same number of electron pairs, and differ only in their arrangement, there is very little difference in their polarizability.

Some bonded phases, such as the very polar tetranitrofluoeno (TENF) and other nitro-group phases, are said to be governed by a charge-transfer mechanism. The moiety's interaction with the LPAHs is not the dipole-dipole or dipole-induced dipole interactions of most normal phases. Instead, these nitro-group-based phases form stronger bonds, analogous to the complexes formed by LPAHs and nitro-group-substituted solvents. This interaction induces a large electron donation by the LPAH, creating a complex in which the LPAH has a slightly positive charge and the nitro solvent that complexes has a slightly negative charge.

8.4.2 Reversed-Phase HPLC

In reversed-phase chromatography, the interaction between LPAHs is strictly a partitioning behavior between the hydrocarbon-like bonded phase and the mobile phase. LPAHs have an affinity for alkane solvents, although their solubilities in such solvents are low. Therefore, they also have an affinity for the long alkyl chains of the bonded phase. The mobile phase must be a better medium for the LPAHs. Thus the aqueous mixtures commonly used for the separation of the smaller PAHs cannot be used because they have almost no affinity for the LPAHs. Mobile phases are nonaqueous, with typical solvents being ethyl acetate, dichloromethane, tetrahydrofuran, and chlorobenzene. Methanol and acetonitrile act as the *weak* solvent in gradients

Certain reversed-phase packings have been found to have a high capability of separating LPAH isomers and are thus widely used in LPAH separations (22–33). It is now necessary to give a short explanation of what these packings are and why their structures are so well suited to separating isomeric LPAHs.

The most commonly used reversed-phase packings have phases derived from reaction of a monofunctional octadecyl (ODS) silane. Monochlorodimethyl-ODS silane is most often used. This moiety attaches to the silica surface at a silanol site, with the moiety replacing the hydrogen atom. Each ODS moiety is bound only at this one point to the surface. A drastically different morphology results, however, if a di- or trifunctional silane is used (for example, dichloromethyl-ODS silane). One of the reactive sites reacts with the surface silanol, as in the case of the monofunctional silane. The moiety still has at least one reactive site. Small amounts of water present in the reaction mixture, either added at appropriate times or present in trace amounts, will turn these reactive sites into silanol groups. These then can react with more of the initial silane. A succession of these reactions can build a "polymeric" bonded phase. This is composed of a siloxane backbone, attached at one end to the silica surface, with an ODS chain branching off at each oligomeric unit.

The moieties in a polymeric bonded phase are very large. Their bulkiness hinders their movement, and they tend to orient in sheetlike structures because of the steric effects. Carbon-13 studies of monomeric and polymeric ODS phases show that the polymeric structure is much more rigid and the long carbon chain is fully extended in organic solvents. LPAHs that are planar can diffuse easily between these sheets and are exposed to more bonded phase than nonplanar LPAHs. Therefore, nonplanar LPAHs have shorter retention times because of this exclusion effect. Wise and Sander have labeled this shape-selective, permeation-dominated mechanism for planar versus nonplanar PAHs the "slot model" of reversed-phase retention. A conceptual rendering, showing the different retention mode for the planar coronene and nonplanar phenanthro[3,4-*c*]phenanthrene, is shown in Figure 77. Sander and Wise have also developed a standard reference material for use in determining the discrimination of ODS columns for shape selective separation of PAHs (see Ref. 34). This material, Standard Reference Material (SRM) 869, is offered by the National Institute for Standards and Technology as a "column selectivity test mixture for liquid chromatography." It contains three components, the planar benzo[*a*]pyrene and the nonplanar LPAHs phenanthro[3,4-*c*]phenanthrene (*compound 50*) and dibenzo[*g,p*]chrysene (*compound 135*). The separation factors (α values) between pairs of these three components can be used to determine the isomeric separating power of new phases.

The validity of the slot model of PAH retention and isomer separation selectivities has been supported by the custom synthesis of several longer alkyl-chain phases. C-22, C-30, and C-36 phases have also shown this type of behavior (35–38). With these longer chains, the monomeric versions of the phases are much more orderly than the monomeric ODS ones. Neutrin scattering, NMR, and fluorescence probe studies of these phases have shown that the longer chains are sterically inhibited in their movements and show much more orderly

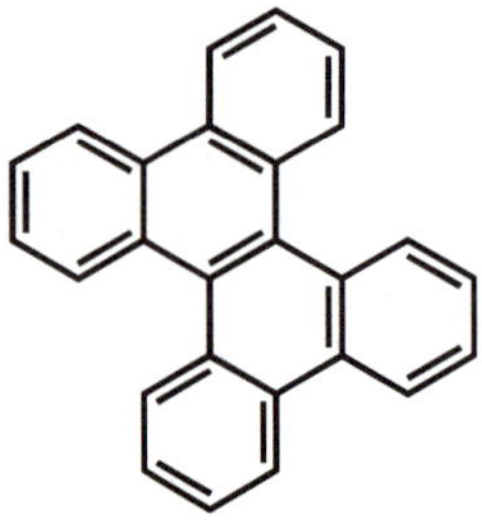

135

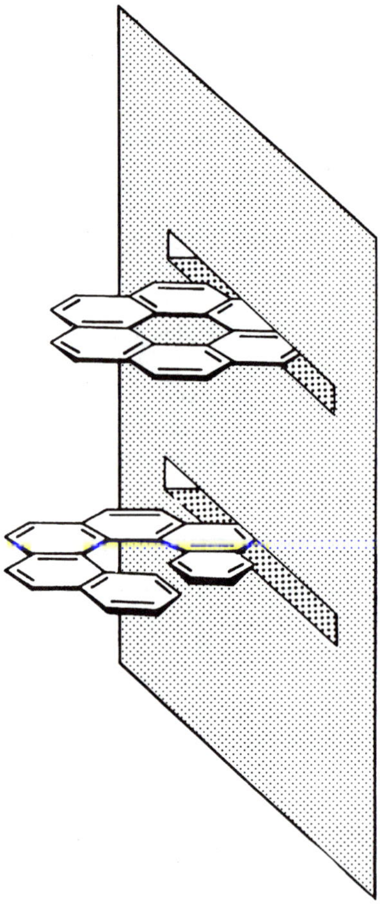

Figure 77. Slot model of Wise and Sander showing the selectivity for planar and nonplanar LPAHs.

arrangements of the alkyl chains (Figs. 78 and 79). The second of these figures shows the variety of bonded phase structures possible and why various reversed-phase columns show different selectivities. The phenomenal difference in separating LPAH isomers was highlighted in a study (38) that examined the separation of the $C_{24}H_{14}$ isomers on monomeric and polymeric phases. Chromatograms of these two separations are shown in Figure 80. The bonding and assembling chemistry of the phases has been thoroughly studied. These columns were used in HPLC separations, and they exceed the selectivities of the polymeric ODS phases for PAHs, carotenoids, and other classes of molecules.

Three other studies have utilized LPAHs as retention probes and the understanding of the shape selectivities of this type of mechanism on very unconventional packing materials. The planar LPAH benzo[1,2,3-*bc*:4,5,6-*b′c′*]dicoronene was packed as a fine powder into a column (39). The large planar structure of this molecule when arrayed in the parallel sheets present in its solid form is analogous to that of the polymeric ODS phases described by the slot model (40). This column showed similar retention behavior for discriminating planar and nonplanar LPAHs, with the nonplanar LPAHs eluting much earlier. The retention orders of several sets of model compounds were the same on both types of columns. In contrast, powdered C-60 fullerene was also packed into a column (41, 42). This perfectly spherical molecule had a stronger retention for nonplanar LPAHs than for planar ones. The nonplanar LPAHs were thought to have more contact with the fullerene surface because they could curl around and fit the curved surface better than the flat planar LPAHs. Thus the slot model was shown to be the probable mechanism in polymeric ODS phases. The planar copper phthalocyanine was bonded to silica and showed the expected shape selectivity.

Another LPAH structural factor that effects retention on polymeric ODS phases is the length-to-breadth ratio (L/B) of the molecule. For planar LPAH isomers, those with longer structures penetrate deeper into the bonded phase. Thus compact isomeric structures like dibenzo[*a,j*]coronene (*compound 58*) elute earlier than longer ones like dinaphtho[8,1,2-*defg*:8′,1′,2′-*opqr*]pentacene (*compound 78*). This can be explained by the slot model. The longer LPAHs can permeate between the large moieties and fit deeper into the polymeric ODS phase. The increased contact of the LPAH with the surface results in a longer retention.

Sander and Wise (23) have calculated L/B values for many PAHs up to 34 carbons. The algorithms they used for deriving the two-dimensional L/B values were modified and expanded to calculate three-dimensional L/B values. They also calculated the maximum value for each of the three dimensions (which Sander and Wise still call L/B values although 3 values are used; perhaps these values should now be called LBD values or some such variation because they also involve the calculated molecular depth). This allows for some estimation of

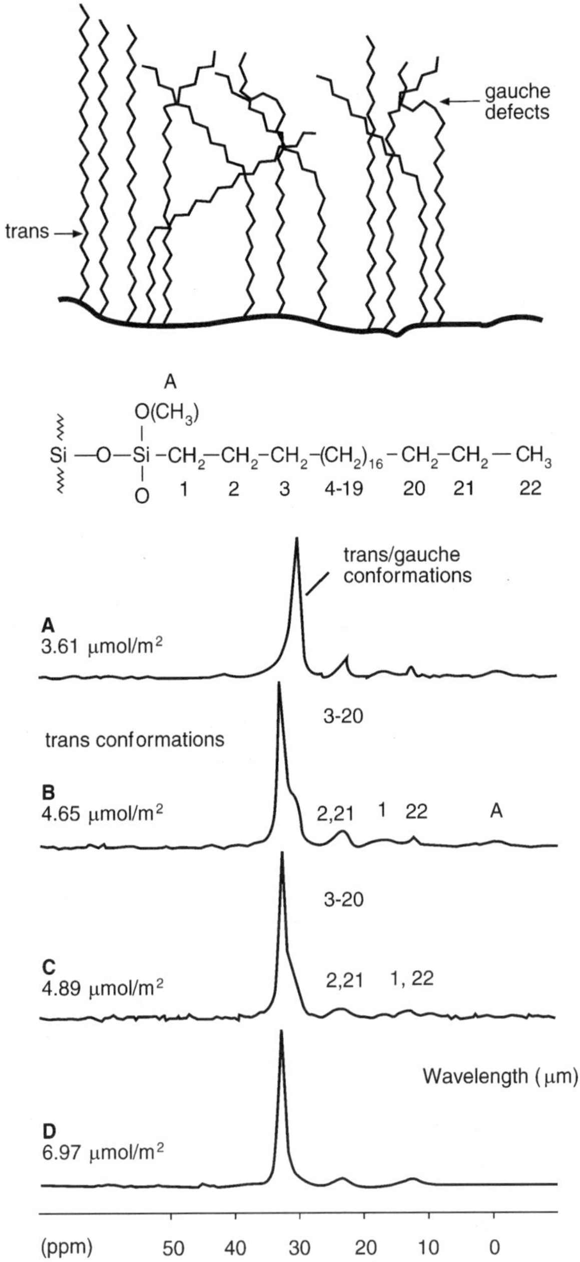

Figure 78. Carbon-13 NMR results showing some of the signals assigned to various structures in the orderly long-carbon-chain bonded phases From M. Pursch, L. C. Sander, H.-J. Egelhaaf, M. Raitza, S. A. Wise, D. Delkrup, and K. Albert, *J. Am. Chem. Soc.* **121** 3205 and 3212 (1999), reproduced with permission from the American Chemical Society

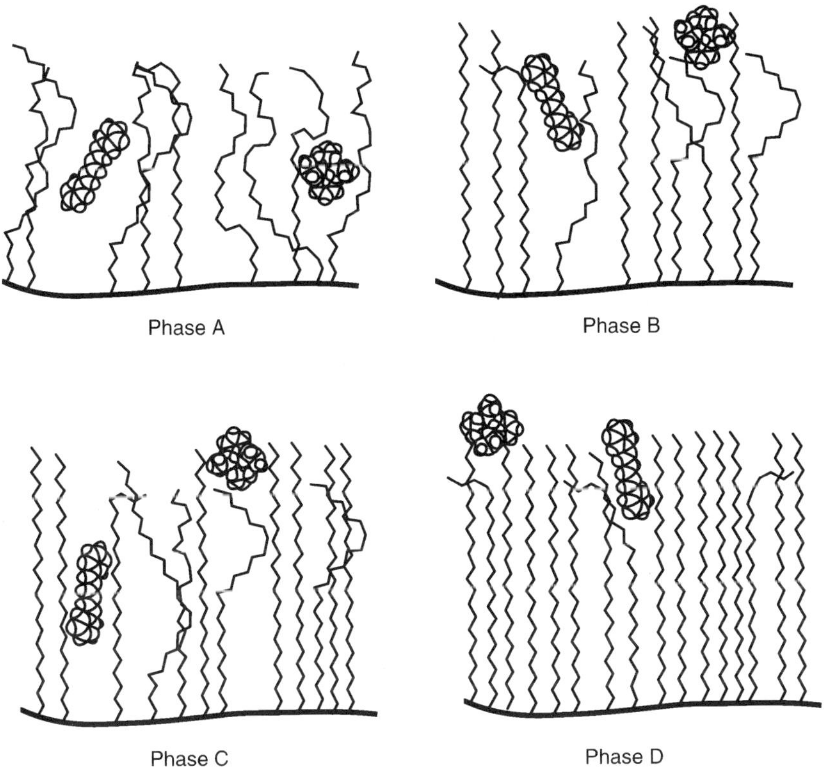

Figure 79. A representation of some of the structures of long-chain alkyl bonded phases and how two differently shaped LPAHs fit within the moieties. From M. Pursch, L. C. Sander, H.-J. Egelhaaf, M. Raitza, S. A. Wise, D. Delkrup, and K. Albert, *J. Am. Chem. Soc.* **121** 3205 and 3212 (1999), reproduced with permission from the American Chemical Society.

the effect of nonplanarity on retention. This publication also gives Chemical Abstracts Service registry numbers, molecular formulae, and IUPAC names for each of the over 900 compounds.

Lai, Ang, and Li (43) studied the retention on polymeric ODS phases of acenaphtheno[1,2-*e*]pyrene (*compound 136*), benzo[*p*]naphtho[1,8,7-*ghi*]chrysene (*compound 137*, which these authors incorrectly name phenanthro[9,10-*e*]pyrene), and dihydro derivatives. All of these standards were newly synthesized in their laboratory. They found that the retention was dominated by the nonplanar structures of these LPAHs, with carbon number and L/B not correlating well with the retention behavior.

The limit for elution with polymeric ODS phases is about 12 rings (44). The very long structure tetrabenzo[*def*,*lm*,*qrs*,*yz*]pyranthrene (*compound 138*) was

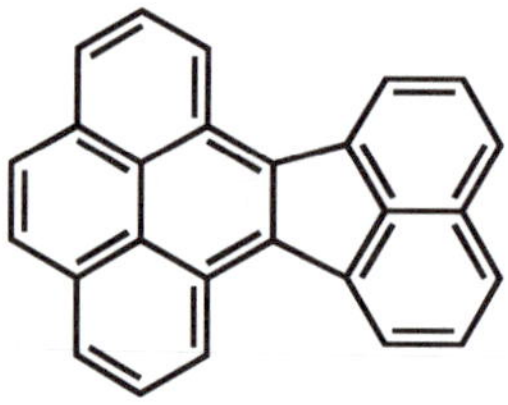

136

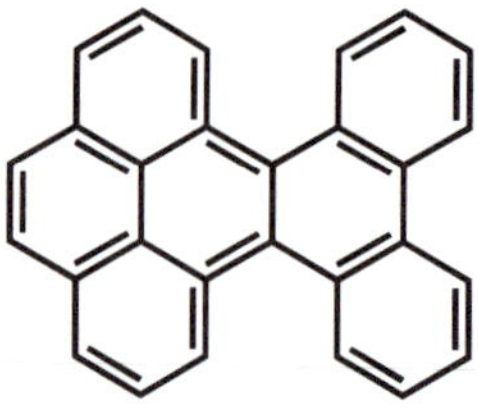

137

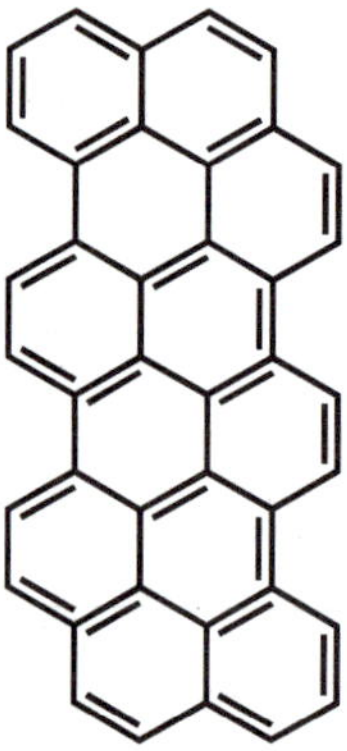

138

found to be elute with extremely long retention times, with relative retention values of >60 when even extremely strong mobile phases, like chlorobenzene, toluene, or 1,2,4-trichlorobenzene, were used.

The solvents used in the mobile phase can also have an effect on resolution, especially in the separation of isomers. With the polymeric ODS phases, using acetonitrile as the base (weaker) solvent gives slightly better separation of isomers than methanol. For the stronger solvent, tetrahydrofuran gives very poor separation between isomers, dichloromethane is very good, and ethyl acetate is even better (but it is somewhat weaker and so can only be used for LPAHs of up to 30 or 32 carbons). It is thought that the different solvents solvate the bonded phase differently. Some solvents give it a more extended and rigid structure that would enhance the isomer separating ability, whereas others collapse the bonded-phase structure into a more amorphous structure.

This increased selectivity was used to isolate a new 28-carbon LPAH, phenanthro[5,4,3-*efghi*]perylene (*compound 48*). This compound is produced along with its two isomers, benzo[*a*]coronene and benzo[*pqr*]naphtho[8,1,2-*bcd*], in petroleum hydrocracking units (45). The latter isomer is easily separated from the other two because it elutes much later, but the first pair have an almost identical overall shape. The retention time in acetonitrile-dichloro-

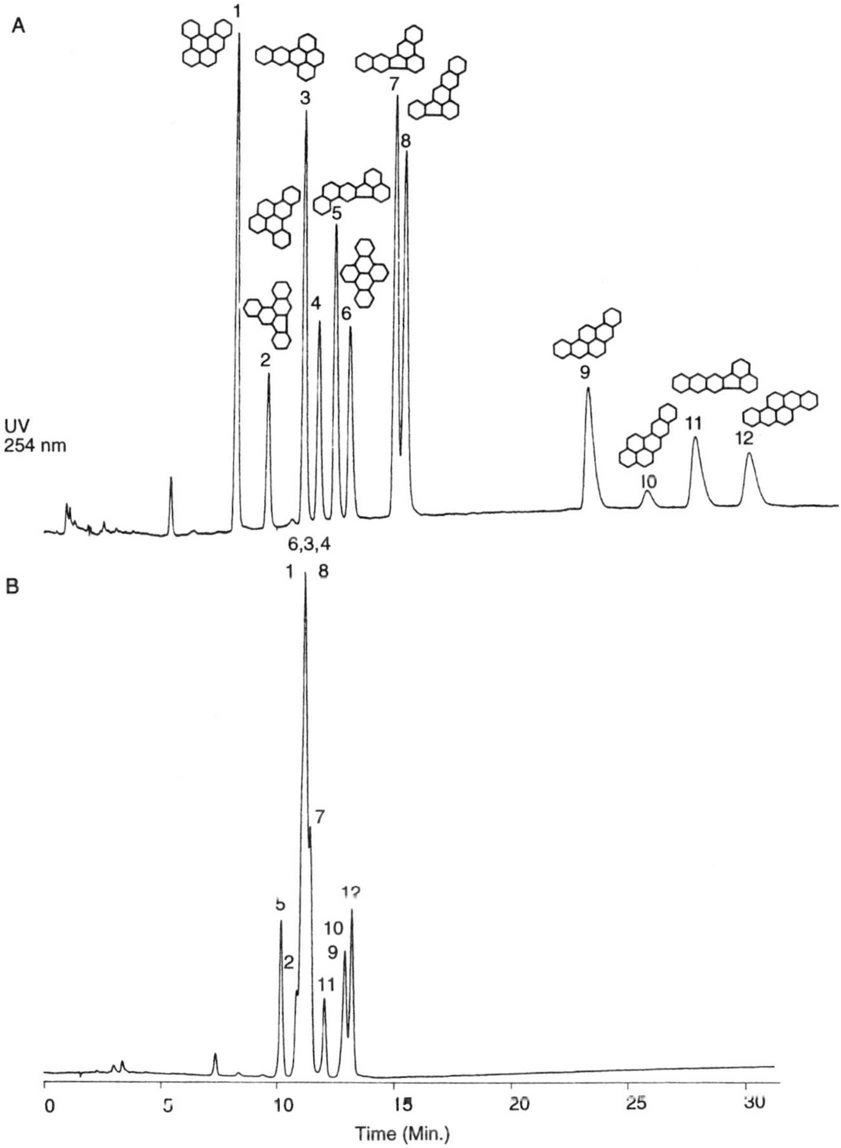

Figure 80. Comparison of the separation of a standard mixture of $C_{24}H_{14}$ isomers on a polymeric ODS and a monomeric ODS column. From L. C. Sander, M. Pursch, and S. A. Wise, *Anal. Chem.* **71** 4824 (1999), reproduced with permission from the American Chemical Society.

methane mobile phases was also almost identical (a separation factor α of 1.050). A change to ethyl acetate increased the separation enough (α of 1.097 in 54 %) that triplicate sequential injections and collections gave a pure sample of this new LPAH.

It is interesting to note the predictive power of the three-dimensional L/B values in this study. For the three isomers, the values are: benzo[*a*]coronene 1.215 (14.23, 11.71, and 3.888), phenanthro[5,4,3,2-*efghi*]perylene 1.216 (14.28, 11.74, and 3.885), and benzo[*pqr*]naphtho[8,1,2-*bcd*]perylene 1.331 (15.96, 10.50, and 4.261). The separation factors and retention orders match these L/B values almost exactly.

In one extreme example of the effect of nonplanarity on retention (27), with identical mobile phases the very nonplanar tribenzo[*a,cd,lm*]perylene (*compound 139*) elutes with a relative retention of 2.5, less than 5% of the value of the structurally-similar planar isomer, benzo[*rst*]naphtho[8,1,2-*cde*]pentaphene (*compound 140*). The separation is so different that in the original synthesis and

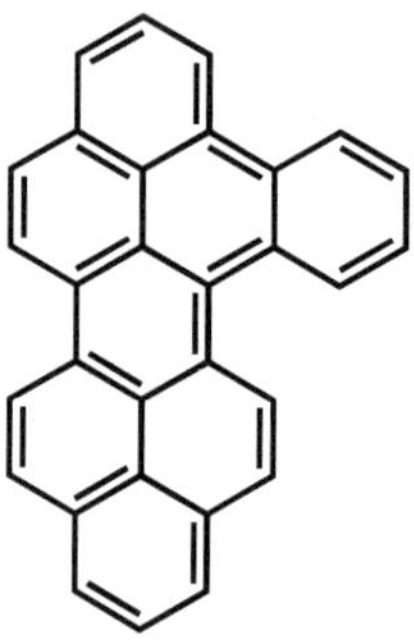

139

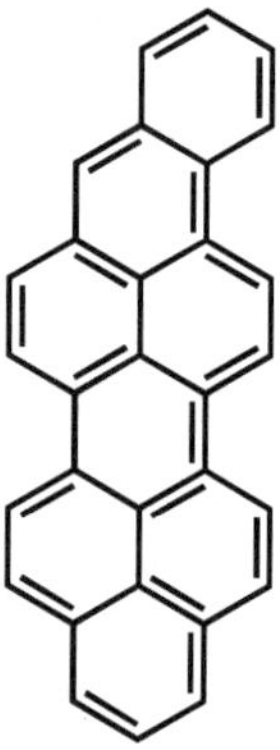

140

isolation, only the nonplanar isomer was seen and collected. The planar one was so strongly retained that it was only observed during the later cleaning of the ODS column by using pure dichloromethane as the mobile phase. The mass spectral analysis of this collected fraction showed this very strongly held LPAH (recognizable as one by the many bands in the absorbance spectrum; Fig. 23) was surprisingly isomeric to the early-eluting peak. The LBD values for these two are 1.354 (15.79, 11.66, and 5.583) and 1.955 (18.20, 9.308, and 3.892). Once again, the slot model of retention and the calculated molecular dimensions predict an extremely wide separation, matching the observed behavior.

The fact that the amount of nonplanarity dominates LPAH separations on polymeric ODS phases can cause some anomalous retention behavior. Fetzer and Biggs and later Jinno, Ibuki, Fetzer, Biggs, Griffiths, and Ollinger (26–28,

46) observed that some LPAHs appear to change and become more nonplanar in stronger mobile phases. The amount of these changes differs for each of the LPAHs studied, so that the plots of the logarithm of retention factor versus the percentage of string solvent have different slopes. For planar LPAHs, these have similar slopes. For the nonplanar LPAHs, the slopes are steeper. This results in intersecting plots, indicating retention order reversals.

The proton NMR spectra of one such LPAH, tetrabenzo[*a,cd,j,lm*]perylene, shows changes due to peak splitting from spatially adjacent protons that reflect changes in their proximities with changes in dichloromethane concentration. Both the retention plots and chromatograms for an extreme example of this type of behavior are shown. Dibenzo[*cd,lm*]perylene is a 26-carbon planar LPAH. Its two naphthologues, benzo[*lm*]phenanthro[4,5,6-*abcd*]perylene and dibenzo[*j-k,uv*]dinaphtho[2,1,8,7-*defg*:2′,1′,8′,7′-*opqr*]pentacene (*compounds 141* and *142*), become increasingly nonplanar as the strong solvent (in this case dichloromethane) percentage increases. At high dichloromethane concentrations the order is completely reversed from the order at low concentrations. In fact, for the preparative-scale isolation of the intermediate-sized LPAH, the best mobile phase was high in dichloromethane. This is contrary to conventional wisdom that stronger mobile phases shorten retention at the cost of lower resolution.

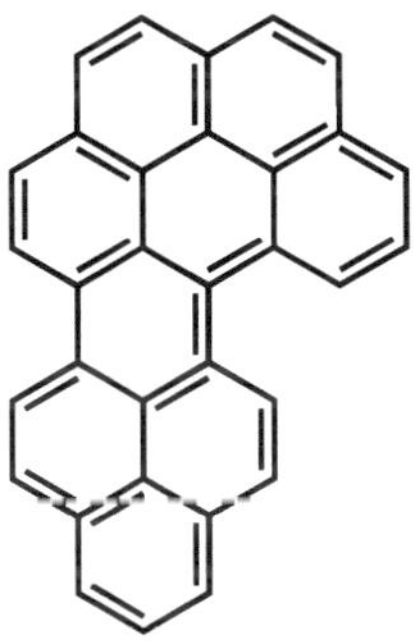

141

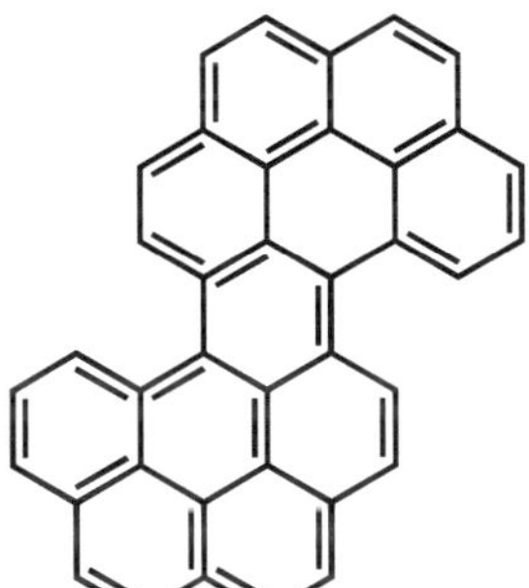

142

The solvation interactions of LPAH molecules in strong mobile phases (and good solvents), like dichloromethane, ethyl acetate, and chlorobenzene, are stronger than in solvents like methanol or acetonitrile. The increased solvation energies must be sufficient to overcome the slight loss in π electron delocalization due to the increased nonplanarity. An example of this is seen in the retention behavior of tribenzo[*a,cd,lm*]perylene (*compound 138*) and the similarly structured benzo[*lm*]phenanthro[4,5,6-*abcd*]perylene (*compound 140*). Figure 81 is a plot of retention behavior as a function of increasing concentration of the strong solvent dichloromethane. At high concentrations, the relative nonplanarities appear to change enough so as to result in a retention reversal.

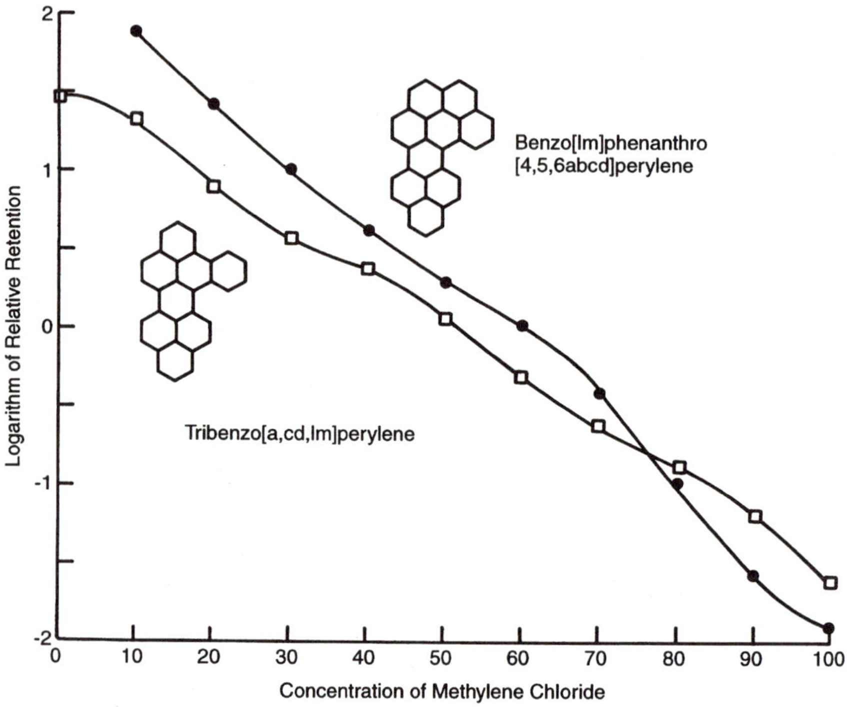

Figure 81. Retention order reversal for two nonplanar LPAHs.

Peaden, Lee, Hirata, and Novotny (47) used a polymeric ODS phase to separate several dozen LPAHs in an extract from an experimental carbon black (the same sample as that Lee and Hites examined by using GC-MS described earlier). Individual peaks were collected and analyzed by off-line fluorescence and mass spectrometries. Although 44 fractions were isolated and their molecular weights and fluorescence excitation and emission maxima were reported, only eleven LPAHs could be unequivocally identified by comparison to known literature compound spectra, mainly from Clar's *Polycyclic Hydrocarbons*. Several others were "identified" as certain proposed structures. Hirose, Wiesler, and Novotny (48) increased the separation efficiency by using packed microcolumns. They observed many more peaks but could not identify any further LPAHs. For example, the first study found seven $C_{26}H_{14}$ peaks and six $C_{28}H_{14}$ peaks. The second study found 16 peaks for each of the molecular formulae. In the second study, the authors classified their peaks only in terms of ring number and degree of condensation.

A later reexamination by Fetzer and Rechsteiner (12) of the data reported in the first study, however, showed that some of those proposed structures were those of known LPAHs. In some cases, the fluorescence excitation and emission maxima reported did not match those of the known compounds, and so these must have been misidentifications. For example, peak 52 is reported by Peaden et al. as dibenzo[*fg,ij*]phenanthro[2,1,10,9,8,7-*pqrstuv*]pentaphene (*compound 143*); the fluorescence wavelengths are much too high to be this compound. In

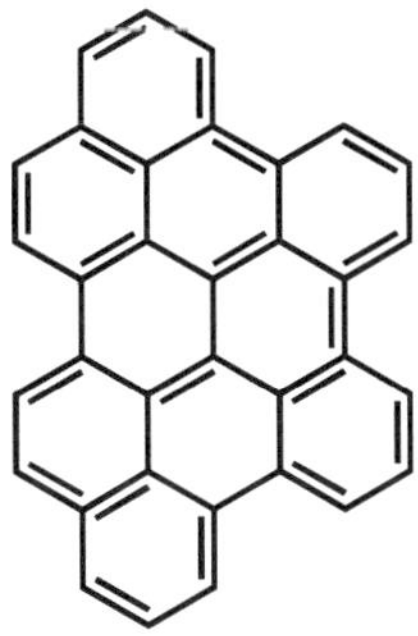

143

the later work another extraction of the same carbon black was performed and some of these earlier identifications were confirmed. Fifteen further LPAHs were identified. This second study used the same type of polymeric ODS phase as in the earlier study but relied on a diode-array absorbance detector (DAD). This type of detector is described in the next section. Figure 82 shows the chromatogram obtained by Peaden et al., but superimposed with the correct identifications confirmed by Fetzer and Rechsteiner. This example also highlights the value of having a large collection of standard compounds that can be compared to components of a complex mixture, because the authors in the later study were able to compare the unknown peaks to several hundred LPAH standards. Still, several dozen peaks could not be identified owing to a lack of standards or reference spectra.

Wise, Benner, Liu, Byrd, and Colmsjo (49) identified many 24-carbon LPAHs isolated from a fraction of SRM1597, a coal-tar pitch extract. They used a normal-phase separation to fractionate the material into "ring number fractions." These were then each separated on a polymeric ODS phase. The final identification used Shpol'skii spectroscopy on individually isolated peaks from the ODS separation. Wise, Deissler, and Sander (50) followed this work up by further refinement of the separations and extending it to three other SRMs. These were two air-particulate matter and a marine sediment samples. They analyzed for both the $C_{22}H_{14}$ and $C_{24}H_{14}$ isomer sets.

8.4.3 Diode-Array and Other Full-Spectrum HPLC Detectors

As discussed in the chapter describing the UV/visible absorbance spectra of LPAHs, the large amount of information found in LPAH absorbance spectra has

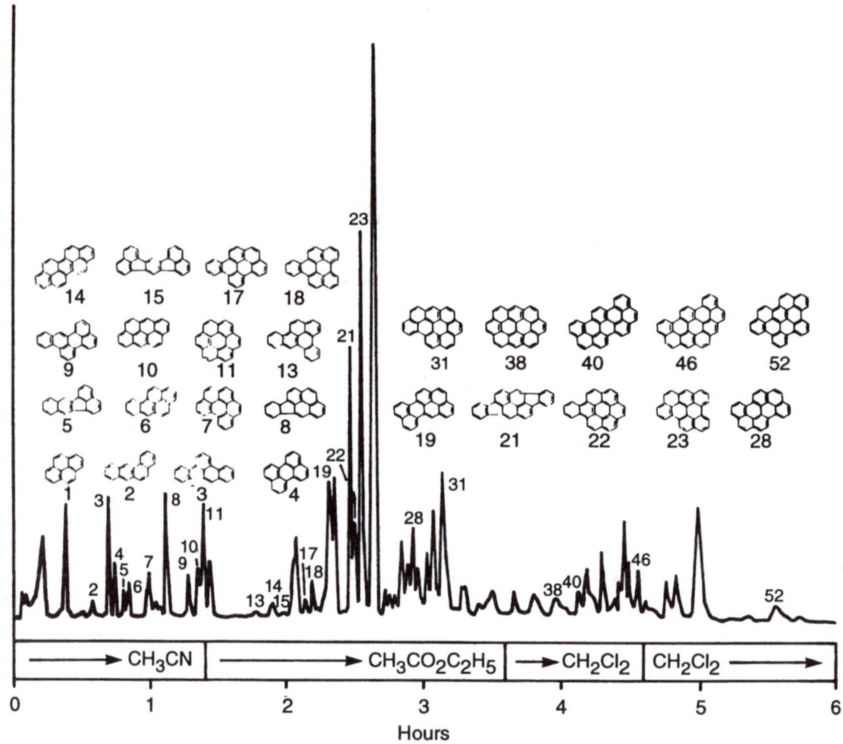

Figure 82. Reversed-phase chromatogram of a carbon black extract with some of the identification of LPAHs found by comparison to standard compounds.

led to the use of the DAD as an instrument that is virtually an ideal HPLC detector for LPAHs. An example of the very spectrally rich and varied spectra, the DAD spectra, is shown of the two "new" isomers tribenzo[*a*,*cd*,*lm*]perylene and benzo[*rst*]naphtho[8,1,2-*cde*]pentaphene discussed in the earlier section on retention on polymeric ODS phases(top of Figure 20 and Figure 23).

The DAD differs fundamentally from the single-wavelength detector that had been the conventional absorbance detectors before the development of DADs. Before the development of DADs, absorbance detectors relied on a monochromator design in which a light source was used to generate a single observation wavelength. The monochromator would only pass a light beam with a narrow wavelength range.

For many single-wavelength detectors, the deuterium lamp was commonly used as the light source in detection for PAHs because of its high intensity at certain wavelengths. One of these wavelengths, 254 nm, was in the intense absorbance region of the π-to-π* transition of benzene. In the spectral regions in which there were no emission bands from the deuterium lamp, no light was emitted, Thus the light intensity was zero and this detector could not be used. Because individual PAHs absorb at many wavelengths, the deuterium source is of limited use if selectivity of detection or sensitivity are issues.

When a wider range of single wavelengths was needed, for example to selectively detect PAHs at their most intense bands, another light source had to be used. Xenon lamps have been used to provide "white light" sources in the UV range, and tungsten lamps are used when wavelengths in the visible range are needed. A monochromator diffuses a single wavelength, which is focused and passed through the sample flow cell. A detection device then monitors the intensity.

The DAD, however, first passes all of the light from the "white light" source through the flow cell. The sample components passing through the flow cell absorb at all wavelengths. The light is then dispersed by a grating and is focused on a series of photo-sensitive semiconductor devices, the diode array. The array is positioned so as be on the optical path. Each diode monitors a segment of the full wavelength range. The total output of the array thus generates an absorbance spectrum.

For LPAHs, the DAD wavelength range needs to include the spectral region of 400–650 nm. Many common LPAHs have all of their most characteristic spectral features in this range. Also, a spectral resolution of <2 nm is desirable to detect the small red shifts due to the number of alkyl substituents. Typically ,each substitution on the aromatic core only shifts the spectrum 1–2 nm higher.

Another useful feature of DAD system is the computer systems necessary for this type of detector. Specifically, the data handling hardware and manipulation software are very important. Complex samples require that many spectra be saved. Often this is accomplished by collecting spectra at preset intervals, sometimes as frequently as each second, throughout the chromatographic run. This may mean thousands of spectra per sample. When stored in the data system, these spectra can be many megabytes of data per sample. Large storage capacities, with quick access to data, become extremely important in data evaluation.

With these large numbers of sample spectra and the many, many possible LPAHs, spectral matching software has also become very important to the analytical chemist. This software allows the comparison of an unknown peak to those of standards, both for spectral patterns and retention times.

For simple compounds, meaning either the unsubstituted LPAHs or their monomethyl derivatives, spectral matching programs usually work well (51, 52). This is not true, on the other hand, for other more substituted LPAHs, where the spectral shift due to substitution is >1–2 nm (53), The currently available software does not work well. These programs generally work by spectral matching done on a point-to-point comparison rather than through looking at the patterns of the spectral bands (comparison of the relative intensities of spectral features and their distances from each other).

With greater substitution, the common programs see the large differences in location, due to the shifting of the spectrum by the substitution, as different peaks, and do not see the similarities in patterns. For example, if the spectra for unsubstituted coronene and for a tetramethylcoronene isomer are compared. The similarity in the peak shapes and relative locations of the absorbance bands is obvious to the naked eye. A spectral matching program, however, cannot recognize the second compound as a coronene-type species because the spectral shift of 5nm is too large for the similarity to be found in a point-to-point comparison.

Because the DAD is a multiple-wavelength detector, there are data manipulations possible with it that are impossible for single-wavelength detection. A simple example is coelution of two components. If the retention times are slightly different, then the spectra from different times across the chromatographic peak will differ and show the coelution. In the case in which there is coelution of a small number of components in a retention range, the overlapping peaks can be deconvoluted if there is any difference in retention times. For each wavelength, the intensity changes as the sum of the individual contributions due to each component (the Beer-Lambert Law). Across the peak of a single component, the spectra will all be proportional to each other. If two components with even slightly different retention times are within the peak, then the relative intensities with change. The contribution of the two components to each wavelength will be their individual absorbances based on their concentration at the retention time of the spectrum. Algorithms can separate the contribution from each and recreate the spectra of each component.

In the simplest case of two coeluting LPAHs, the molar absorptivities need only be very different at two distinct wavelengths for deconvolution to be done. The relative changes in intensity at these two wavelengths are enough to mathematically separate the spectral contributions of each compound at the different retention times. Both the retention times of the peak maxima of each component and their absorbance spectra can be separated from each other by the computer's algorithms.

Ramos, Stewart, and Rohrback (54) first pulled the individual spectra of benzo[*b*]fluoranthene and benzo[*k*]fluoranthene or chrysene and benz[*a*]anthracene from each other in a purposely created coeluting peak of mixtures of each of the two pairs of PAHs (Fig. 83). They then were able to deconvolute the individual spectra from a mixture of the isomers benzo[*e*]pyrene, benzo[*b*]fluoranthene, and benzo[*k*]fluoranthene in a similar fashion. In all cases they purposely generated peaks with severe overlap ($>90\%$ of each peak coeluting with the other components) to show the power of deconvolution. Tauler, Durand,

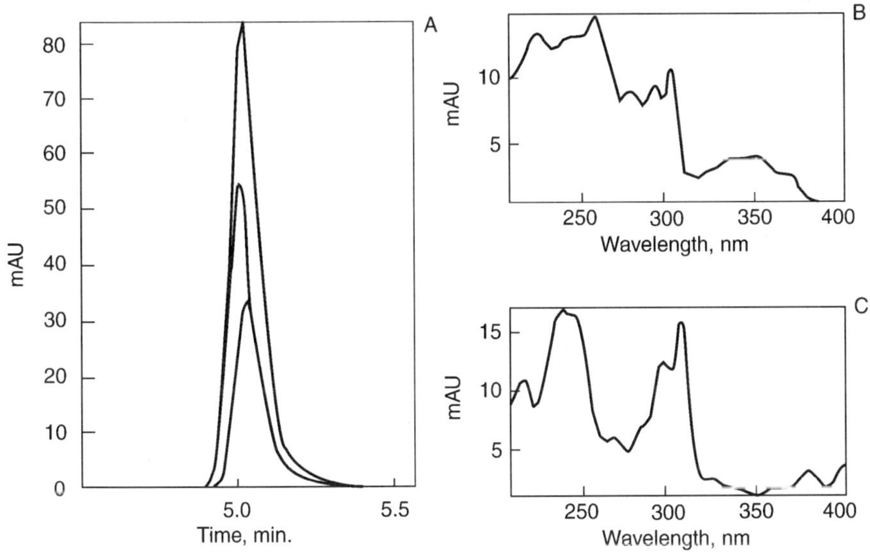

Figure 83. Deconvolution of DAD spectra from a mixture of benzo[*b*]fluoranthene and benzo[*k*]fluoranthene, where the chromatography was done to make their peaks overlap. From L. S. Ramos, R. J. Stewart, and B. G. Rohrback, *Chromatogr. (LC/GC)* **2** 98 (1987), reproduced with permission from Cahners Publishing.

and Barcelo (55) described another algorithm to deconvolute the individual spectra in coeluting peaks and reviewed similar efforts by others.

In a series of papers, Sim, Quilliam, and various co-workers (56–58) coupled a mass spectrometer to the outlet of the DAD. The mode was atmospheric pressure chemical ionization, where the effluent was volatilized in a heated nebulizer. The mobile phase was preferentially removed, but the remaining solvent also acted as the chemical-ionization reagent. Both reversed-phase HPLC and supercritical fluid chromatography were used as the separation methods. This allowed separation of LPAHs of >500 daltons. This group was only able to identify a few of the LPAH peaks through the use of standards or reference UV absorbance spectra. These included coronene, benzo[*pqr*]naphtho-[8,1,2-*bcd*]perylene, naphtho[8,1,2-*abc*]coronene, and ovalene. Larger PAHs were analyzed by direct introduction of the samples into the mass spectrometer.

There is one major drawback in the DAD. The much more complex optical system and the associated data system make the DAD a very expensive HPLC detector. DADs can be many more times expensive than the single-wavelength detectors, with some HPLC-DAD-computer systems costing well over $50,000. This makes their purchase, and eventual replacement, more difficult to budget.

There have been a few reports on the use of full-spectrum fluorescence emission detectors, some of which were used in PAH analyses. Although many of these presented no examples of an application to LPAHs, the instrumentation would be directly useful with different chromatographic conditions. Jadamec, Saner, and Talmi (59) used a vidicon (television type) camera to collect spectra from an oscilloscope. They separated crude oil fractions and identified naphthalene and fluorene. Some of the later-eluting peaks were described as possible "polyphenyls."

Gluckman and Novotny (60, 61) described a diode array-based emission monitor. They examined a carbon-black extract, identical to that used for the earlier studies described above. Only a few LPAHs were identified because of a lack of standard reference compounds or spectra, but two peaks were ascribed to LPAHs similar to rubrene and decacyclene. Chromatograms collected with this detector are shown in Figure 84. The selectivity of emission wavelengths is highlighted.

To increase the sensitivity of these detectors high-intensity light sources, such as lasers or flash lamps, have been used. Also, cooling of the diode array to reduce the background dark currents has also been done. Even though most of these set-ups for doing full fluorescence emission spectrum collection are currently not commercially available, the hardware for a complete instrument can be readily purchased from several spectrometer manufacturers and assembled. The lack of published reports of full-spectrum fluorescence detection with HPLC has been mainly due to the lack of a readily available commercial instrument. One such instrument has recently been introduced (62), so this situation may soon change and this type of detection may become common for LPAHs. The initial product booklet for this instrument shows a reversed-phase chromatogram of the USEPA 16 priority-pollutant PAHs in the full-emission mode. The detector optics rely on rapid scanning technology, which can yield full spectra in either excitation or emission with only a single wavelength used for the alternative mode.

Cecil (63) examined the corrections that must be made in the fluorescence spectra gathered by diode arrays when there are differences in mobile-phase solvents. He found that both the wavelength shifts and band height changes are significant (see the section in the chapter on fluorescence for more detail). For example, perylene has a 6-nm red shift when the solvent is changed from pure methanol to 20% water in methanol. The collection of a standard spectral library under set conditions was recommended, with algorithms dealing with the wavelength shifts. The second issue could not be adequately addressed.

8.5 SUPERCRITICAL FLUID CHROMATOGRAPHY

There has been little work on the supercritical fluid chromatography (SFC) of LPAHs. A recent review by Sharif, Robson, and Bartle (64) does describe the uses of SFC in the separation of the PAHs. Jackson and Lee (65, 66) examined

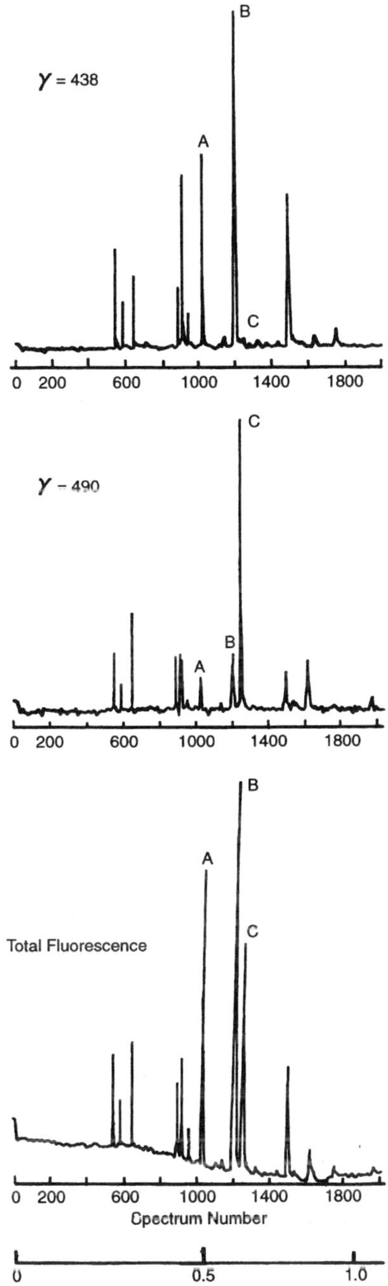

Figure 84. Chromatograms of a carbon black extract collected at different wavelengths simultaneously on a DAD fluorescence detector. From J. C. Gluckman and M. Novotny, *H. R. C. and C. C.* **8** 672–677 (1985), reproduced with permission of Wiley-VCH.

standard mixtures, the LPAH synthesis products from the Clar condensation of perinaphenone and benzanthrone, and a carbon black extract. They used *n*-pentane as the fluid, with up to 10 % benzene, isopropanol, or dichloromethane as the modifier. Capillary polysiloxane-coated columns were used, with the stationary phases polydimethyl, polymethylphenyl-, and polymethyl-n-octyl-siloxanes. A scanning fluorometer with a high-pressure cell was the detector. No appreciable differences were observed in the detection limit or excitation spectrum of coronene when the system was operated at lower pressure in a liquid-phase mode or in the SFC mode. They identified four of the synthesis product peaks and the coronene and ovalene peaks in the carbon black extract chromatogram.

8.6 CHROMATOGRAPHIC ANALYSIS FOR LPAHs

The LPAHs in a wide variety of materials have been analyzed for. Many of these analyses used the powerful combination of HPLC-DAD, because of the ability of this detector to identify in-line LPAHs. Not long after the commercialization of DAD detectors, their power in PAH analysis was realized. DAD has now become the UV detector preferred by most chemists for these reason (52, 67–70). One of the earliest works using HPLC-DAD was identification of the LPAHs in a diesel particulate extract (71–73). About a dozen LPAHs were found, including the never-before identified nine-ring naphtho[8,1,2-*abc*]coronene. A carbon black extract, obtained from the same carbon black as that used in the work of Lee, Novotny, and co-workers cited earlier, was examined by HPLC-DAD (12). The DAD, as well as a much larger collection of standard compounds, allowed the identification of around 20 more LPAHs than in the earlier study.

McCarry et al. (74–80) performed a similar series of HPLC fractionations to determine the PAHs in sediment samples. After drying, the sediments (from a harbor in western Lake Ontario and one on the Atlantic coast of Nova Scotia) were extracted either with ultrasound or sequentially with dichloromethane and methanol in a Soxhlet apparatus. Each fraction was further separated by either a propyl-amino-cyano normal phase or a sequence of alumina and resin-packed adsorption columns. The mobile phase in the normal-phase HPLC separation was step gradients of *n*-hexane and dichloromethane. This is a mixed-moiety phase that has a nitrile group at the terminus and an amino group on the β carbon. Each of these fractions was then analyzed by polymeric ODS phase using an acetonitrile to 1:1 acetonitrile-dichloromethane gradient. McCarry observed $C_{24}H_{14}$ LPAHs similar to those found by in a coal tar SRM. In later work, LPAHs of 26, 28, 30, and 32 carbons were found. For the reversed-phase HPLC both a DAD and direct atmospheric-pressure chemical-ionization mass

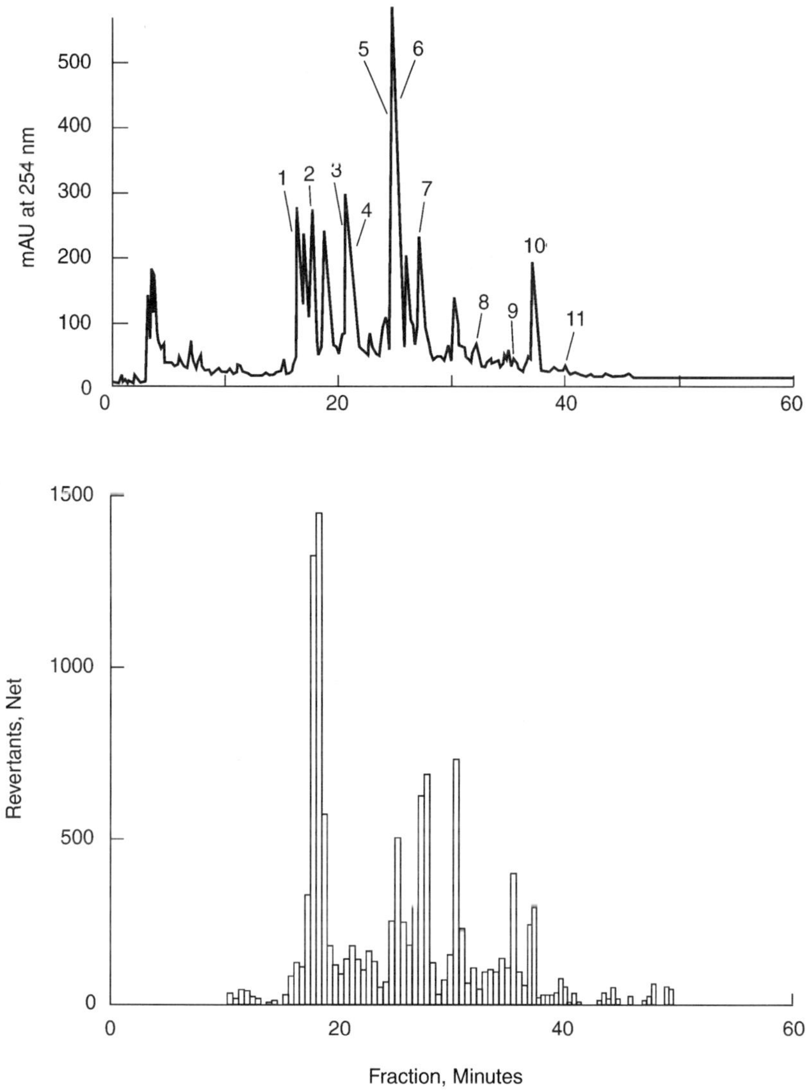

Figure 85. A sediment fraction containing the 24-carbon LPAHs versus fraction mutagenicity. From C. H. Marvin, B. E. McCarry, J. A. Lundrigan, K. Roberts, and D. W. Bryant, *Sci. Total Environ.* **231** 140 (1999), reproduced with permission from Elsevier Science Publishers BV.

spectrometry were used for detection. To make the DAD less specific and more universal, the average response from 250 to 370 nm was collected. Fractions were also collected and the mutagenicity of each was measured. Figure 85 is a comparison of DAD and mutagenicity responses for a sediment sample.

A coal tar pitch was separated and DAD spectra were used to identify the LPAHs (81). Several LPAHs, such as benzo[*a*]perylene and dibenzo[*a*, *j*]perylene, were found in this sample. The presence of these structures implied that the formation was caused by condensation reactions. These types of LPAHs are much less thermodynamically favored than other isomers but can be made through Scholl condensation of smaller PAHs (the Scholl condensation is a type of Friedel-Crafts reaction in which 2 small PAHs react to form a bridging ring, such as a naphthalene molecule and an anthracene molecule forming benzo[*a*]perylene).

Some studies have focused on the formation of LPAHs as a route to soots (82, 83). HPLC-DAD analysis of the products of pyrolysis of smaller PAHs have found numerous LPAHs. Many can be ascribed to direct condensations, but other LPAHs found can only result from drastic rearrangements of the carbon skeleton. Similar results were found in the soots and tars (extractable material from the soots) obtained during flame combustion of ethylene, acetylene, benzene, and other light hydrocarbons. Much of this work has been marked by the use of advanced analytical methods, but the key to identification of individual LPAHs has still been the synthesis of numerous new LPAHs to be available for use as standards. Comparisons of retention times and absorbance spectra made identification possible in these complex mixtures.

Pyrolysates of naphthalene and anthracene were separated and analyzed by HPLC-DAD. The combination of separation, selective detection, and a large number of standard compounds led to the identification of four main classes of LPAH products. These groups of compounds were partially condensed structures containing small aromatic parts joined by single bond bridges, condensed species with five-member ring bridges, condensed species with six-member bridges, and structures that can result from complex rearrangements of the carbon skeletons. The first three groups are made up of recognizable "oligomeric units" of the starting small aromatics.

Deep sea hydrothermal vents have been shown to produce PAH containing material. Hydrothermal vents are geologically active spots where heat, in the form of magma or superheated steam, escapes and comes in contact with water. Many hydrothermal vents are found in the deep ocean. In these deep ocean trenches, associated with tectonic fault lines, the contact is usually directly with the very hot magma. In these areas, numerous unique physical and chemical phenomena are found. These include "smoking chimneys" where the buildup of dissolved minerals results in deposited tubes of venting superheated gases and steam, nodules of very pure metals such as manganese, and organic bitumens.

These bitumens are separated into fractions containing saturated hydrocarbons, PAHs, LPAHs, and polar heteroatom-containing molecules. GC-MS analysis of these bitumen fractions showed that they have saturated hydrocarbon compositions that closely resemble heavy petroleums. The bitumens are, however, are extremely young. Carbon-14 dating shows them to be only a few thousand years old or less, in contrast to heavy petroleum, which is many tens or hundreds of millions of years old. It is thought that the heat, combined with the extreme pressures of the deep sea waters, is sufficient to make the water a supercritical fluid. This highly reactive water then reacts with the organic detritus (the organic bottom silt resulting from algae, plankton, and other life forms) on the sea floor to form the bitumen.

Interestingly, at least in the context of this book, several LPAHs were found by HPLC-DAD analyses in certain of these samples (84, 85), including coronene, benzo[*b*]picene, dibenzo[*cd,lm*]perylene, naphtho[8,1,2-*abc*]coronene, and ovalene. This variety of LPAH structures indicates the probability of a variety of reaction mechanisms in the formation. Preparative-scale HPLC was used to isolate sufficient amounts of each LPAH to allow Shpol'skii spectroscopy, which gave unequivocal identifications. Figure 86 shows the portion of the 3-dimensional chromatogram containing the 28- and 30-carbon LPAHs.

McKay and Latham (86) found several LPAHs in a Wyoming crude oil. The HPLC separation was gel-permeation (size exclusion) chromatography instead of adsorbent or bonded-phase HPLC. The separation mechanism, however, was not by molecular size because the large PAHs eluted later than the smaller ones. Similar behavior has been seen for fullerenes and has been ascribed to interactions with the polystyrene-divinylbenzene matrix of the GPC column packing. TLC of the fractions from GPC showed six LPAHs. Figures 87–89 show comparison spectra of isolated fractions and standard LPAHs. Desideri, Lepri, and Checchini (87) analyzed eight crude oils from the Middle East, Africa, Russia, and Venezuela. They used sequential reversed-phase and alumina-adsorbent HPLC to find coronene and two other LPAHs of 24 carbons.

McKinney, Clifford, Hou, Bogdan, and Hatcher (88) used a normal-phase HPLC separation and DAD to determine the PAHs in several coal liquefaction products. They found coronene and two LPAHs that they called "bisanthene" and "benzobisanthene." These names are trivial ones used by Clar for *compounds 144* and *145*, phenanthro[1,10,9,8-*pqra*]perylene and dibenzo[*bc,ef*]cor-

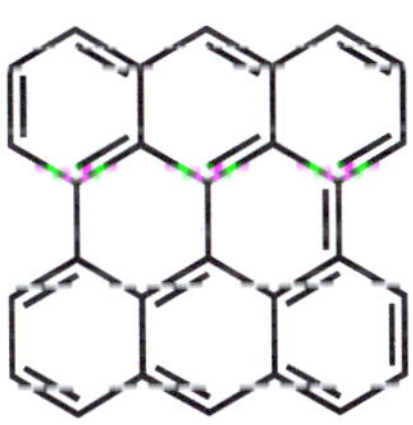

144

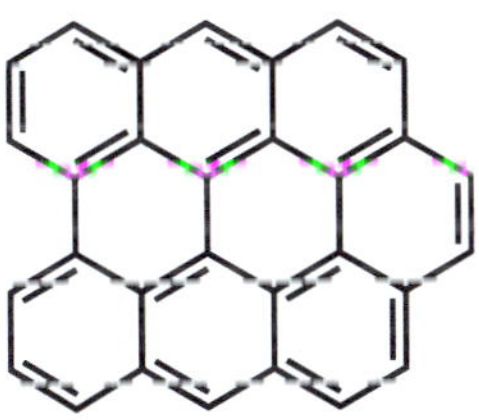

145

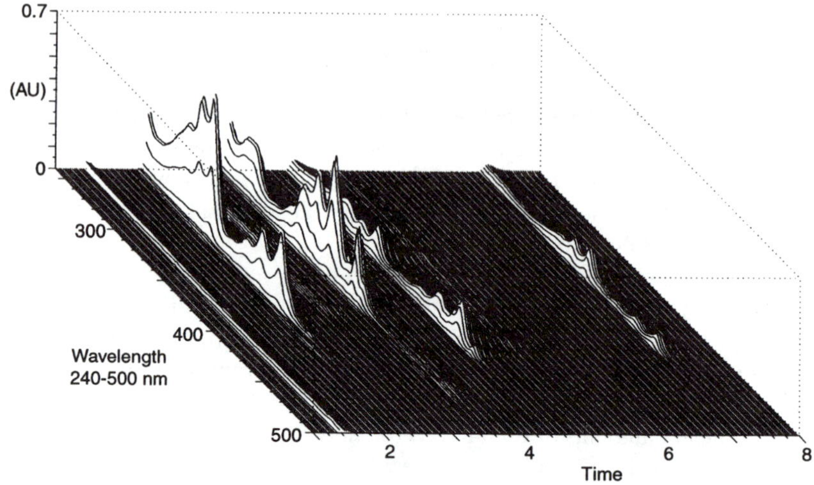

Figure 86. Hydrothermal vent 3-D chromatogram showing dibenzo[*b,pqr*]perylene, phenanthro[5,4,3,2-*efghi*]perylene, benzo[*pqr*]naphtho[8,1,2-*bcd*]perylene, and naphtho[8,1,2-*abc*]elution. From B. R. T. Simoneit and J. C. Fetzer, in *Organic Geochemistry: Developments and Applications to Energy, Climate, Environment and Human History*, J. O. Grimalt and C. Dorronsoro, Eds., A. I. G.

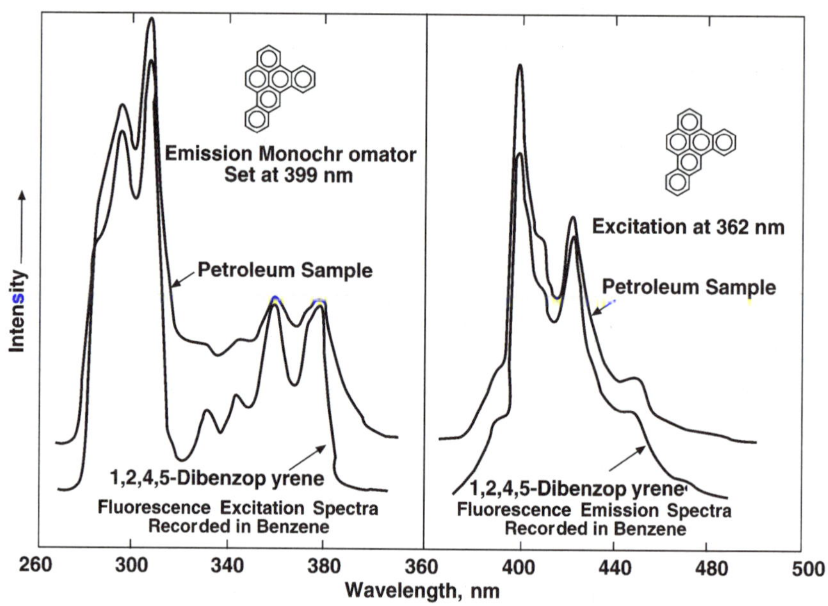

Figure 87. Identification of a fraction as naphtho[1,2,3,4-*def*]chrysene. From J. F. McKay and D. R. Latham, *Anal. Chem.* **45** 1052 (1973), reproduced with permission from the American Chemical Society.

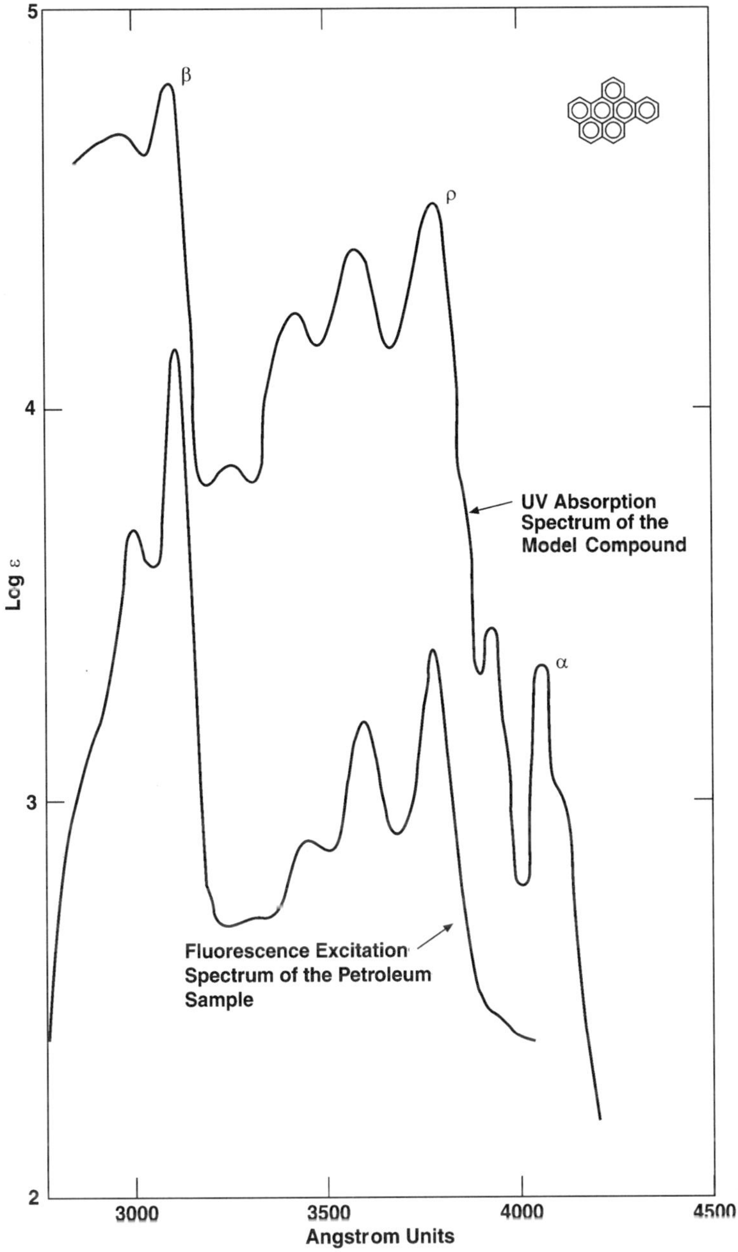

Figure 88. Identification of a fraction as dibenzo[*b,pqr*]perylene. From J. F. McKay and D. R. Latham, *Anal. Chem.* **45** 1053 (1973), reproduced with permission from the American Chemical Society.

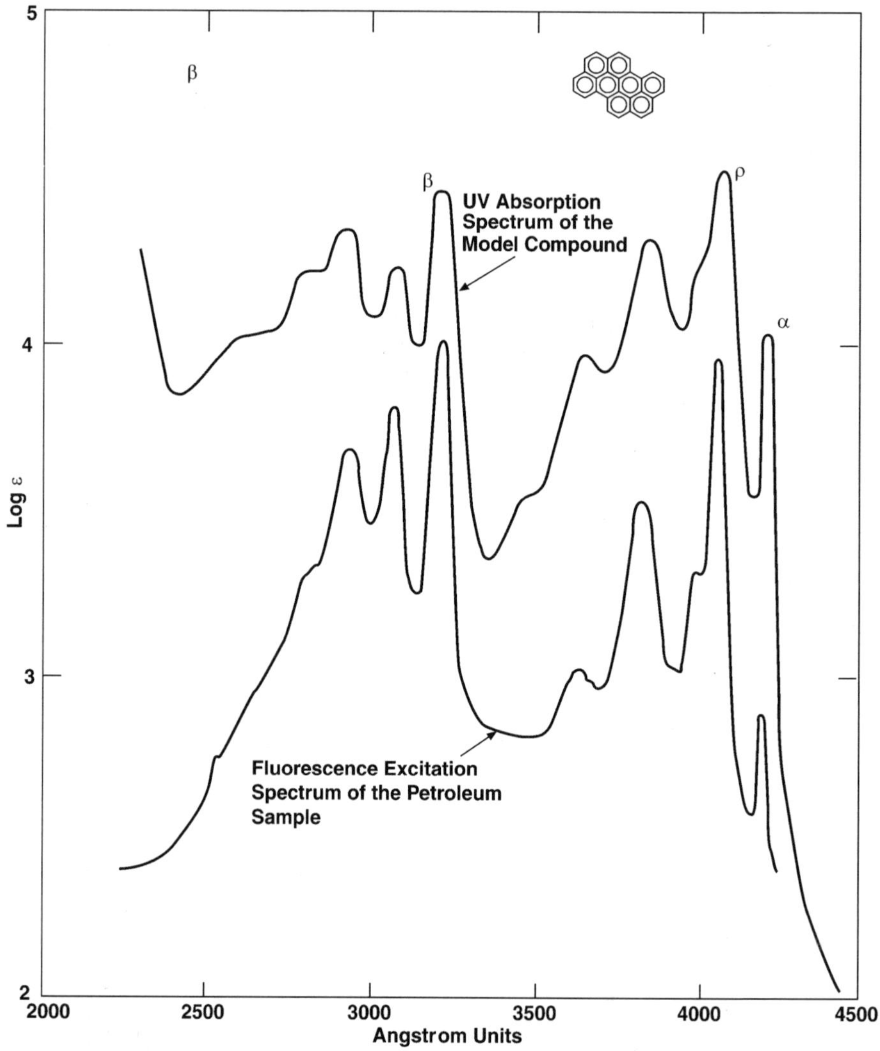

Figure 89. Identification of a fraction as benzo[*pqr*]naphtho[8,1,2-*bcd*]perylene. From J. F. McKay and D. R. Latham, *Anal. Chem.* **45** 1053 (1973), reproduced with permission from the American Chemical Society.

onene (89). Because the authors did not have authentic standards and did not describe how they obtained these identifications, the results must only be assumed to show a $C_{28}H_{14}$ LPAH and a $C_{30}H_{14}$ LPAH. The reason for the ambiguity is that these two LPAHs have very high wavelength bands (663 nm and 535 nm are the highest bands for each) that should be definitive if these were the LPAHs in these samples. The band shapes in one of these spectra are also characteristic for the first in having the broad bands and shallow valleys of a nonplanar LPAH. These two LPAHs, however, are less energetically-favored isomers than alternatives such as benzo[*a*]coronene or benzo[*pqr*]naphtho[8,1,2-*bcd*]perylene for the lower-mass LPAH and naphtho[8,1,2-*abc*]coronene for the heavier LPAH.

Fetzer et al. (90, 91) have used HPLC-DAD to analyze the oils produced in the hydrocracking processes used in petroleum refining. They found that the LPAHs are produced by one of two routes. The first is a series of one-ring additions of either two- or four-carbon additions. This route exactly matches that calculated by Stein (92) as the most thermodynamically favored buildup route, but it continues well past coronene, which his calculations only went to. The second reaction was a Scholl reaction of pyrene to form two 9-ring, 32-carbon isomers. These are then partially hydrogenated. The resulting saturated rings are cracked off to form two 25-carbon LPAHs. Example chromatograms are shown in Figure 90.

An example of the higher power of HPLC-DAD to separate and identify LPAHs is a comparison of the results in the separation of the reaction mixtures from certain LPAH syntheses. In separate experiments, Clar and McKay (93) condensed the two cyclic ketones benzanthrone and naphthanthrone using the zinc dust fusion reaction. Fetzer and Biggs repeated these same syntheses under identical conditions and on the same scale over 20 years later. In the first reaction product mixture, Clar identified five isomers with the formula $C_{34}H_{18}$ and two more-condensed isomers of $C_{34}H_{16}$. One of the latter pair could not be separated in pure form, and its UV absorbance spectrum could only be obtained by subtraction of the interferences of the other isomer (which was obtained in pure form).

Recent studies (94–96) found, through the use of HPLC-DAD, another new isomer of the first set in the reaction product, and it was at very high concentrations (as high as 3 of Clar's isomers and much higher than the other 2). Fetzer and Biggs also collected a pure sample of the $C_{34}H_{16}$ compound that previously had only been separated in impure form. They also measured its UV absorbance spectrum directly (which varied little from Clar's, except that his molar absorptivity values were slightly smaller, 8% too low, because of the presence of impurities). This second LPAH was also really found to be a pale orange color, whereas Clar had reported it to be dark red due to the presence of one of the impurities.

In the second (naphthanthrone) reaction, Clar found four isomers of $C_{38}H_{18}$; *compounds 142*, *146*, and *147* are the three of the isomers. Fetzer and Biggs (95) easily separated a fifth isomer, dianthra[1,9,8,7-*cdefg*:1′,9′,8′,7′-*nopqr*]pentacene (*compound 126*). This new isomer had been unobserved in one of Clar's

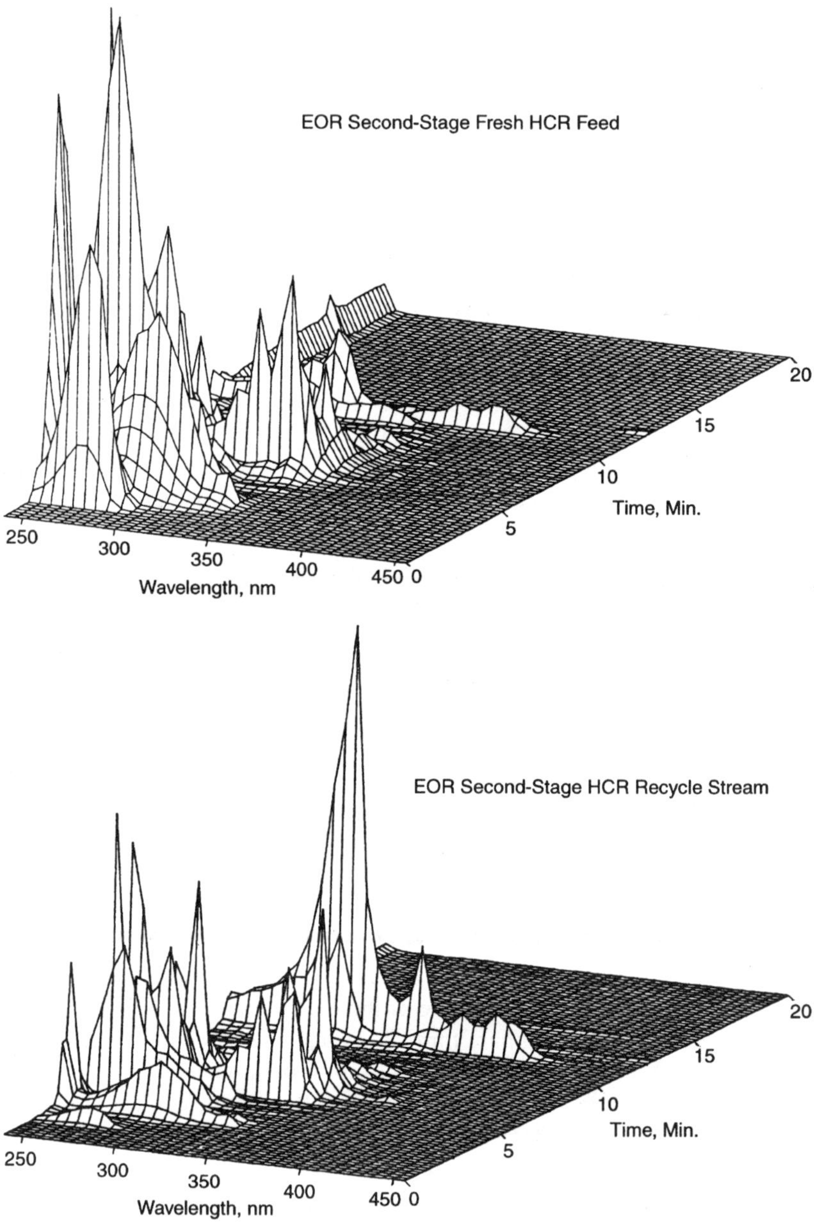

Figure 90. 3-D DAD chromatograms for the PAHs in two hydrocracker recycle oil samples. The top chromatogram shows end-of-run (EOR) conditions for feed to the process' second stage and the bottom chromatogram shows the material recycled back as feed after the light products have been distilled away. From R. F. Sullivan, M. M. Boduszynski, and J. C. Fetzer, *Energy Fuels* **3** 603–612 (1989), reproduced with permission from the American Chemical Society.

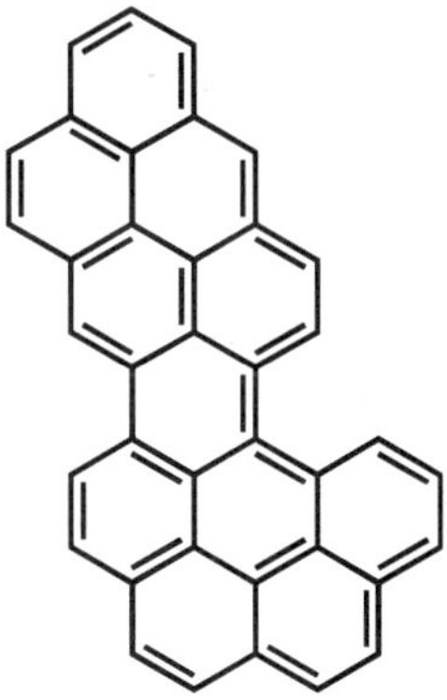

146

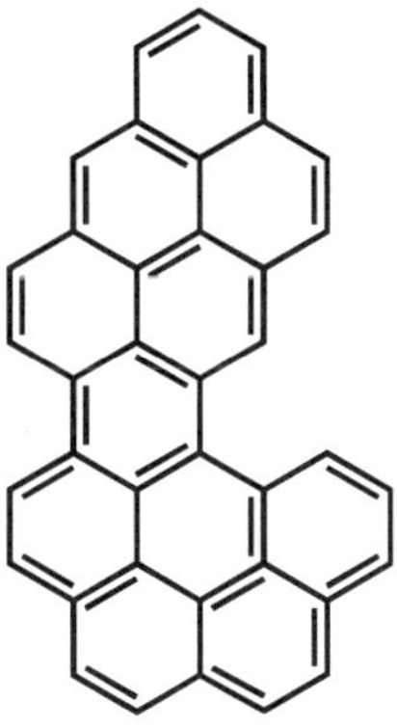

147

four fractions {a sample of the fourth of these, with the isomer naphth[3′,2′,1′,8′,7′-*vwxyz*]anthra[2,1,9,8-*klmn*]hexaphene (*compound 127*), was obtained by Fetzer and Biggs later for comparison. The two isomers were both very long and narrow in shape and could not be separated by adsorption chromatography and vacuum sublimation, the only techniques available to Clar. The subtle separation of these two very similarly shaped isomers was unattainable until many years later.

Clar did not even see the second isomer as an impurity because of a spectral coincidence. This fifth isomer was present in Clar's fraction at ~25%. The major isomer in that fraction has the higher wavelength bands and its second and third highest wavelength bands lie within a few nanometers of the highest two of the less prevalent isomer, completely masking them. Clar's isomer had UV absorbance bands at 566, 537, and 495 nm. The bands at 535, 494, and 456 nm in the UV absorbance spectrum of this new isomer overlapped those of this other isomer, and so it went unseen by being buried under the overlapping bands of the much more prevalent isomer. Another unavailable analytical tool, three-dimensional fluorescence (see Chapter 4) also readily showed both species as being present in Clar's fraction. The coincidence that these two structurally similar LPAHs could not be separated and had severely overlapping absorbance bands led to Clar not finding the isomer with the approaches available.

With the additional separation and characterization capabilities of HPLC-DAD, Fetzer and Biggs also extended Clar's work by reacting mixtures of benzanthrone and naphthanthrone. In addition to the many products found from the two reactions with only one reacting ketone, this reaction mixture also contained five $C_{36}H_{18}$ isomers and one $C_{36}H_{16}$ LPAH. All of these were new LPAHs never produced before.

REFERENCES

1. M. L. Lee and R. A. Hites, *Anal. Chem.* **48** 1890–1893 (1978).
2. W. Schmidt, G. Grimmer, J. Jacob, and G. Dettbarn, *Toxicol. Environ. Chem.* **13** 1–16 (1986).
3. W. Schmidt, G. Grimmer, J. Jacob, G. Dettbarn, and K. W. Naujack, *Fresenius Z. Anal. Chem.* **326** 401–413 (1987).
4. S. A. Wise, B. A. Benner, H. Liu, G. D. Byrd, and A. Colmsjo, *Anal. Chem.* **60** 630–637 (1988).
5. J. M. Gieskes, B. R. T. Simoneit, A. J. Magenheim, and R. N. Leif, *Appl. Geochem.* **5** 93–101 (1990).
6. B. R. T. Simoneit, W. D. Goodfellow, and J. M. Franklin, *Appl. Geochem.* **7** 257–264 (1992).
7. O. E. Kawka and B. R. T. Simoneit, *Org. Geochem.* **22** 947–978 (1994).
8. A. Bemgard, B.-O. Lundmark, and A. Colmsjo, in *Polycyclic Aromatic Compounds: Synthesis, Properties, Analytical Measurements, Occurrence, and Biological Effects*, P. Garrigues and M. Lamotte, Eds., Gordon and Breach Science, Yverdon, Switzerland, 1993, p. 603–610.
9. A. Bemgard, A. Colmsjo, and B.-O. Lundmark, *J. Chromatogr.* **595** 247–258 (1992).
10. A. Bemgard, A. Colmsjo, and B.-O. Lundmark, *J. Chromatogr.* **630** 287–295 (1993).
11. A. Bemgard, A. Colmsjo, and K. Wrangskog, *Anal. Chem.* **66** 4288–4294 (1994).
12. J. C. Fetzer and C. E. Rechsteiner, in *Polycyclic Aromatic Hydrocarbons: Eleventh International Symposium*, M. Cooke, K. Loening, and J. Merritt, Eds., Battelle, Columbus, OH, 1991, p. 259–270.
13. M. L. Lee, D. L. Vassilaros, C. M. White, and M. Novotny, *Anal. Chem.* **51** 768–773 (1979).
14. D. L. Vassilaros, R. C. Kong, D. W. Later, and M. L. Lee, *J. Chromatogr.* **252** 1–20 (1982).
15. W. J. Simonsick and R. A. Hites, *Anal. Chem.* **58** 2114–2121 (1986).
16. R. A. Brown and T. D. Searl, in *Chromatography in Petroleum Analysis*, K. H. Altgelt and T. H. Gouw, Eds., Dekker, New York, 1979, p. 377–384.
17. H. Matsushita, Y. Esmi, A. Suzuki, and T. Handa, *Bunseki Kagaku* **21** 1471 (1972).
18. R. C. Pierce and M. Katz, *Anal. Chem.* **47** 1743–1748 (1975).
19. M. Katz and R. C. Pierce, in *Carcinogenesis—A Comprehensive Study*, R. I. Freudenthal and P. W. Jones, Eds., Raven, New York, 1976, Vol. 1, p. 413.
20. J. F. McKay and D. R. Latham, *Anal. Chem.* **45** 1050–1055 (1973).
21. J. C. Fetzer and W. R. Biggs, *J. Chromatogr.* **346** 81–92 (1985).
22. L. C. Sander and S. A. Wise, in *Advances in Chromatography*, J. C. Giddings, E. Grushka, J. Cazes, and P. R. Brown, Eds., Dekker, New York, 1986, Vol. 25, p 139–218.

23. L. C. Sander and S. A. Wise, "NIST Special Publication 922: Polycyclic Aromatic Hydrocarbon Structure Index", National Institute of Standards and Technology, 1997.
24. K. Jinno, in *Advances in Chromatography*, J. C. Giddings, E. Grushka, and P. R. Brown, Eds., Dekker, 1989, Vol. 30, p. 123–165.
25. P. Shah, L. B. Rogers, and J. C. Fetzer, *J. Chromatogr.* **388** 411–419 (1987).
26. J. C. Fetzer and W. R. Biggs, *J. Chromatogr.* **295** 161–169 (1984).
27. J. C. Fetzer and W. R. Biggs, *J. Chromatogr.* **322** 275–286 (1985).
28. J. C. Fetzer and W. R. Biggs, *J. Chromatogr.* **386** 87–101 (1987).
29. K. Jinno, T. Nagoshi, N. Tanaka, M. Okamoto, J. C. Fetzer, and W. R. Biggs, *J. Chromatogr.* **386** 123–135 (1987).
30. K. Jinno, N. Tanaka, M. Okamoto, J. C. Fetzer, and W. R. Biggs, *J. Chromatogr.* **392** 75–82 (1987).
31. K. Jinno, N. Anazawa, M. Okamoto, N. Tanaka, J. C. Fetzer, and W. R. Biggs, *J. Chromatogr.* **402** 173–178 (1987).
32. J. C. Fetzer, in *Polynuclear Aromatic Hydrocarbons*, L. Ebert, Advances in Chemistry 217, American Chemical Society, Washington, DC, 1988, Chapt. 18, p. 309–331.
33. H. Ohta, K. Jinno, Y. Saito, J. C. Fetzer, W. R. Biggs, J. J. Pesek, M. T. Maryska, and Y.-L. Chen, *Chromatographia* **42** 56–62 (1996).
34. Certificate for Standard Reference Material 869, National Institute for Standards and Technology, Gaithersburg MD, 1998.
35. L. C. Sander, K. E. Sharpless, N. E. Craft, and S. A. Wise, *Anal. Chem.* **66** 1667–1674 (1994).
36. C. M. Bell, L. C. Sander, J. C. Fetzer, and S. A. Wise, *J. Chromatogr.* **753** 37–46 (1996).
37. M. Pursch, L. C. Sander, H.-J. Egelhaaf, M. Raitza, S. A. Wise, D. Delkrup, and K. Albert, *J. Am. Chem. Soc.* **121** 3201–3213 (1999).
38. L. C. Sander, M. Pursch, and S. A. Wise, *Anal. Chem.* **71** 4821–4830 (1999).
39. K. Jinno, S. Shimura, J. C. Fetzer, and W. R. Biggs, *Polycyclic Aromat. Compd.* **1** 151–159 (1990).
40. R. Goddard, M. W. Haenel, W. C. Herndon, C. Krueger, and M. Zander, *J. Am. Chem. Soc.* **117** 30–41 (1995).
41. K. Jinno, K. Yamamoto, J. C. Fetzer, and W. R. Biggs, *J. Microcolumn Chromatogr.* **4** 187–190 (1992).
42. K. Jinno and T. Uemura, *J. Microcolumn Sep.* **4** 325–329 (1993).
43. Y.-H. Lai, S.-G. Ang, and H.-C. Li, *J. Chromatogr.* **606** 251–254 (1992).
44. J. C. Fetzer and W. R. Biggs, *Chromatographia* **27** 118–122 (1989).
45. J. C. Fetzer and W. R. Biggs, *Polycyclic Arom. Cmpds.* **5** 193–199 (1994).
46. K. Jinno, T. Ibuki, J. C. Fetzer, W. R. Biggs, P. R. Griffiths, and J. M. Ollinger, *J. Chromatogr.* **461** 209–227 (1989).
47. P. A. Peaden, M. L. Lee, Y. Hirata, and M. Novotny, *Anal. Chem.* **52** 2268–2271 (1980).
48. A. Hirose, D. Wiesler, and M. Novotny, *Chromatographia* **18** 239–242 (1984).

49. S. A. Wise, B. A. Benner, H. Liu, G. D. Byrd, and A. Colmsjo, *Anal. Chem.* **60** 630–637 (1988).
50. S. A. Wise, A. Deissler, and L. C. Sander, *Polycyclic Arom. Cmpds.* **3** 169–184 (1993).
51. H. J. Boessenkoof, P. Cleij, C. E. Goewie, H. H. van den Broek, and H. A. van't Klooster, *Mikrochim. Acta* **2** 71–92 (1986).
52. D. M. Demorest, J. C. Fetzer, I. S. Lurie, S. M. Carr, and K. B. Chatson, *LC/ GC Magazine* **5** 128–142 (1987).
53. J. C. Fetzer and W. R. Biggs, *J. Chromatogr.* **642** 319–327 (1993).
54. L. S. Ramos, R. J. Stewart, and B. G. Rohrback, *Chromatogr. (LC/GC)* **2** 95–102 (1987).
55. R. Tauler, G. Durand, and D. Barcelo, *Chromatographia* **33** 244–254 (1992).
56. P. G. Sim, M. A. Quilliam, S. Pleasance, and F. M. Benoit, *Polycyclic Arom. Cmpds. Suppl.* **3** 321–328 (1993).
57. J. F. Anacleto, L. Ramaley, R. K. Boyd, S. Pleasance, M. A. Quilliam, G. P. Sim, and F. M. Benoit, *Can. Rapid Commun. Mass Spectrom.* **5** 149–155 (1991).
58. A. L. LaFleur, K. Taghizadeh, J. B. Howard, J. E. Anacleto, and M. A. Quilliam, *Am.* Soc. *Mass Spectrom.* **7** 276–286 (1996).
59. J. R. Jadamec, W. A. Saner, and Y. Talmi, *Anal. Chem.* **49** 1316–1321 (1977).
60. J. C. Gluckman and M. Novotny, *High Resolution Chromatogr. and C. C.* **8** 672–677 (1985).
61. J. C. Gluckman and M. Novotny, *J. Chromatogr. Sci.* **45** 145 (1989).
62. R. Schuster and H. Schulenberg-Schell, *A New Approach to Lower Limits of Detection and Easy Spectral Analysis*, Hewlett Packard Instruments (Agilent Technology Co.), Waldbronn, Germany, 1998.
63. T. L. Cecil, PhD. Thesis, Virginia Commonwealth Univ., 1990.
64. S. M. Sharif, M. M. Robson, and K. D. Bartle, *Polycyclic Arom. Cmpds.* **12** 147–185 (1997).
65. W. P. Jackson, PhD. Thesis, Brigham Young Univ., 1985.
66. W. P. Jackson, R. C. Kong, and M. L. Lee, in *Polynuclear Aromatic Compounds: Mechanisms, Methods, and Metabolism*, M. Cooke, and A. J. Dennis, Eds., Batelle, Columbus, OH, 1985, p. 609–625.
67. M. W. Dong and A. Greenberg, *J. Liq. Chromatogr.* **11** 1887–1905 (1988).
68. J. Wegrzyn, G. Patonay, M. Ford, and I. Warner, *Anal. Chem.* **62** 1754 (1990).
69. R. J. van de Nesse, C. Gooijer, G. P. Hoornweg, U. A. T. Brinkman, N. H. Velthorst, and S. J. Van der Brent, *Anal. Lett.* **24** 1235 (1990).
70. T. L. Cecil and S. C. Rutan, *J. Chromatogr.* **556** 495 (1991).
71. K. Jinno, J. C. Fetzer, and W. R. Biggs, *Chromatographia* **21** 274–276 (1986).
72. J. C. Fetzer, W. R. Biggs, and K. Jinno, *Chromatographia* **21** 439–442 (1986).
73. K. Jinno, Y. Miyashita, S. Sasaki, J. C. Fetzer, and W. R. Biggs, *Envir. Monit. Assess.* **19** 13–25 (1991).
74. C. H. Marvin, L. Allan, B. E. McCarry, and D. W. Bryant, *Environ. Mol. Mutagenesis* **22** 61–70 (1993).

75. C. H. Marvin, M. Tassaro, B. E. McCarry, and D. W. Bryant, *Sci. Total Environ.* **156** 119–131 (1994).
76. A. E. Legzdins, B. E. McCarry, C. H. Marvin, and D. W. Bryant, *J. Environ. Anal. Chem.* **60** 79–94 (1995).
77. C. H. Marvin, J. A. Lundrigan, B. E. McCarry, and D. W. Bryant, *Environ. Toxicol. Chem.* **14** 2059–2066 (1995).
78. C. H. Marvin, B. E. McCarry, J. Villella, D. W. Bryant, and R. W. Smith, *Polycyclic Arom. Cmpds.* **9** 193–200 (1996).
79. C. H. Marvin, B. E. McCarry, J. A. Lundrigan, K. Roberts, and D. W. Bryant, *Sci. Total Environ.* **231** 135–144 (1999).
80. C. H. Marvin, R. W. Smith, D. W. Bryant, and B. E. McCarry, *J. Chromatogr.* **863** 13–24 (1999).
81. J. C. Fetzer and J. R. Kershaw, *Fuel* **74** 1533–1536 (1995).
82. M. J. Wornat, B. A. Vernaglia, A. L. LaFleur, E. F. Plummer, K. Taghizadeh, P. F. Nelson, C.-Z. Li, A. Necula, and L. T. Scott, *Twenty-Seventh International Symposium on Combustion*, The Combustion Institute, Boulder, Colorado, 1998, p. 1677–1686.
83. M. J. Wornat, F. J. J. Vriesendorp, A. L. LaFleur, E. F. Plummer, A. Necula, and L. T. Scott, *Polycyclic Arom. Cmpds.* (1999).
84. J. C. Fetzer, B. R. T. Simoneit, P. Garrigues, and H. Budzinski, *Polycyclic Arom. Compds.* **9** 109–120 (1996).
85. B. R. T. Simoneit and J. C. Fetzer, in *Organic Geochemistry: Developments and Applications to Energy, Climate, Environment and Human History*, J. O. Grimalt and C. Dorronsoro, Eds., A. I. G. O. A., Donostia, San Sebastian, Spain, 1995, p. 414–417.
86. J. F. McKay and D. R. Latham, *Anal. Chem.* **45** 1050–1055 (1973).
87. P. G. Desideri, L. Lepri, and L. Checchini, *J. Chromatogr.* **369** 117–124 (1986).
88. D. E. McKinney, D. J. Clifford, L. Hou, M. R. Bogdan, and P. G. Hatcher, *Energy Fuels* **9** 90–96 (1996).
89. E. Clar, in *Polycyclic Hydrocarbons*, John Wiley and Sons, New York, 1964, Vol. 2, p. 98–102.
90. R. F. Sullivan, M. M. Boduszynski, and J. C. Fetzer, Energy Fuels **3** 603–612 (1989).
91. J. C. Fetzer, in *Catalysts in Petroleum and Petrochemical Processing 1995*, M. Absi-Halabi, J. Beshārah, H. Qabazard, and A. Stanislaus, Eds., Elsevier, Amsterdam, 1996, p. 263-271.
92. S. Stein, *J. Phys. Chem.* **82** 866–871 (1978).
93. E. Clar and C. C. MacKay, *Tetrahedron* **28** 6041–6047 (1972).
94. J. C. Fetzer and W. R. Biggs, *J. Chromatogr.* **346** 81–92 (1985).
95. J. C. Fetzer and W. R. Biggs, *Org. Prep. Proc. Int.* **20** 223–230 (1988).
96. J. C. Fetzer, *Polycyclic Arom. Cmpds.* **14** 1–10 (1999).

CHAPTER

9

LPAH OCCURRENCE

Now that the many techniques for LPAH analysis have been reviewed, as well as many examples of their use, it is necessary to comprehensively review the occurrence of the LPAHs. There are two ways in which this can be done, either by the material in which they are found or by the specific LPAH, then listing in what variety of materials it is found. This section will use the first approach because most researchers are working on the characterization of a particular material.

One caveat must be again mentioned here. Analysis for LPAHs depends heavily on the availability of standard compounds. These are not commonly or universally available at reasonable costs. Thus, in general, these analyses had a limited set of standards and the compositional information is not complete. The absence of any LPAH in a certain type of material may only indicate that the workers who performed the analyses did not have the standard compound.

The types of LPAHs found in materials parallels the conditions of formation of the materials. For similar materials, the most complex mixtures are those formed under very high-energy conditions, where many isomers of alternant and nonalternant structures can be made. Conversely, the simplest are formed under low-energy conditions, where only alternant structures of higher stability (those with lower formation and higher resonance energies).

9.1 HYDROTHERMAL VENT BITUMENS AND PAH MINERALS

An example of LPAH-containing-material was recently found. Hydrocarbon-containing samples from several deep sea hydrothermal vents all contained over a dozen LPAHs, including both ortho- and peri-fused structures (1, 2). The vents were located in the Guaymas Basin in the Gulf of California area, the Escanaba Trough near the Gorda Ridge off the northern coast of California, and the Middle Valley near the Juan de Fuca Ridge area of the north Pacific Ocean. The peri-fused LPAHs produced include highly-condensed ones like coronene and ovalene, which can be produced through the most thermodynamically favored

Large ($C>=24$) Polycyclic Aromatic Hydrocarbons: Chemistry and Analysis, by John C. Fetzer.
ISBN 0-471-36354-5

route of one-ring additions. The largest LPAH found that follows this path had 34 carbons. Another production route, however, must produce the dibenzo[*cd,lm*]perylene found because it cannot be built up from one-ring additions without going through some higher-energy PAHs. In contrast to the deep sea-generated materials, those from the cooler waters of the hot springs of Yellowstone Park, Wyoming, and in Lake Tanganyika only make the highly condensed and very stable LPAH coronene.

There are certain minerals that contain inclusions of LPAHs. Pendletonite, curtisite, karpatite (carpathite), and idrialite are among the more common of the many of these odd and generally rare minerals. Pendletonite (1) is a mercury ore that is found in San Benito County, California, and contains flakes of virtually pure coronene (99 + %, with the remainder being methyl-coronene). Karpatite (2) is a very similar material that may also contain inclusions of coronene. Curtisite is a sandstone found in the area of Skaggs Springs, Sonoma County, California. Idrialite (3, 4) is also a mercury ore that is found in the Transcarpathia region of the Czech Republic and in the Dalmatian coastal region of Croatia. It contains a mixture of LPAHs of up to 26 carbons, including benzo[*b*]picene (*compound 148*), benzo[*c*]picene (*compound 128*), and naphtho[2,1-*b*]chrysene (*compound 149*) and six nonalternant $C_{24}H_{16}$ isomers

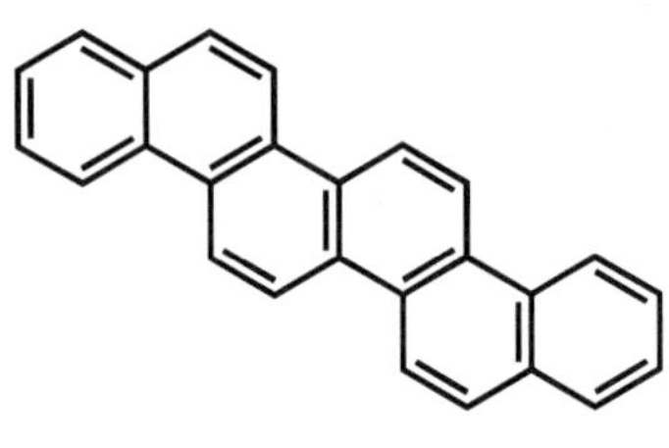

148

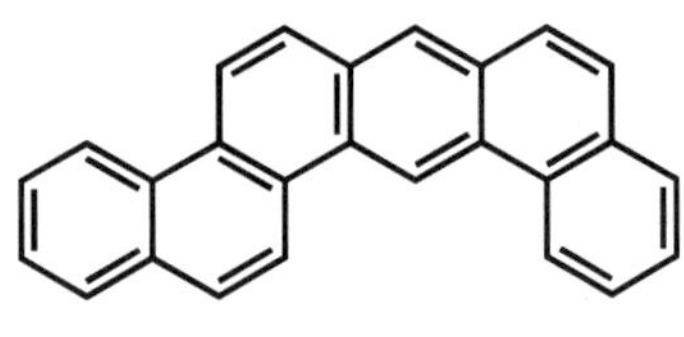

149

(which must have at least two 5-member rings). There are similarities between these organic minerals and the composition of hydrothermal vent LPAHs. It has been proposed that a possible source for those PAHs found in these minerals is the high-temperature aqueous conditions similar to those in hydrothermal areas (7).

9.2 COAL TAR AND PYROLYTIC PITCHES

Coal tar pitches are the most complex and have the highest concentrations of LPAHs of any material. Zander and Collin (8), and Zander in a later article (9), reviewed the significance of PAHs, and especially LPAHs, to coal tar pitch chemistry and properties. One presumption about coal tar pitch composition is that the LPAH distribution is neither kinetically or thermodynamically limited. Therefore, the number of LPAHs is defined by the statistics involving enumeration of the isomers. For a given carbon number, the distribution of any LPAH is $1/n$, where n is the number of isomers for that carbon number.

After chemical class fractionation by adsorption chromatography, MS of the fractions reveals that the LPAHs extend up to 2,500 daltons (10–12). Little alkylation is found in coal tar pitches, so these very large molecules consist of fused ring systems. Many of these LPAHs consist of networks of aromatic cores linked through single bonds, as well as bridging five- and six-member rings. These pitches contain many extended ortho-fused LPAHs. Of the peri-fused LPAHs, some thermodynamically unfavored species are found, benzo[*a*]perylene and dibenzo[*a,j*]perylene (*compounds 150* and *151*) being examples. These probably are formed by pyrolytic condensation of smaller PAHs, such as naph-

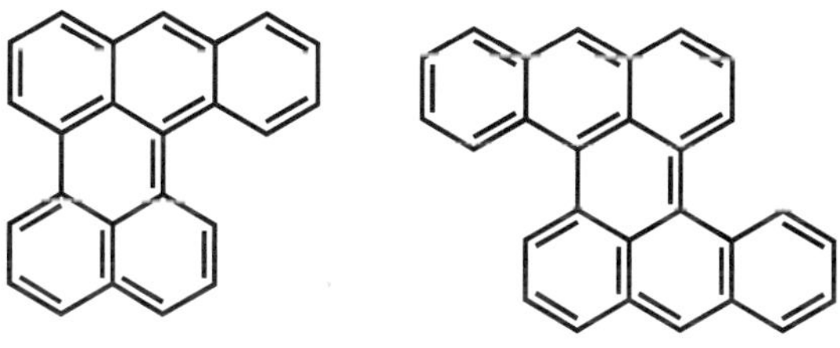

150 **151**

thalene, phenanthrene, anthracene, and other small PAHs. Coal tars also contain many nonalternant LPAHs. Analogous structures with either five- or six-member bridging rings are common, as well as the heteroatom-containing analogues. Irwin and Lewis report an unidentified $C_{68}H_{30}$ LPAH in a pitch, and Fetzer and Rechsteiner were able to identify benzo[1,2,3-*bc*:4,5,6-*b′c′*]dicoronene (*compound 53*) in another sample.

Suzuki, Kaneko, and Tsuchiya (13) used chemical-ionization MS to characterize a coal tar. They assigned the masses they found to the LPAHs benzo[*b,pqr*]perylene, anthra[2,1,9-*qra*]tetracene, dibenzo[*a,c*]tetracene, dibenzo[*fgh,pqr*]trinaphthylene, benzo[*pqr*]naphtho[8,1,2-*cde*]picene (*compounds 152–156*), and others. Because this technique does not differentiate isomers and this group did not possess authentic standard LPAHs, these assignments can be taken as only representative examples. Sabor, Smidl, and Mitera (14) used a wide variety of separation techniques to generate fractions from a coal tar. Dialysis, extraction, adsorption chromatography, and size-exclusion chromatography in various combinations and sequences gave fractions containing PAHs, LPAHs, and other heteroatom-containing polycyclic aromatic compounds. GC-MS was used for the final analysis and showed several $C_{24}H_{14}$ isomers, coronene, and benzo[*a*]coronene.

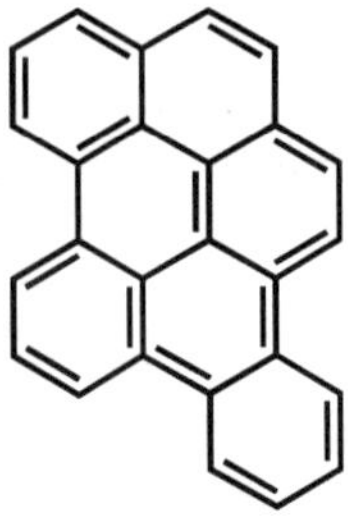

152

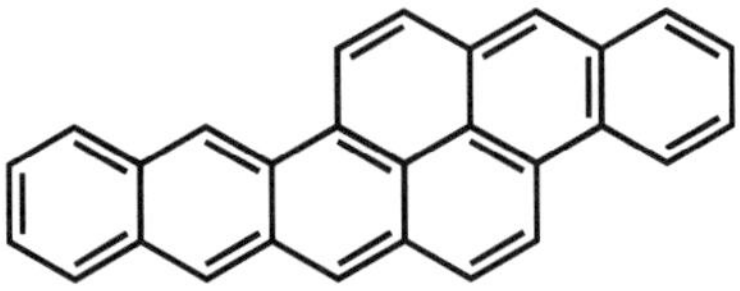

153

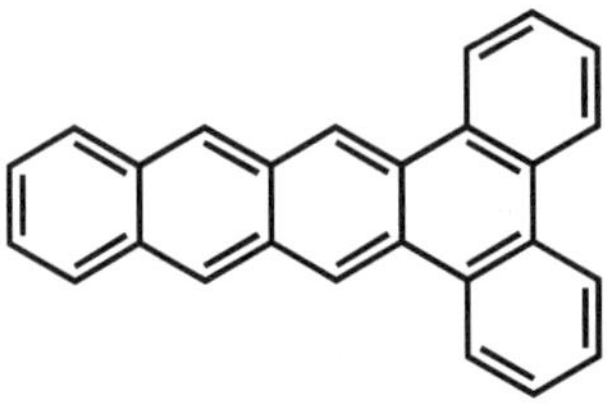

154

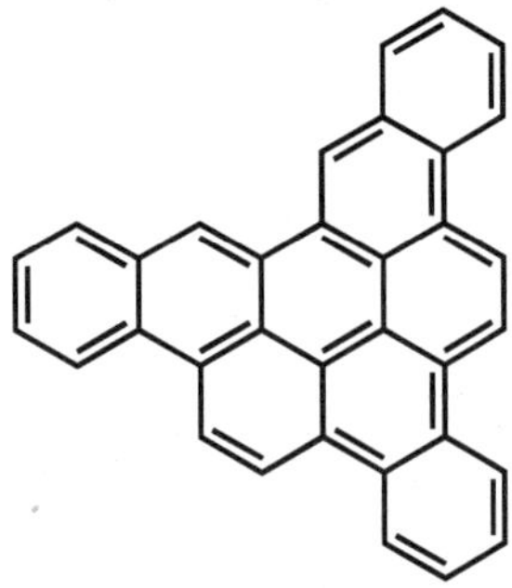

155

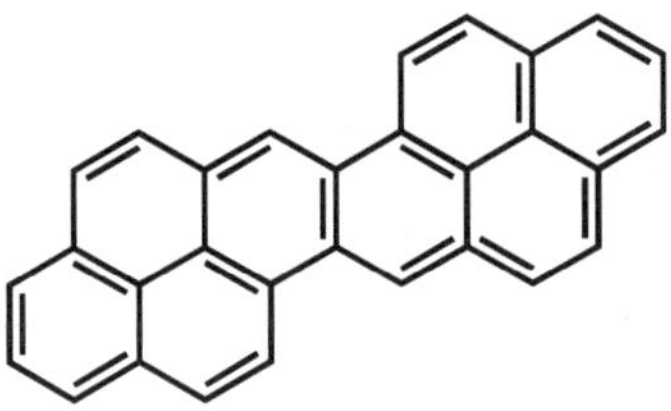

156

The numerous research efforts have found that the condensation reaction route is the most prevalent in the formation of pitches. Senthilnathan and Stein (15) followed the reactions of small PAHs and modeled the reactions. Formation of single-bond linkages, followed by formation of bridging rings, is common and both kinetically and thermodynamically favored. This was confirmed by analytical work on the real materials made from condensation of model compounds, such as naphthalene, anthracene, and pyrene, and in the production of coal tar pitches.

Colmsjo, Wise, and their co-workers (16–18) and Kershaw and Fetzer (12) found these types of condensed LPAHs in coal tar pitch, as well as many large

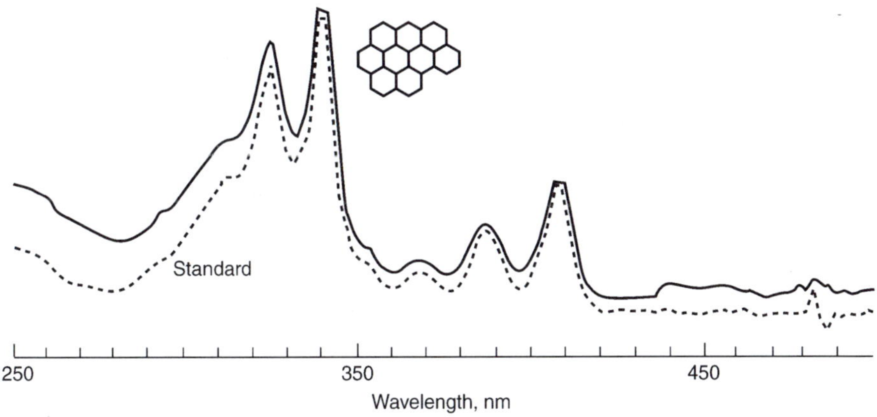

Figure 91. Comparison of the UV spectra of a coal tar pitch component and naphtho[8,1,2-*abc*]coronene. From J. C. Fetzer and J. R. Kershaw, *Fuels* **74** 1533–1536 (1995).

peri-fused structures that do not result from direct condensation of smaller PAHs. Figure 91 shows their spectra of a coal tar pitch component and a standard of naphtho[8,1,2-*abc*]coronene. Wise and co-workers found a wide variety of 24-carbon LPAHs in a similar coal tar material that is used as a standard reference material for PAH analysis. They separated a coal tar into carbon-number fractions, and analyzed specifically for LPAHs of this carbon number. This was part of a comprehensive program to identify as many as possible of the PAHs of up to 24 carbons. They identified many LPAHs that had never been seen in a coal tar before because they had made the considerable effort to obtain all 34 of the possible alternant and nonalternant fused $C_{24}H_{14}$ isomers. With this large standard collection they identified several minor components of their fractions.

Chen and Qian (19) used field-ionization and field-desorption MS techniques to look at the composition of solvent-refined coal extracts. These materials are very complex. LPAH masses corresponding to $C_{24}H_{14}$, $C_{26}H_{14}$, $C_{26}H_{16}$, $C_{28}H_{14}$, $C_{28}H_{16}$, $C_{30}H_{16}$, $C_{32}H_{14}$, and $C_{32}H_{16}$ were seen. They postulated that some of the structures were caused by cyclization or condensation of aromatized steranes or hopanes also seen. These types of hydrocarbons are known to arise from biological materials and are commonly found as the aromatized forms in coal and petroleum.

There are also numerous LPAHs in the effluent from hard (anthracite) coal combustion. Schmidt, Grimmer, Jacob, Dettbarn, and Naujack (20, 21) have found many of either 300 or 302 daltons (24-ring carbons). They found that the PAH and LPAH levels in these smokestack effluents were much higher than those from burning either fuel oil or brown (lignite) coal. For example, coronene was 24 more times abundant. There were also many nonalternant species found.

There were four LPAH of $C_{24}H_{12}$, and coronene is the only alternant structure of that molecular weight. Interestingly, some alternant 24-carbon LPAHs were not found. These were dibenzo[*b,def*]chrysene (*compound 130*), dibenzo[*def,p*]-chrysene (*compound 37*), and naphtho[8,1,2-*cde*]tetracene (*compound 157*). It was thought that specific degradation routes present in the combustion zone resulted in transformation of these to other structures (one that would be one of the 300-dalton nonalternant LPAHs).

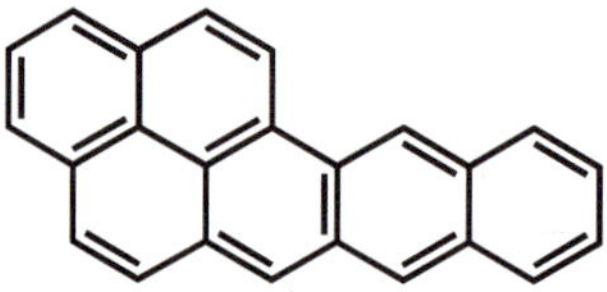

157

Ruckmick, Hurtubise, and Silver (22) used a sequence of separation methods to fractionate a coal liquid. Field-ionization MS showed many high-molecular-weight components in the fractions. The even-mass ones were assumed to be aromatic hydrocarbons. These authors drew several postulated structures of between 30 and 40 carbons, but they did not identify any specific LPAHs.

Pierce and Katz (23, 24) extracted an air particulate sample. Two-dimensional TLC separated the LPAHs. Fluorescence was used to identify 12 LPAHs. They found naphtho[1,2,3,4-*def*]chrysene (*compound 158*), benzo[*rst*]pentaphene, dibenzo[*b,def*]chrysene, and naphtho[2,1,8-*qra*]chrysene (*compound 159*). Wise

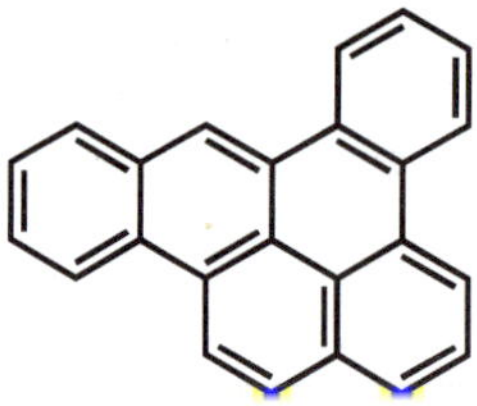

158

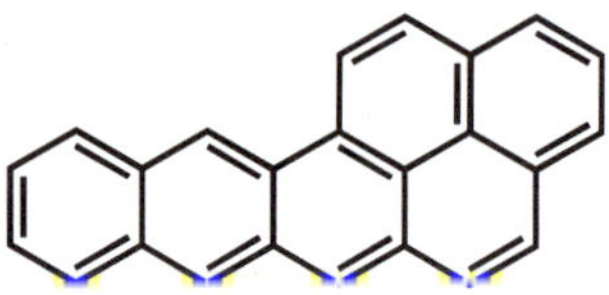

159

and co-workers (25, 26) examined two SRM air particulate and one marine sediment sample for the $C_{24}H_{12}$ and $C_{24}H_{14}$ isomer sets using a sequence of normal- and reversed-phase separations. They fractionated extracts of SRMs 1648 and 1649 by HPLC and used GC-MS for specific compound determination. They found nine LPAH isomers of the latter group. The distribution of these isomers was similar in all three SRMs, in contrast to a fourth SRM that was examined. A coal tar pitch material showed a very different distribution. This hints that air particulates might be the major source of PAHs in marine sediments.

Diesel exhaust particulate contains many LPAHs at low concentrations (27–29). The alternant peri-fused LPAHs dominate. Figure 92 shows an HPLC

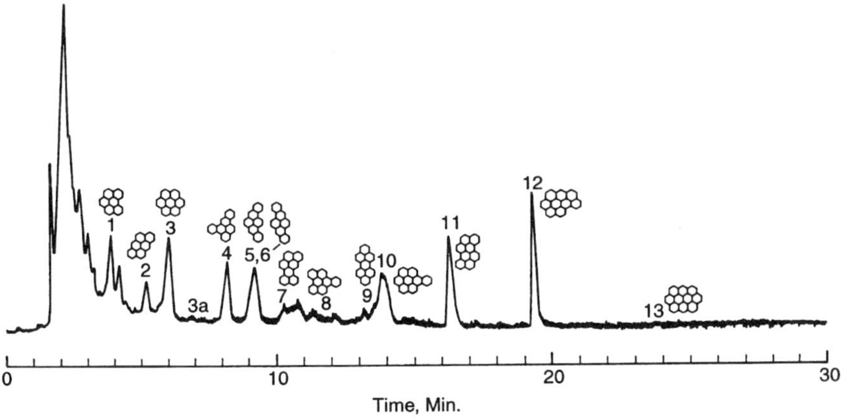

Figure 92. Reversed-phase HPLC separation of DIESEL particulate LPAHs. From J. C. Fetzer, W. R. Biggs, and K. Jinno, *Chromatographia* **21** 439–442 (1986).

chromatogram of the LPAH fraction of a diesel particulate extract. The highly-condensed species, such as naphtho[8,1,2-*abc*]coronene and ovalene, are major components of the LPAH fraction. The effluent from used biomass gasifiers also contains PAHs up to at least coronene. They saw coronene and 9 other $C_{24}H_{12}$ isomers and dibenzo[*b*,*k*]fluoranthene, naphtho[1,2,3,4-*def*]chrysene, and 13 other $C_{24}H_{14}$ isomers (30).

Blumer, Blumer, and Reich (31) observed similar LPAHs in extracts of soil samples collected from roadsides in Switzerland. They ascribed the presence of these compounds to the exhaust from automobiles. The concentrations of these LPAHs were much lower in areas away from heavily-used roads.

McCarry et al. (32–38), found LPAHs of 24–32 carbons in sediments from harbors on Lake Ontario and the Atlantic coast. They also used a microbial bioassay to examine the mutagenic potency of 50 fractions from a polymeric reversed-phase HPLC separation. They observed that several $C_{24}H_{14}$ fractions were highly active. A fraction that had naphtho[2,1,8-*qra*]tetracene (*compound 159*) and benzo[*pqr*]picene (*compound 160*) was the most potent one found. Several of these LPAHs were identified through comparison of standard

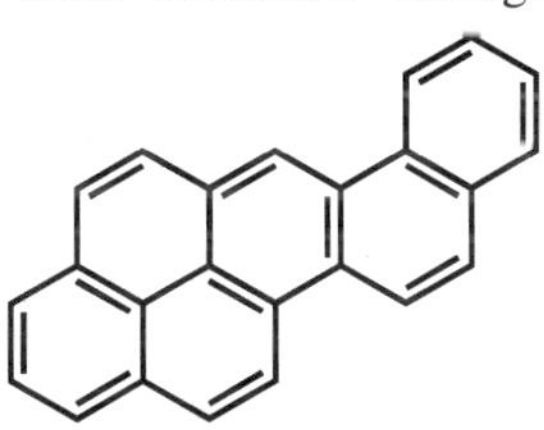

160

compounds (36, 37), including dibenzo[*cd,lm*]perylene, benzo[*a*]coronene, benzo[*pqr*]naphtho[8,1,2-*bcd*]perylene, and naphtho[8,1,2-*abc*]coronene.

9.3 PETROLEUM AND ITS PROCESSED PRODUCTS

Petroleum has been found to contain ultra-trace levels of several LPAHs. McKay and Latham (39) found coronene, naphtho[1,2,3,4-*def*]chrysene, naphtho[1,2,3.4-*ghi*]perylene, dibenzo[*fg,ij*]pentaphene (*compound 161*), dibenzo[*b,pqr*]perylene, and benzo[*pqr*]naphtho[8,1,2-*bcd*]perylene using a series of separations and comparison to UV absorbance and fluorescence spectra. Desideri, Lepri, and Checchini (40) found the first two LPAHs and dibenzo[*b,def*]chrysene in eight crude oils from Saudi Arabia, Dubai, Qatar, Libya, Congo, Gabon, Russia, and Venezuela. The smaller aromatic molecules in petroleums are highly alkylated. This is not true for those PAHs of 20 or more ring carbons. McKay and Latham inferred this from UV and fluorescence shifts relative to the unsubstituted model compounds. Boduszynski (41, 42) later showed through the use of HPLC fractionation followed by field-ionization MS that the substituents are predominantly methyl groups. Even the four- and five-carbon substituted LPAHs appear to be polymethylated.

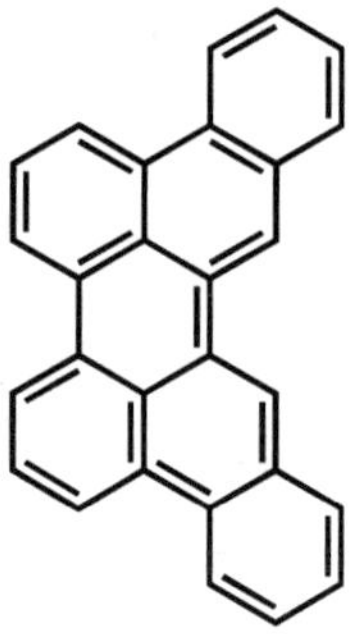

161

Petroleum hydrocracking is a process whereby partially processed oils that have been treated catalytically to remove the heteroatomic molecules, are further treated with other catalysts under high hydrogen pressures to produce motor and aviation fuels. An early study (43) used MS and identified the LPAHs produced in this process as "benzocoronenes." The mechanism was thought to be a wide variety of ring-addition reaction mechanisms. In these, many isomers are formed as the product LPAHs. Later work by another research group (44–47) has shown, instead, that there are two rather simple and elegant LPAH production routes. In the first, a series of one-ring additions occurs through the condensation reaction of two or four carbons to the PAH core (Fig. 93). This ultimately leads

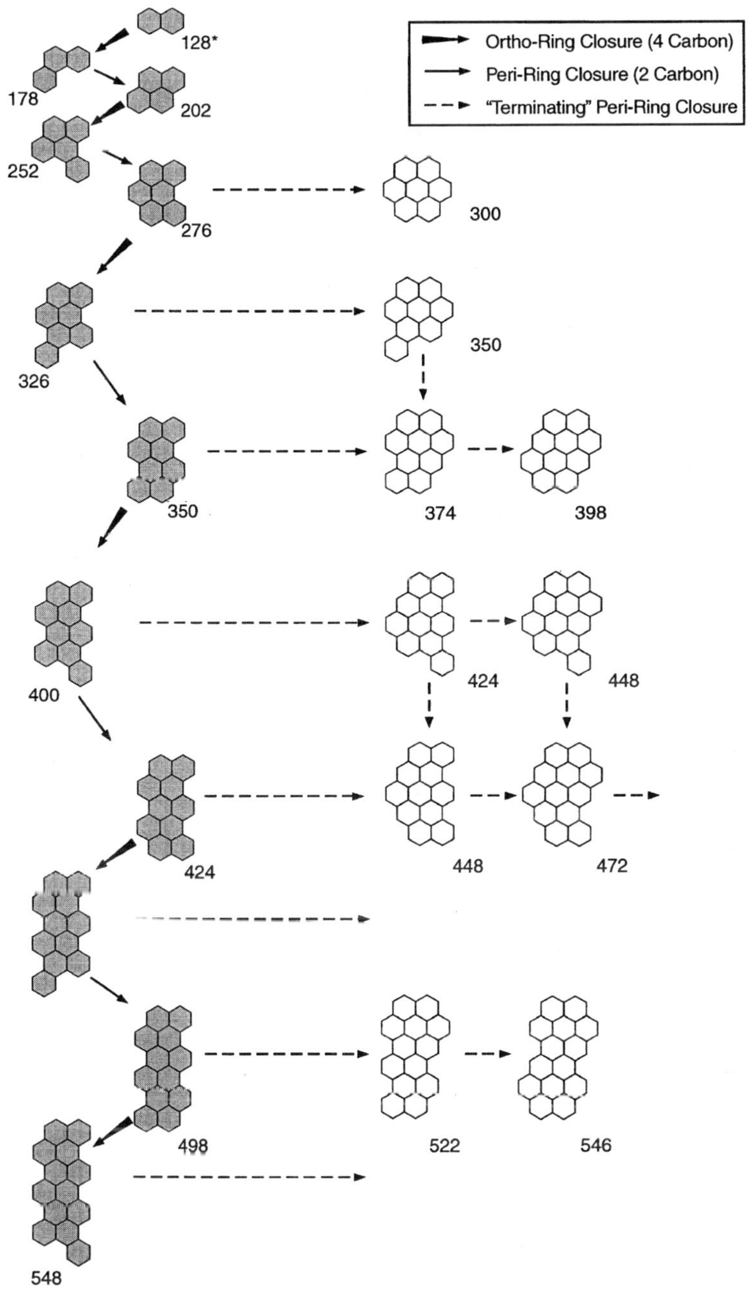

Figure 93. Production route of LPAHs by one-ring build-up in catalytic hydrocracking during petroleum processing. From J. C. Fetzer, in *Catalysts in Petroleum and Petrochemical Processing 1995*, M. Absi-Halabi, J. Besharah, H. Qabazard, and A. Stanislaus, Eds., Elsevier, 1996, p. 263–271.

to LPAHs such as coronene, benzo[*a*]coronene, phenanthro[5,4,3,2-*efghi*]perylene, benzo[*pqr*]naphtho[8,1,2-*bcd*]perylene, naphtho[8,1,2-*abc*]coronene, and ovalene (Fig. 94, a close-up view of one section of the reactions in Fig. 93). The second reaction type is the Scholl condensation of two pyrene molecules, or two coronene molecules, or a coronene molecule with an ovalene molecule. These result in dinaphtho[2,1,8,7-*defg*:2′,1′,8′,7′-*ijkl*]pentaphene and dinaphtho-[2,1,8,7-*defg*:2′,1′,8′,7′-*opqr*]pentacene, or benzo[1,2,3-*bc*:4,5,6-*b*′*c*′]-dicoronene, or benzo[10,11]phenanthro[2′,3′,4′,5′,6′:4,5,6,7]chryseno[1,2,3-*hi*]ovalene. The first three are orange or reddish in color, but the last 56-carbon LPAH is purple and gives that very distinct color to the deposits in which it is present in greater than 5%.

These LPAHs often precipitate out from the process stream, and during shutdowns these are found as yellow, orange, or red deposits. After removal of the residual oil by extraction with an alkane solvent, the LPAHs of up to ~14 rings can be extracted with dichloromethane. It was from such an extract that the first pure, macroscopic samples of phenanthro[5,4,3,2-*efghi*]perylene and naphtho[8,1,2-*abc*]coronene were obtained. Figure 95 shows a detailed reaction scheme that highlights the specificity in this production route and explains the presence of the three $C_{28}H_{14}$ isomers found. The very large PAHs resulting from Scholl condensation of coronene and ovalene are soluble in 1,2,4-trichlorobenzene and can be isolated in pure form from the deposits (usually leaving inorganic material, catalyst fines and ferric oxide, as the extraction residue). Because only alternant LPAHs are produced in hydrocracking, the coronene dimerization product is only the alternant structure. The aluminum(III) chloride synthetic route produces some nonalternant dimerization product. Figures 96 and 97 are the mass spectrum and fluorescence excitation spectrum of isolated hydrocracker deposit material, showing that it is benzo[1,2,3-*bc*:4,5,6-*b*′*c*′]dicoronene.

The production of those 24-carbon and larger LPAHs that are produced by the one-ring addition route sometimes results in their precipitation as process deposits or as coprecipitants with the much larger ones that result from the Scholl condensation. The first pure samples of phenanthro[5,4,3,2-*efghi*]perylene and the first macroscopic samples of naphtho[8,1,2-*abc*]coronene were isolated from petroleum hydrocracker deposits. These two LPAHs, and samples of the commercially unavailable benzo[*pqr*]naphtho[8,1,2-*bcd*]perylene, then were used as standards in much of the analytical work that has involved these three compounds.

It is interesting to note that benzo[1,2,3-*bc*:4,5,6-*b*′*c*′]dicoronene, with an empirical formula of $C_{48}H_{20}$, may be one of the most abundant LPAHs because of its formation in these processes. This is counter to conventional thinking that believes there is a monotonic decrease in LPAH concentration as the carbon number increases in any material. The dimerization reaction results in very, very

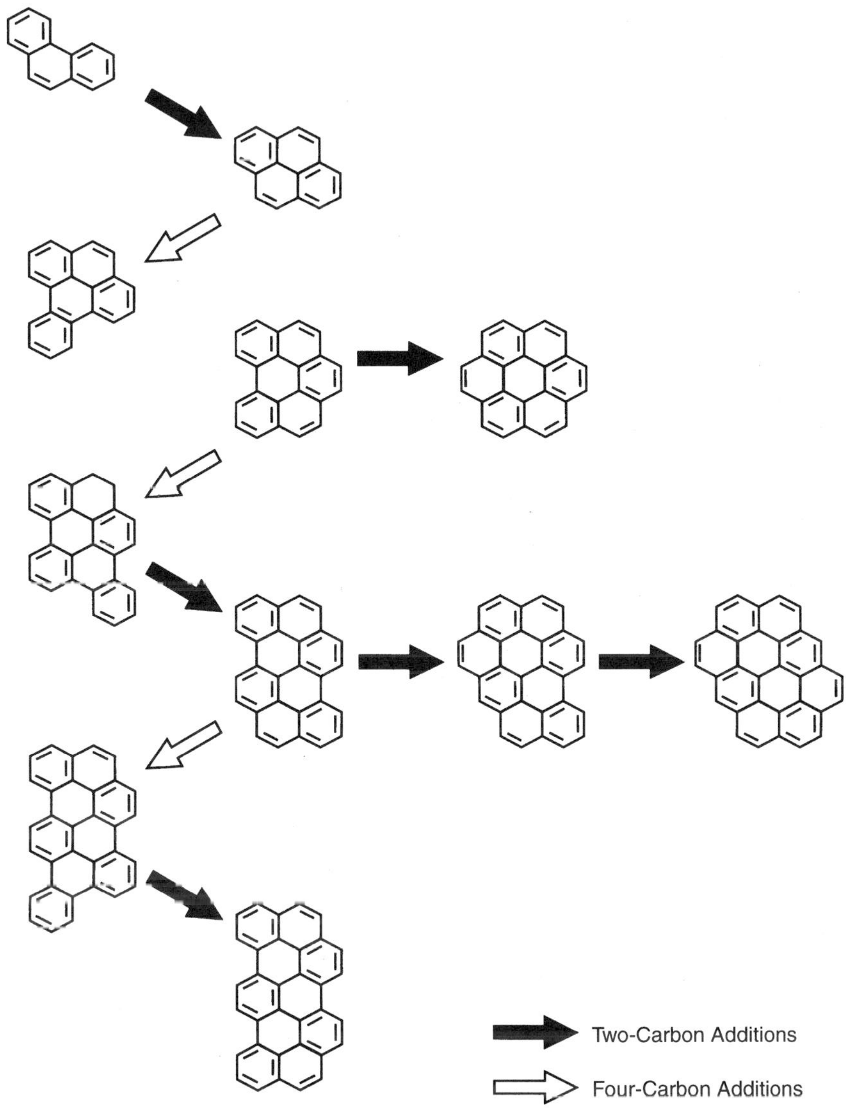

Figure 94. Close-up of a portion of Figure 93 showing the specific addition pathways. From R. F. Sullivan, M. M. Boduszynski, and J. C. Fetzer, *Energy Fuels* **3** 603–612 (1989), reproduced with permission from the American Chemical Society.

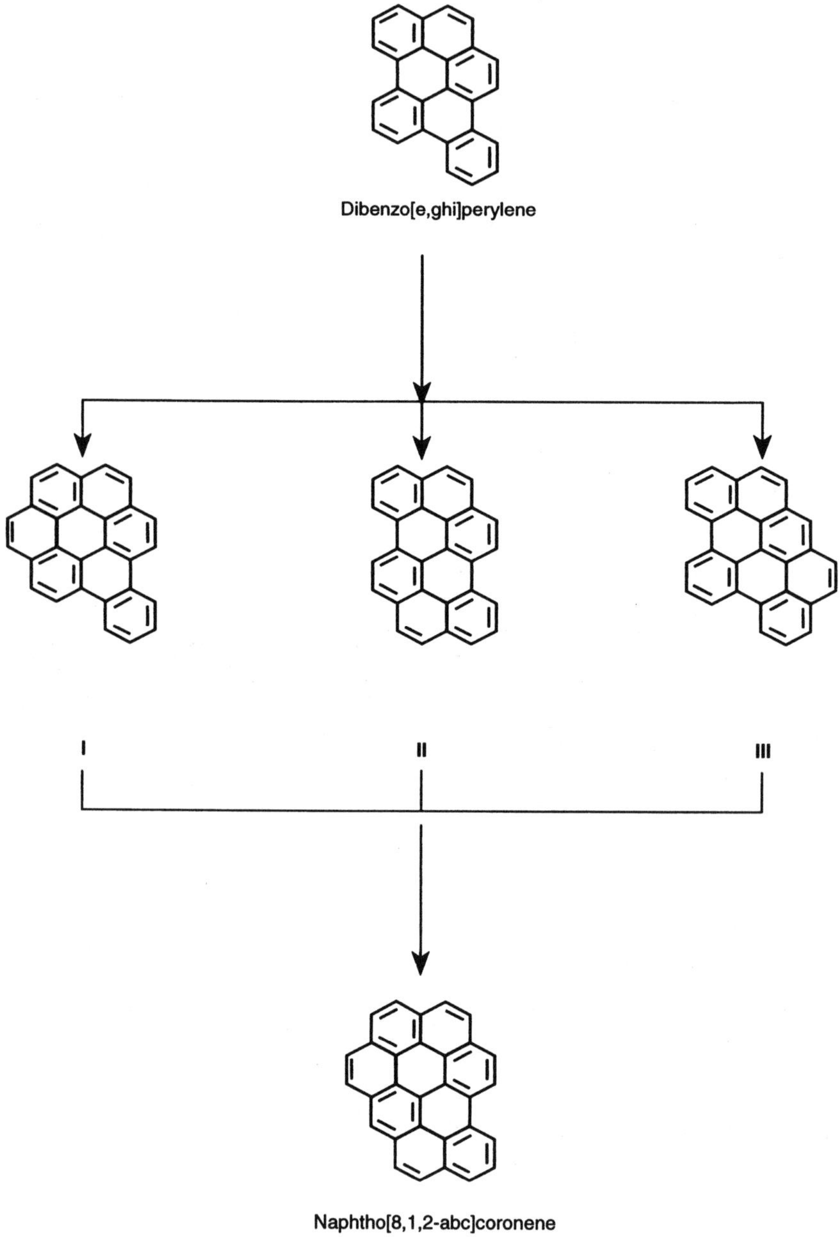

Figure 95. Specific pathways to the previously unknown LPAH phenanthro[5,4,3,2-*efghi*]perylene, as found in isolated fractions from a catalytic hydrocracking deposit. From J. C. Fetzer and W. R. Biggs, *Polycyclic Arom. Cmpds.* **5** 193–199 (1994), reproduced with permission of Gordon and Breach Science Publishers.

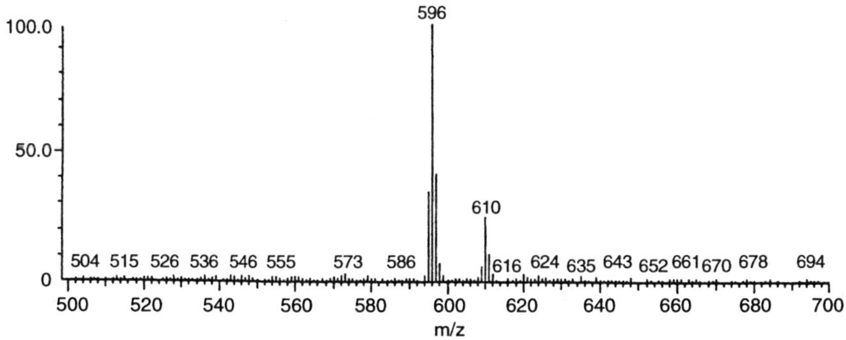

Figure 96. Mass spectrum of isolated hydrocracker deposit showing the molecular ion indicative of $C_{48}H_{20}$. From J. C. Fetzer, in *Catalysts in Petroleum and Petrochemical Processing 1995*, M. Absi-Halabi, J. Besharah, H. Qabazard, and A. Stanislaus, Eds., Elsevier, 1996, p. 263–271.

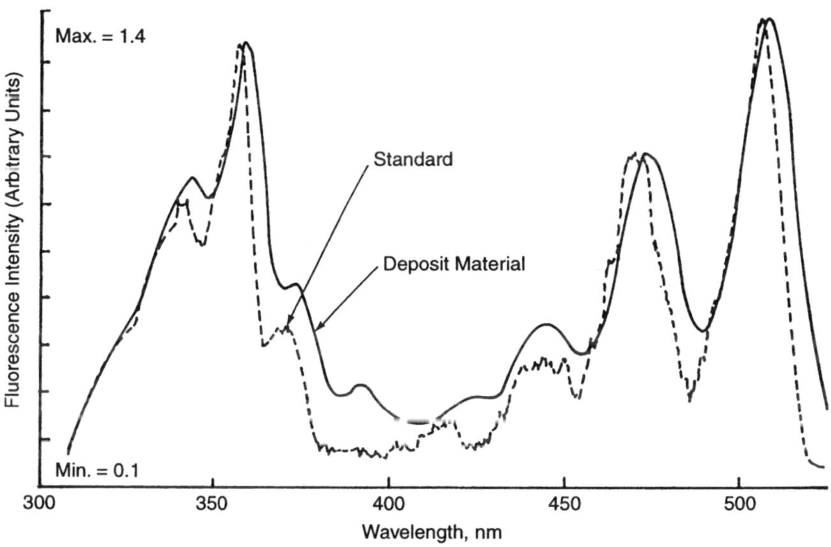

Figure 97. Fluorescence emission spectrum of the 1,2,4-trichlorobenzene solution of the same hydrocracker deposit as in Figure 96, showing it to be benzo[1,2,3-*bc*:4,5,6-*bc*]dicoronene. From J. C. Fetzer, in *Catalysts in Petroleum and Petrochemical Processing 1995*, M. Absi-Halabi, J. Besharah, H. Qabazard, and A. Stanislaus, Eds., Elsevier, 1996, p. 263–271.

little of the LPAHs of 34- to 46-ring carbons. There are many dozens of these processing units throughout the world. Each hydrocracking unit may produce several dozens of kilograms of this particular LPAH (or the other LPAH product) per year. Gu, to confirm that this chemistry was more universal, performed sequential Soxhlet extractions of hydrocracker deposits from several Chinese refineries (48). Fluorescence and mass spectra analyses showed the deposit to be the same Scholl condensation products as found by Fetzer.

Most carbon blacks are very low in LPAHs. One, however, has spawned a large number of analytical papers. An experimental carbon black that was manufactured by Cabot Corporation was first studied by Lee and Hites (49). It has much higher levels of PAHs than usually seen. Further studies (50, 51), using reversed-phase HPLC and off-line fluorescence and mass spectroscopies, showed many LPAHs. Fetzer and Rechsteiner (11) used a similar HPLC method with a DAD and identified many of these by comparison to standards.

9.4 OTHER COMBUSTED MATERIAL

Commercial coronene was fractionated by Schmidt using vacuum sublimation (52). A fraction of higher subliming LPAHs, that was ~1% of the sample charge, was isolated. Fluorescence analysis of the fractions showed benzo[*pqr*]-naphtho[8,1,2-*bcd*]perylene, naphtho[8,1,2-*abc*]coronene, ovalene, and benzo[4,10]anthra[1,9,8-*abcd*]coronene (*compound 162*). Mass spectral analysis showed lower levels of LPAHs with 38, 40, and 42 carbons. Schmidt concluded that this material must have been produced during the conversion of coal to motor fuels. Boente had earlier studied this material (53) and tentatively identified dinaphtho[8,1,2-*abc*:8′,1′,2′-*jkl*]coronene (*compound 106*) as a low-level impurity.

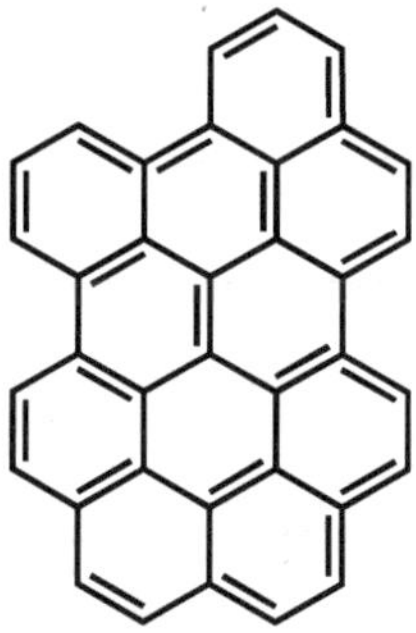

162

The research efforts of LaFleur, Wornat, et al. have focused on the formation of PAHs in flame combustion (54, 55). In material generated from the flame combustion of an ethylene-naphthalene mixture, they found coronene, benzo[*pqr*]naphtho[8,1,2-*bcd*]perylene, naphtho[8,1,2-*abc*]coronene, and ovalene. They used HPLC-DAD-MS as the analytical method. Direct introduction of these same samples into the mass spectrometer showed LPAHs as large as 160 carbons. They also looked at the LPAHs in the effluent from coal pyrolysis. They also used anthracene as the model compound to aid in understanding of the mechanisms. They found the expected structures of anthracene condensed through single bond and ring formation (leading to both alternant and nonalternant products). Interestingly, many LPAHs were found that could only occur through rearrangements of the carbon skeletons. These are the isomers naphtho[1,2,3,4-*rst*]pentaphene (*compound 85*), dibenzo[*de*,*uv*]pentacene (*compound 163*), and benzo[*vwx*]hexaphene (*compound 164*), as well as naphtho[8,1,2-*abc*]coronene. Figure 98 shows a distribution graph from this work, which displays the relative abundance of several $C_{28}H_{16}$ isomers. Previous work on pitches by Lewis and co-workers (56, 57) had not shown such a high degree of rearrangement, most probably because the bulk of the material is large aromatic hydrocarbons formed by linkage of the starting material by single bonds and bridging rings. The condensed LPAHs, and in particular those made through carbon skeletal rearrangements, are minor components compared to these "PAH polymeric species."

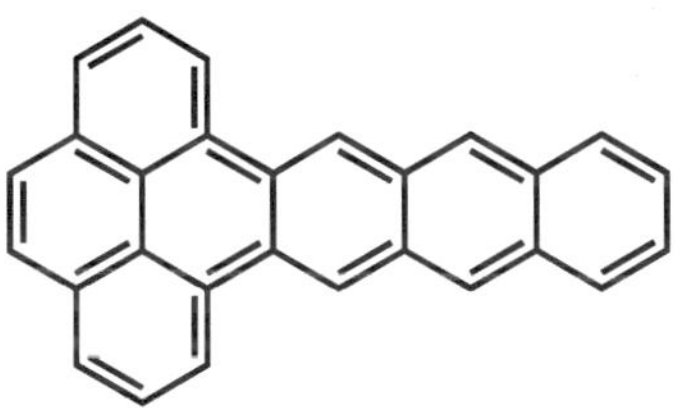

163

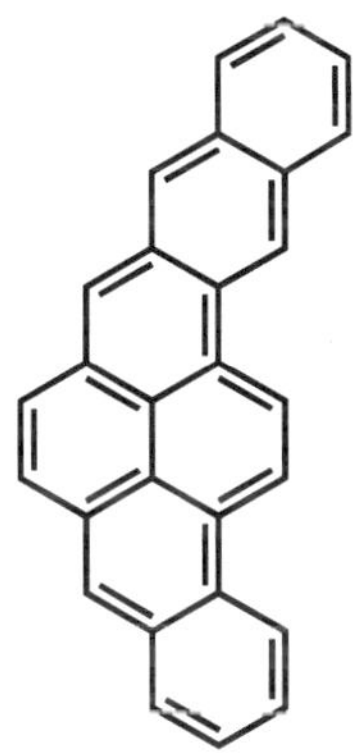

164

When the research group of Lafleur, and co-workers (58) pyrolyzed brown coal under similar conditions to the naphthalene pyrolysis, they found many fused cyclopenta-substituted PAHs, including many with more than one substitution. The two largest compounds were the two LPAHs benzo[*ghi*]cyclopenta[*cd*]perylene and cyclopenta[*bc*]coronene (*compounds 165* and *166*). They compared to newly synthesized standards and compared the HPLC-DAD spectra from the samples to reference UV absorbance spectra. In the ethylene-rich flame studies (59) many species were found with peripheral condensation of an ethylene to form an "ace" ring, the cyclopenta-fused PAHs. This observation was specifically aided by synthesis of these specific LPAH model compounds to

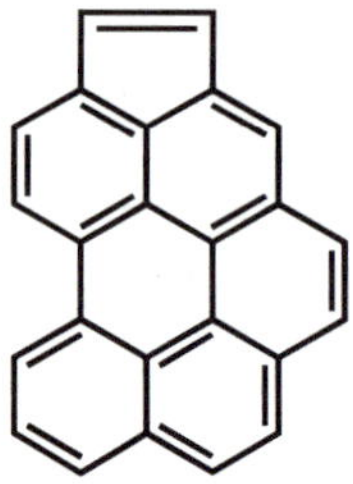

165

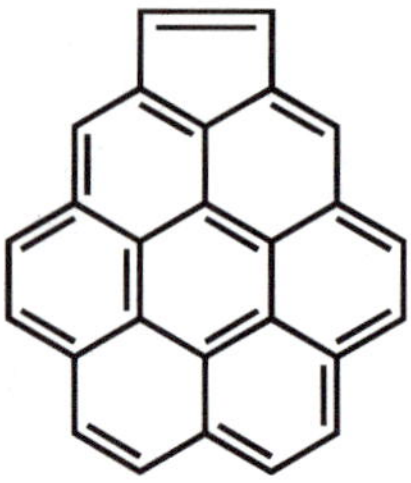

166

be compared to the chromatographic retention times and UV absorbance spectra of the many unknown peaks in their chromatograms.

They found similar structures in their anthracene pyrolysis products (60). In the anthracene pyrolysis there was definite evidence of molecular rearrangements, as shown by the presence of the several $C_{28}H_{16}$ isomers and naphtho[8,1,2-*abc*]coronene that were found. In both the coal and model compound studies, they found that condensation to larger structures was very common. Ring closure to form both five- and six-member bridging rings was seen, as well as drastic rearrangements. They observed, not unexpectedly, that these reactions became more and more prevalent as the pyrolysis temperature was increased.

Pyrolysis of *o*-terphenyl (61) yields highly-condensed LPAHs. Among the products were dibenzo[*fg*,*op*]tetracene (104), tribenzo[*fg*,*ij*,*rst*]pentaphene (*compound 167*), and phenanthro[9,10-*b*]triphenylene (*compound 168*). Rearrangements similar to those seen in the naphthalene and anthracene experiments also occurred, as shown by benzo[*rst*]pentaphene, naphtho[2,3-*e*]acephenanthrylene (*compound 169*), and naphtho[8,1,2-*abc*]coronene. Some of the condensed structures that were found hint that cleavage of the single bonds in *o*-terphenyl occurs. This would give reactive phenyl and biphenyl species that

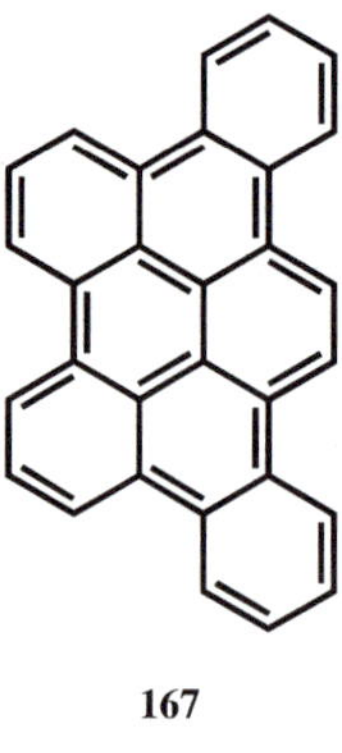

167

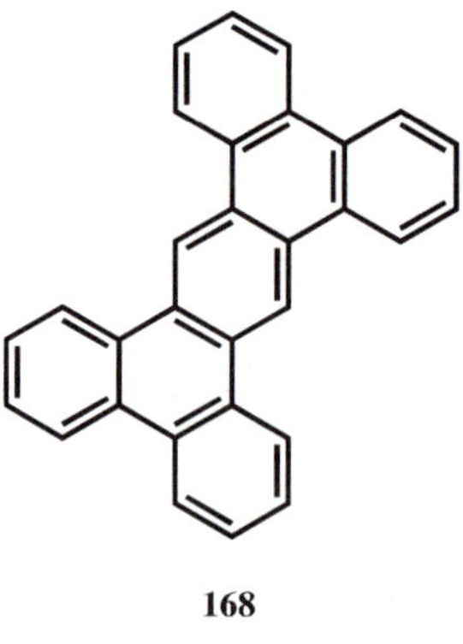

168

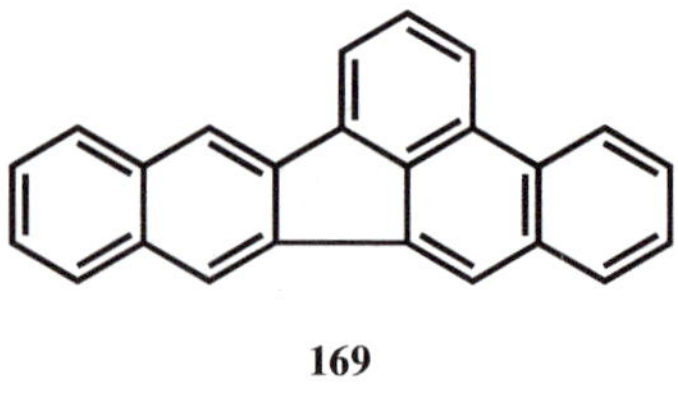

169

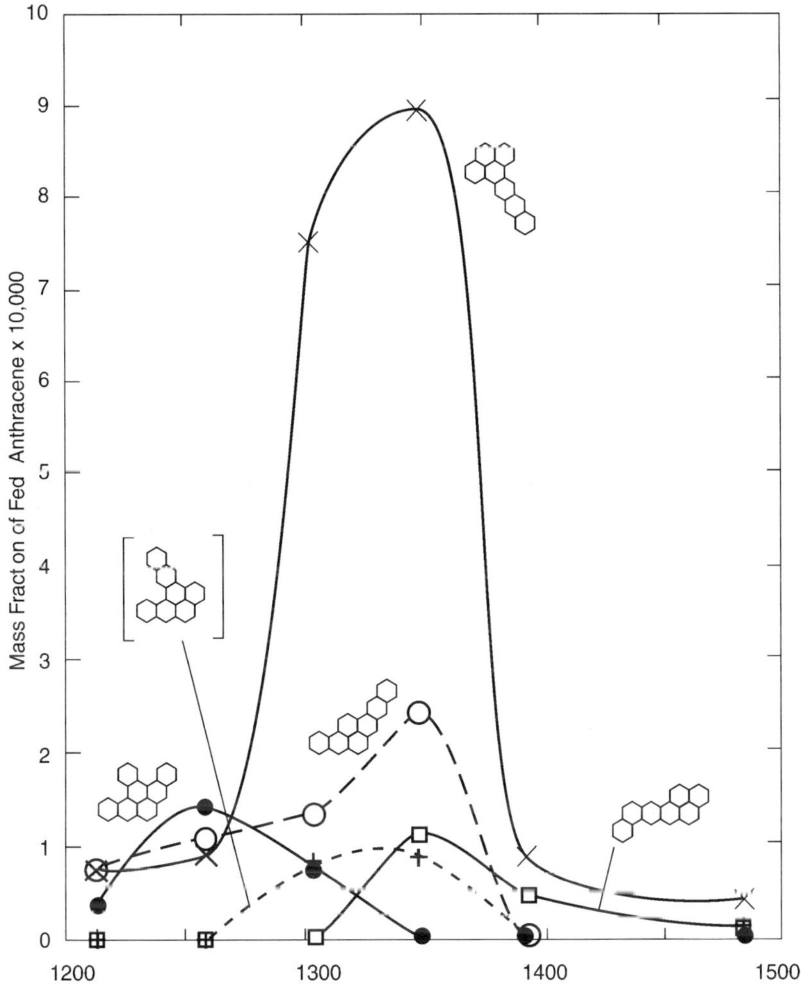

Figure 98. Temperature-dependent distribution of several isomeric LPAHs produced in anthracene pyrolysis anthracene pyrolysis yields of mass 352 isomers M. J. Wornat, A. F. Sarofim, and A. L. LaFleur, *Twenty-Fourth International Symposium on Combustion*, The Combustion Institute, 1995, p. 959, reproduced with permission from Dr. W. J. Wornat and The Combustion Institute.

would explain the fact that the products were found with four and five fully-benzenoid ring structures.

In another study (62) by LaFleur and co-workers, pyrene was pyrolyzed. Besides numerous bipyrenyl isomers, two alternant and three nonalternant $C_{32}H_{16}$ isomers were found. The yields of these five products varies with temperature, reflecting the kinetic, steric, and thermodynamic factors that control the formation of the six bipyrenyl isomers.

Kubica and Jastrzebski (63) pyrolyzed phenanthrene, anthracene, and mixtures of the two at temperatures from 475°C to 600°C. They found the expected dibenzoperylene iusomers and rearrangement products like rubicene (*compound 112*) and dibenzo[*cd,lm*]perylene (*compound 38*).

The use of sequential separations with both UV absorbance and MS being used for identification, has led to a better understanding of the formation of LPAHs during flame-combustion as the route to soot (64–66). The products obtained during flame combustion of ethylene, acetylene, benzene, and other light hydrocarbons are very similar to each other. The new analytical methods have allowed separation of the many individual species present. This, coupled with the use of newly-synthesized individual LPAHs as comparison standards, has led to a better understanding of the build-up mechanisms. Comparisons of retention times and absorbance spectra made identification possible in these complex mixtures.

The efforts of Homann et al. have extended the understanding of soot formation even further (67–71). They used GC-MS to study the PAHs of up to 24 carbons, but molecular beam and resonance enhanced multi-photon ionization MS have allowed them to observe species in fuel-rich flames of several hundred carbon atoms. These include neutral LPAHs, LPAH radicals, and LPAH ions, as well as the fullerenes and their corresponding radicals and ions. Because the mass spectral techniques cannot differentiate isomers, only specific very selective masses can be ascribed to a certain structure, whereas the rest are only given an empirical formula. For example, this group observed both $C_{294}H_{42}$ and $C_{294}H_{46}$ ions. The first has only one possible structure, and this can be assigned to this species, whereas the second has numerous isomeric possibilities.

Boehm, Boenig, Feldermann, Jander, Rudolph, and Wagner (72) studied the products of combustion in ethylene-fueled flames. They found LPAHs such as naphtho[8,1,2-*abc*]coronene and 1,1′-bicoronene (and many other molecules formed by single bonds joining two PAHs). They observed a pressure dependence on the amount of LPAHs and soot produced. At lower pressures, this was found to correlate with P^2, but was approximately linear at higher pressures.

REFERENCES

1. J. C. Fetzer, B. R. T. Simoneit, P. Garrigues, and H. Budzinski, *Polycyclic Arom. Compds.* **9** 109–120 (1996).
2. B. R. T. Simoneit and J. C. Fetzer, in *Organic Geochemistry: Developments and Applications to Energy, Climate, Environment and Human History*, J. O. Grimalt and C. Dorronsoro, Eds., A. I. G. O. A., Donostia, San Sebastian, Spain, 1995, p. 414–417.
3. J. Murdoch and T. A. Geissman, *Am. Mineral.* **52** 11–16 (1967).

4. M. Blumer, *Chem. Geol.* **16** 245 (1975).
5. W. R. West, S. A. Wise, R. M. Campbell, K. D. Bartle, and M. L. Lee, in *Polycyclic Aromatic Hydrocarbons: Chemistry, Characterization, and Carcinogenesis.* M. Cooke and A. J. Dennis, Eds., Batelle, Columbus, OH, 1986, p. 995–1009.
6. S. A. Wise, R. M. Campbell, W. R. West, M. L. Lee, and K. D. Bartle, *Chem. Geol.* **54** 339–357 (1986).
7. T. M. McCollom, B. R. T. Simoneit, and E. L. Stock, *Energy Fuels* **13** 401–410 (1999).
8. M. Zander and G. Collin, *Fuel* **72** 1281–1285 (1993).
9. M. Zander, *Polycyclic Arom. Cmpds.* **7** 209–221 (1995).
10. I. C. Lewis and B. A. Petro, *J. Polym. Sci.* **14** 1975 (1976).
11. J. C. Fetzer and C. E. Rechsteiner, in *Polycyclic Aromatic Hydrocarbons: Eleventh International Symposium*, M. Cooke, K. Loening, and J. Merritt, Eds., Battelle, Columbus, OH, 1991, p. 259–270.
12. J. C. Fetzer and J. R. Kershaw, *Fuel* **74** 1533–1536 (1995).
13. S. Suzuki, T. Kaneko, and M. Tsuchiya, *Kankyo Kagaku* **6** 511–520 (1996).
14. G. Sabor, P. Smidl, and J. Mitera, *Fuel* **68** 1285–1289 (1989).
15. V. P. Senthilnathan and S. E. Stein, *J. Org. Chem.* **53** 3000–3007 (1986).
16. A. Colmsjo, *Anal. Chim. Acta* **197** 65–70 (1987).
17. A. Colmsjo and C. Ostman, in *Polynuclear Aromatic Hydrocarbons: Physical and Biological Chemistry*, M. Cooke, A. J. Dennis, and G. L. Fisher, Eds., Batelle, Columbus, OH, 1982, p. 201–210.
18. S. A. Wise, B. A. Benner, H. Liu, G. D. Byrd, and A. Colmsjo, *Anal. Chem.* **60** 630–637 (1988).
19. J. Chen and S. Qian, *Fuel Sci. Technol. Int.* **6** 687–722 (1988).
20. W. Schmidt, G. Grimmer, J. Jacob, and G. Dettbarn, *Toxicol. Environ. Chem.* **13** 1–16 (1986).
21. W. Schmidt, G. Grimmer, J. Jacob, G. Dettbarn, and K. W. Naujack, *Fresenius Z. Anal. Chem.* **326** 401–413 (1987).
22. S. C. Ruckmick, R. J. Hurtubise, and H. F. Silver, *Fuel* **66** 1677–1683 (1986).
23. R. C. Pierce and M. Katz, *Anal. Chem.* **47** 1743–1748 (1975).
24. M. Katz and R. C. Pierce, in *Carcinogenesis—A Comprehensive Study*, R. I. Freudenthal and P. W. Jones, Eds., Raven, New York, 1976, Vol. 1, p. 413.
25. S. A. Wise, B. A. Benner, S. N. Chesler, L. R. Hilpert, C. R. Vogt, W. E. May, *Anal. Chem.* **58** 3067–3077 (1986).
26. S. A. Wise, A. Deissler, and L. C. Sander, *Polycyclic Arom. Cmpds.* **3** 169–184 (1993).
27. K. Jinno, J. C. Fetzer, and W. R. Biggs, *Chromatographia* **21** 274–276 (1986).
28. J. C. Fetzer, W. R. Biggs, and K. Jinno, *Chromatographia* **21** 439–442 (1986).
29. K. Jinno, Y. Miyashita, S. Sasaki, J. C. Fetzer, and W. R. Biggs, *Envir. Monit. Assess.* **19** 13–25 (1991).

30. D. J. Desilets, P. T. Kissinger, F. E. Lytle, M. A. Horne, M. S. Ludwiczak, and R. B. Jacko, *Environ. Sci. Technol.* **18** 386–391 (1984).
31. M. Blumer, W. Blumer, and T. Reich, *Environ. Sci. Technol.* **11** 1082–1084 (1977).
32. C. H. Marvin, L. Allan, B. E. McCarry, and D. W. Bryant, *Environ. Mol. Mutagenesis* **22** 61–70 (1993).
33. C. H. Marvin, M. Tassaro, B. E. McCarry, and D. W. Bryant, *Sci. Total Environ.* **156** 119–131 (1994).
34. A. E. Legzdins, B. E. McCarry, C. H. Marvin, and D. W. Bryant, *J. Environ. Anal. Chem.* **60** 79–94 (1995).
35. C. H. Marvin, J. A. Lundrigan, B. E. McCarry, and D. W. Bryant, *Environ. Toxicol. Chem.* **14** 2059–2066 (1995).
36. C. H. Marvin, B. E. McCarry, J. Villella, D. W. Bryant, and R. W. Smith, *Polycyclic Arom. Cmpds.* **9** 193–200 (1996).
37. C. H. Marvin, B. E. McCarry, J. A. Lundrigan, K. Roberts, and D. W. Bryant, *Sci. Total Environ.* **231** 135–144 (1999).
38. C. H. Marvin, R. W. Smith, D. W. Bryant, and B. E. McCarry, *J. Chromatogr.* **863** 13–24 (1999).
39. J. F. McKay and D. R. Latham, *Anal. Chem.* **45** 1050–1055 (1973).
40. P. G. Desideri, L. Lepri, and L. Checchini, *J. Chromatogr.* **369** 117–124 (1986).
41. M. M. Boduszynski, *Energy Fuels* **1** 2–11 (1987).
42. M. M. Boduszynski, *Energy Fuels* **2** 597–613 (1988).
43. U. S. patent 3,619,407, November 9, 1971, G. W. Hendricks, E. C. Attane, and J. W. Wildon (to Union Oil).
44. R. F. Sullivan, M. M. Boduszynski, and J. C. Fetzer, *Energy Fuels* **3** 603–612 (1989).
45. U. S. patent 5,190,633, March 3, 1993, J. C. Fetzer and D. G. Lammel (to Chevron Research and Technology).
46. J. C. Fetzer, in *Catalysts in Petroleum and Petrochemical* Processing 1995, M. Absi-Halabi, J. Besharah, H. Qabazard, and A. Stanislaus, Eds., Elsevier, 1996, p. 263–271.
47. J. C. Fetzer and W. R. Biggs, *Polycyclic Arom. Cmpds.* **5** 193–199 (1994).
48. K. Gu, *Shiyou Lianzhi Yu Huagong* **29** 43–45 (1998), in Chinese.
49. M. L. Lee and R. A. Hites, *Anal. Chem.* **48** 1890–1893 (1978).
50. P. A. Peaden, M. L. Lee, Y. Hirata, and M. Novotny, *Anal. Chem.* **52** 2268–2271 (1980).
51. A. Hirose, D. Wiesler, and M. Novotny, *Chromatographia* **18** 239–242 (1984).
52. W. Schmidt, in *Polycyclic Aromatic Hydrocarbons and Astrophysics*, A. Leger, L. d'Hendecourt, and B. Boccara, Eds., Reidel, Dordrecht, The Netherlands, 1987, p. 149–164.
53. L. Boente, *Brennstoff Chem.* **36** 210–214 (1955).
54. M. J. Wornat, A. F. Sarofim, and J. P. Longwell, *Twenty-Second International Symposium on Combustion*, The Combustion Institute, 1988, p. 135–145.

55. M. J. Wornat, A. F. Sarofim, and A. L. LaFleur, *Twenty-Fourth International Symposium on Combustion*, The Combustion Institute, 1995, p. 955–963.
56. R. A. Greinke and I. C. Lewis, *Carbon* **22** 305–314 (1984).
57. I. C. Lewis and L. S. Singer, in *Polynuclear Aromatic Hydrocarbons*, L. Ebert, Ed., Advances in Chemistry 217, American Chemical Society, Washington, DC, 1988, p. 269–285.
58. A. L. LaFleur, K. Taghizadeh, J. B. Howard, J. E. Anacleto, and M. A. Quilliam, *Am. Soc. Mass Spectrom.* **7** 276–286 (1996).
59. M. J. Wornat, B. A. Vernaglia, A. L. LaFleur, E. F. Plummer, K. Taghizadeh, P. F. Nelson, C.-Z. Li, A. Necula, and L. T. Scott, *Twenty-Seventh International Symposium on Combustion*, The Combustion Institute, Boulder, Colorado, 1998, p. 1677–1686.
60. M. J. Wornat, F. J. J. Vriesendorp, A. L. LaFleur, E. F. Plummer, A. Necula, and L. T. Scott, *Polycyclic Arom. Cmpds.* (in press).
61. J. Mukherjee, A. F. Sarofim, and J. P. Longwell, *Combust. Flame* **96** 191–200 (1994).
62. M. C. Masonjones, J. Mukherjee, A. F. Sarofim, K. Taghizadeh, and A. L. LaFleur, *Polycyclic Arom. Cmpds.* **8** 229–242 (1996).
63. K. Kubica and J. Jastrzebski, *Koks Smola Gaz* **32** 211–217 (1987), in Polish.
64. P. G. Sim, M. A. Quilliam, S. Pleasance, and F. M. Benoit, *Polycyclic Arom. Cmpds. Suppl.* **3** 321–328 (1993).
65. J. F. Anacleto, L. Ramaley, R. K. Boyd, S. Pleasance, M. A. Quilliam, G. P. Sim, and F. M. Benoit, *Can. Rapid Commun. Mass Spectrom.* **5** 149–155 (1991).
66. A. L. LaFleur, K. Taghizadeh, J. B. Howard, J. E. Anacleto, and M. A. Quilliam, *Am. Soc. Mass Spectrom.* **7** 276–286 (1996).
67. M. Hausmann and K.-H. Homann, *Ber. Bunsen-Ges. Phys. Chem.* **101** 651–667 (1997).
68. P. Weilmuenster, A. Keller, and K.-H. Homann, *Combust. Flame* **116** 62–83 (1998).
69. J. Griesheimer and K.-H. Homann, *Twenty-Seventh International Symposium on Combustion*, The Combustion Institute, Boulder, Colorado, 1998, p. 1753–1759.
70. J. Ahrens, A. Keller, R. Kovacs, and K.-H. Homann, *Ber. Bunsen-Ges. Phys. Chem.* **102** 1823–1839 (1998).
71. K.-H. Homann, *Angew. Chem. Int. Ed. Engl.* **37** 2434–2451 (1998).
72. H. Boehm, M. Boenig, C. Feldermann, H. Jander, G. Rudolph, and H. G. Wagner, *Springer Ser. Chem. Phys.* **59** 145–164 (1994) (Soot Formation and Combustion).

CHAPTER

10

ANALYTICAL TECHNIQUES FOR STRUCTURE ELUCIDATION

An area in which no comprehensive review has ever been written for the LPAHs is one that describes those techniques that are useful for structural determination. After isolation of synthetic products, or other LPAH fractions, the specific structure must be determined. The specific properties of isomeric LPAHs, and the specific structures of these isomers, is key to assessing formation mechanisms. For many of the condensation reactions described in Chapter 1, the mechanisms are not known. Isomer distributions can hint at a mechanism that can then be used to predict the products of other similar reactions. This, for example, was useful in the various new reactions of mixed ketones that were reacted in the Clar condensation to produce several dozen new LPAHs.

There are many specific techniques that can be useful in helping to determine the structure of newly synthesized products or LPAHs that have been isolated from complex mixtures. Several of these do not fall within the topical areas in the previous chapters. Because there are so many possible isomers of a particular carbon number and degree of condensation, this is a critical step for the synthetic organic chemist or the analytical chemist trying to characterize their materials.

Zander (1) has reviewed the specific case of how to determine the structures of the LPAHs that have been isolated as fractions from coal tar pitches. He describes the relative strengths and weaknesses of many techniques, including some of the same ones that are described below. The effect of the complex original matrix involved and the probable presence of impurities is a dominating factor in this type of work. This changes how many of the approaches should be used, however, and lessens or strengthens the relative merits of each compared to the use of the same techniques for characterizing fractions from other types of materials or from synthesis products.

10.1 X-RAY CRYSTALLOGRAPHY

This is a brief description of this most widely used technique for structure determination. and how it has been specifically used for LPAHs. There are

---

Large (C>=24) Polycyclic Aromatic Hydrocarbons: Chemistry and Analysis, by John C. Fetzer.
ISBN 0-471-36354-5

several good books on the basic principles and application of X-ray crystallography, and the reader is directed to them for more in-depth descriptions (2–5).

The scattering of X rays by crystals produces a pattern of intensities, called the diffraction pattern. This pattern is determined by the location of the atomic arrays in the crystal structure, relative to the direction of the incident X-ray beam.

For LPAHs, one of the most difficult steps in crystallography is the obtaining of a sufficiently large enough crystal. The limited solubility and volatility of LPAHs are often high barriers to easily producing crystals. The purity of LPAH samples used in crystal formation must be greater than 99%. This is also a severe limitation because purification on a large scale is time consuming.

The first problem can sometimes be overcome if the right combinations of temperature and solvent mixtures are used for crystal formation in solution. Often mixtures of a very good solvent and a poor one are utilized. Crystal formation and precipitation can be caused by a careful adding of the poorer solvent or slow removal of the better one. For example, a solution of a LPAH may be dissolved in dichloromethane, a very good, but volatile, solvent. If this solution is slowly diluted with a miscible poor LPAH solvent, like methanol or acetonitrile, crystals may form. Alternatively, when some of the poorer solvent is added to the dichloromethane solution, but not enough to induce precipitation, it can be left to stand or gently heated. Some of the dichloromethane will evaporate. The ensuing solvent composition becomes gradually less able to keep the LPAH in solution. Crystals may slowly form. Often, the right combinations of solvents and temperature cannot be found and amorphous solids result. There are times when these can even appear to be crystalline, but microscopy reveals a polymorphic structure unsuitable for crystallography.

To overcome the second limitation, the right type of sublimation apparatus is used for vacuum or gas-phase formation of crystals. This apparatus must have either controllable pressure and temperature or a set vacuum pressure and very tightly controlled temperature gradients.

The research group of Aoki and Fujisawa and their co-workers (6–11) has used X-ray crystallography to determine the structure of several newly synthesized "peropyrene-type" LPAHs produced through variations on the Clar reaction involving condensation of perinaphthenone-type ketones. Their initial work involved several LPAHs of that class that had been first synthesized by Clar. They studied the effect of spatially adjacent rings on the nonplanarity of these LPAHs by obtaining crystal structures tetrabenzo[*a*,*cd*,*j*,*lm*]perylene (*compound 121*), benzo[*rst*]phenanthro[1,10,9-*cde*]pentaphene (*compound 170*), dibenzo[*a*,*rst*]naphtho[8,1,2-*cde*]pentaphene (*compound 171*), dibenzo[*jk*,*uv*]dinaphtho[2,1,8,7-*defg*:2′,1′,8′,7′-*opqr*]pentacene (*compound 141*), benzo[*rs*]dinaphtho[2,1,8,7-*klmn*:3′,2′,1′,8′,7′-*uvwxy*]hexaphene (*compound 147*), and

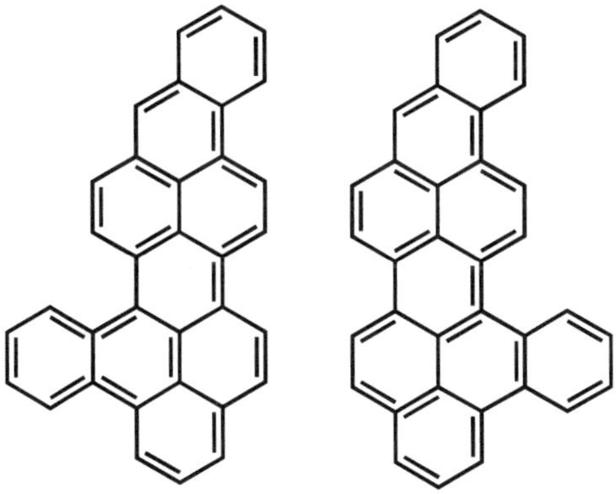

170 **171**

anthra[2,1,9,8-*hijkl*]benzo[*de*]naphtho[2′,1′,8′,7′-*stuv*]pentacene (*compound 146*). Figure 99 shows the long-axis view of dibenzo[*jk,uv*]dinaphtho[2,1,8,7-*defg*:2′,1′,8′,7′-*opqr*]pentacene (*compound 141*), as determined by X-ray crystallography and highlights the highly nonplanar structure.

After this work, this Japanese group developed syntheses for several different new LPAHs using a variety of new cyclic aromatic ketones of five or six rings. They obtained the crystal structure of the new LPAHs tribenzo[*a,h,rst*]phenanthro[1,2,10-*cde*]pentaphene and dibenzo[*def,i*]naphtho[8,1,2-*vwx*]pyranthrene (*compounds 67 and 68*). Figure 100 shows the nonplanar structure of the first of these, as found by its crystallography.

In general, they found that the presence of bay and cove regions in these molecules result in nonplanarity. This occurs through twisting and distortion of the central bridging ring of those structures with only a single bridging ring. The crystal structures revealed that the strain in these molecules was generally

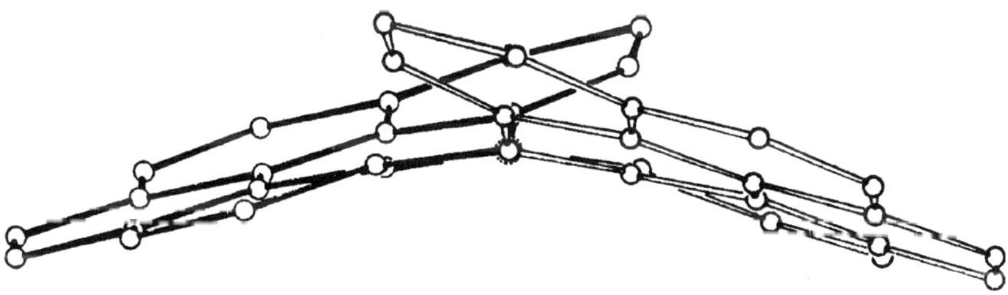

A perspective drawing of DPP

Figure 99. 3-D rendering of the structures of dibenzo[*jk,uv*]dinaphtho[2,1,8,7-*defg*:2′,1′,8′,7′-*opqr*]pentacene (*compound 141*) as determined by x-ray crystallography. From I. Oonishi, S. Fujisawa, J. Aoki, and T. Danno, *Bull. Chem. Soc. Jpn.* **51** 2259 (1978), reproduced with permission from the Chemical Society of Japan.

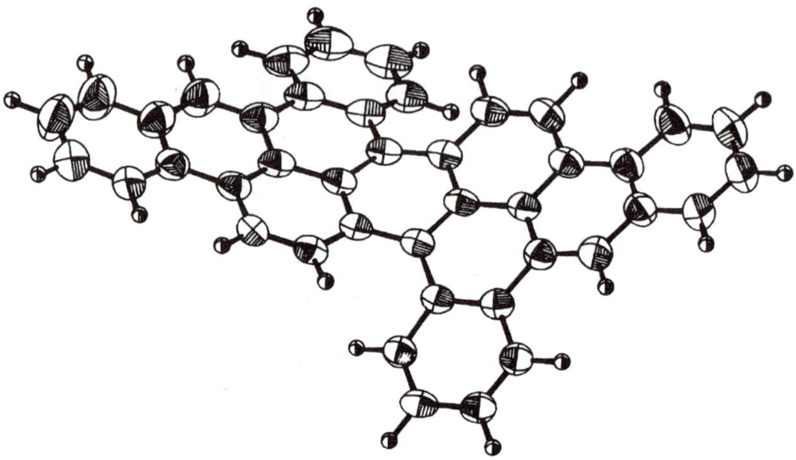

A perspective drawing of tribenzo [a,h,rst]phenanthra[1,2,10-cde]pentaphene.

Figure 100. From I. Oonishi, S. Ohshima, S. Fujisawa, J. Aoki, Y. Ohashi, and T. M. Krygowski, *J. Mol. Struct.* **265** 288 (1992), reproduced with permission from Elsevier Science Publishers BV.

reduced through twisting across the central bridging ring, with the portions on either side twisting in opposite orientation (so that the end on view of the PAH looked like an "X"). They termed this in-plane distortion because the two portions were still somewhat planar. The distortion around the central ring was termed out-of-plane distortion because the ring and parts of the adjacent rings contorted and were not planar. For the structure with two bridging rings and substantial steric strain, dibenzo[*def,i*]naphtho[1,8,7-*vwx*]pyranthrene, the distortion occurs through some twisting of the bridging rings and through a bending or flexing of the overall structure to reduce the steric strain. There is much more of the second effect than in molecules with only one bridging ring. This is the only example of this deformation in a LPAH structure, as all of the previously known products of the Clar reaction that possessed two-bridging rings only had a bay region on either side. These are all planar structures. The additional ring in this structure that resulted in a cove region thus has a great effect on the planarity (12).

Goddard, Haenel, Herndon, Krueger, and Zander (13) determined the structure of the planar PAHs hexabenzo[*bc,de,gh,jk,mn,pq*]coronene and benzo[1,2,3-*ab*:4,5,6-*a′b′*]dicoronene (*compounds 46 and 53*). They obtained suitable crystals by using boiling pyrene as the solvent. To prevent cocrystallization of the LPAH and pyrene, mesitylene (1,3,5-trimethylbenzene) was added as a cosolvent. Figure 101 shows the two crystal array patterns found. This offset pattern of the intersecting structures is common for planar PAHs.

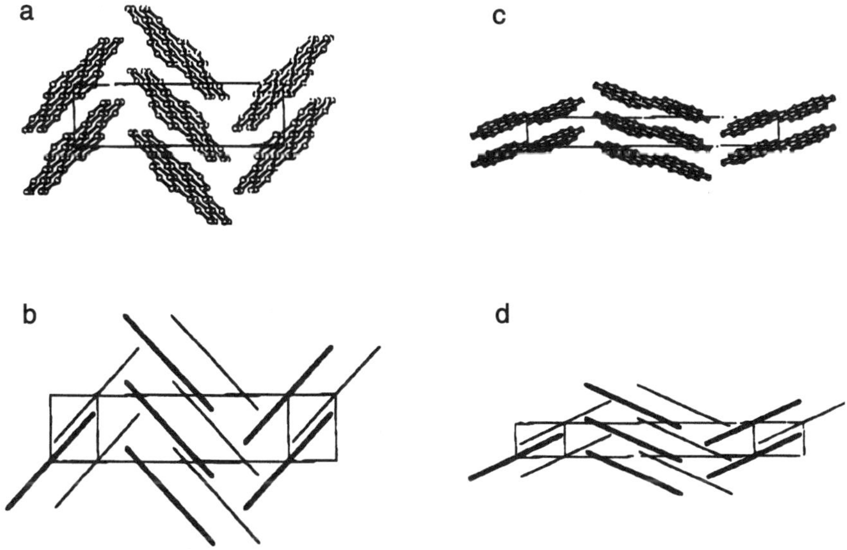

Figure 101. R. Goddard, M. W. Haenel, W. C. Herndon, C. Krueger, and M. Zander, *J. Am. Chem. Soc.* **117** 35 (1995), reproduced with permission from the American Chemical Society.

Barnett, Ho, Baldridge, and Pascal (14) studied the very nonplanar hexabenzotriphenylene (dibenzo[*f*,*j*]phenanthro[9,10-*r*]picene; *compound 97*). They were able to grow crystals from slowly cooling a saturated, hot nitrobenzene solution.

10.2 ANNELLATION THEORY

Clar was the predominant synthetic organic chemist for PAHs for over four decades (15). In the course of this work, he and his numerous collaborators developed many syntheses for new PAHs. They also copied the syntheses of other research groups to obtain PAHs that he did not have in his collection. From the collection of absorbance spectra from this vast number (several hundred) of compounds, he looked at comparisons of the structural features of the PAHs and the absorbance spectra. Through this comparison, he saw patterns between the two. These trends led to the development of a systematic interpretation of PAH absorbance spectra. The theory depended on trends seen in the absorbance spectra, in the arrangement of the aromatic rings in the structures, and in the resonance structures for the PAHs. Clar called this annellation theory. The patterns in UV absorbance band wavelengths and structure were described in many papers and summarized in Clar's book *The Aromatic Sextet* (16).

The first step in annellation theory is looking at the LPAH structure and drawing its Kekule structures. Fully-aromatic rings were defined as those in which there are three π bonds (as in a benzene molecule). Clar called one of these rings an aromatic sextet. The concept of an aromatic sextet was first proposed by Armitt and Robinson (17) in 1925 to describe and designate the aromaticity shared through the six π electrons in benzene. The number of these sextets in the LPAH was found and counted. Annellated rings were those that cannot not be drawn as a "fully benzenoid" sextet. These rings are drawn with double bonds to show the positions of the π electron pair that are not placed into fully benzenoid rings. These double bonds extended the π cloud of the adjacent aromatic sextets through partial localization of the π bonds.

In drawing an LPAH, it is correct only to use a hexagon with an incised circle to designate a fully benzenoid sextet. The frequent practice of drawing all of the rings in this fashion is both incorrect and uninformative because it tells nothing about the distribution of the π electrons throughout the structure. It is also good to keep in mind that these Clar structures are useful in predicting UV absorbance bands, but in reality the π electrons are localized neither in double bonds nor fully benzenoid sextets. If all the Kekule structures of an LPAH are drawn, the π electron distribution will coincide with the Clar structure as the most probable configuration of the electrons, but the electrons are delocalized throughout the molecule. Thus the middle ring(s) in molecules such as perylene, coronene, or ovalene has some π aromatic character.

There are many striking examples of the trends predicted by annellation theory. Two 5-ring structures result when an additional ring is ortho-fused to pyrene (*compound 20*), benzo[*a*]pyrene and benzo[*e*]pyrene (*compounds 22* and *24*). The highest-wavelength major absorbance maximum for pyrene is around 335 nm (depending on the solvent); The one for benzo[*a*]pyrene is at 385 nm and that for benzo[*e*]pyrene is at 331 nm. What causes this divergence?

The Kekule structures for the three PAHs shows why. Pyrene can be drawn with two fully-benzenoid rings that have three π bonds each, and there are two other π bonds. Benzo[*a*]pyrene also has two fully benzenoid rings, this time with four other π bonds. Benzo[*e*]pyrene, on the other hand, can be drawn with three fully benzenoid rings and only one other π bond. The additional ring in benzo[*e*]pyrene results only in an additional fully-aromatic ring where one of the isolated π bonds in the pyrene structure had been. This does not effect the UV absorbance band wavelength. The additional ring in benzo[*a*]pyrene results in more isolated π bonds in the Clar structure, and shifts the wavelength 50 nm higher.

This pattern holds for multiple substitutions on an aromatic core. If another ring is added to benzo[*e*]pyrene on its isolated π bond, the resulting dibenzo[*fg,op*]tetracene (*compound 104*) has a UV maxima at 328 nm. Addition of a ring to any of the faces that are fully benzenoid rings gives LPAHs with much higher absorbances. As examples, dibenzo[*def,p*]chrysene (*compound 37*) has its band at 391 nm and naphtho[1,2,3,4-*def*]chrysene (*compound 158*) has its band at 396 nm.

As another variation on extension of the π electron structure of a PAH, the addition of a peri-fused ring that also increases the number of isolated π bonds will result in higher-wavelength absorbance bands. This is the case for going from benzo[*e*]pyrene to benzo[*ghi*]perylene, where the maximum is at 385 nm (an upward shift of 50 nm). The two structures both have three fully benzenoid rings, but the additional peri-fused ring gives benzo[*ghi*]perylene one more isolated π bond than benzo[*e*]pyrene. Conversely, if rings are added through peri-fusion to give the resulting structure a greater number of fully benzenoid rings, the UV absorbance maximum wavelength will go down. The addition of such a peri-fused ring to perylene, with a maximum wavelength of 435 nm, also results in benzo[*ghi*]perylene (a downward shift of 50 nm).

The benzologues of dibenzo[*cd,lm*]perylene provide a perfect example of the effect of additional rings onto a LPAH structure (18–20). Many of these peropyrene-type LPAHs, using Clar's designation for this 26-carbon structure, have been synthesized. Over the past 15 years, several new additions to this class of LPAHs have been made. This was mainly through the reaction known (coincidentally) as the Clar reaction. These give a more in-depth view of the variations in ring addition onto the aromatic core and the resulting effects on the UV spectrum. The structures and UV absorbance spectra are shown in Figures 20–22. Dibenzo[*cd,lm*]perylene has three fully-benzenoid rings and four isolated π bonds. Its highest-wavelength UV absorbance band is at 444 nm. Both isomers with one additional ring have been recently synthesized. Tribenzo[*a,cd,lm*]perylene has the additional ring where one of the isolated π bonds was in the Clar structure of peropyrene. Analogously to pyrene and benzo[*e*]pyrene, this LPAH has a very similar UV wavelength to the parent compound, being at 445 nm. The other isomer, benzo[*rst*]naphtho-[8,1,2-*cde*]pentaphene, analogous to benzo[*a*]pyrene, shifts the wavelength to 473 nm.

The other structures that add rings onto the isolated π bonds of dibenzo[*cd,lm*]perylene do not shift the UV maximum significantly. Pentabenzo[*a,cd,f,j,lm*]perylene (*compound 172*) and hexabenzo[*a,cd,f,j,lm,o*]perylene (*compound 47*), with three and four rings in such positions, have much lower wavelengths than benzo[*rst*]naphtho[8,1,2-*cde*]pentaphene, with only one additional ring in the positions that extend the π conjugation. Its maximum is at 473 nm in dichloromethane. Even for hexabenzo[*a,cd,f,j,lm,o*]perylene, with four rings added onto each of the four isolated π bonds of dibenzo[*cd,lm*]perylene, the shift is only a little less than 20 nm. Their spectra are all very similar

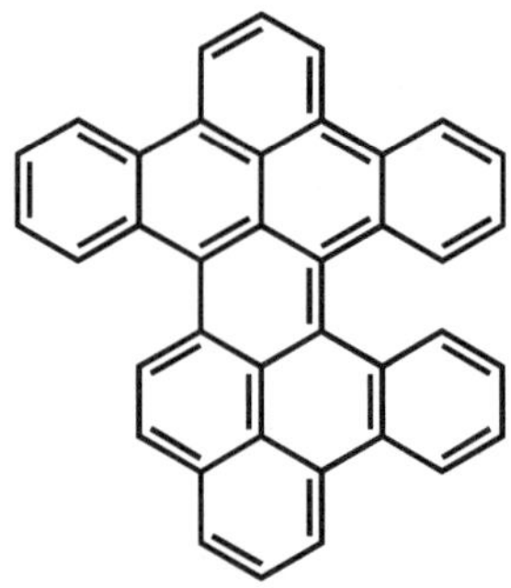

172

in pattern with clusters that have similar numbers of bands and similar wavelength locations.

The one LPAH, benzo[*rst*]naphtho[8,1,2-*cde*]pentaphene (*compound 139*) that has a ring on the long end of the core structure absorbs at much higher wavelength than its isomer tribenzo[*a,cd,lm*]perylene (with the added ring on the side). The others, which also have one or two rings added to the "ends" of the structure, all shift the wavelength upward. The other two structures with only one ring on the end (annellating the spectra) and one that is at one of the side faces have maxima at 472 and 482 nm. Those that have two rings substituted in this fashion all have absorbances much higher than any of the others. Benzo[*a,rst*]naphtho[8,1,2-*cde*]pentaphene and benzo[*rst*]phenanthro[8,1,2-*cde*]-pentaphene (*compounds 170* and *171*) have maxima that are shifted by around 50 and 70 nm higher than the core structure, to 496 and 508 nm in dichloromethane. The new LPAH synthesized by Fujisawa and co-workers, tribenzo-[*a,h,rst*]phenanthro[1,2,10-*cde*]pentaphene (*compound 67*), has a maxima even higher at 509 nm in chlorobenzene. The overall spectral patterns of these LPAHs are also similar to each other, having many more lower intensity bands of intermediate wavelengths.

Two other of the products of one of the Clar reactions give another dramatic example of the differences in UV absorbance wavelengths and ring arrangements. The two isomers that result from benzanthrone through formation of two bridging rings are dibenzo[*fg,ij*]phenanthro[2,1,10,9,8,7-*pqrstuv*]pentaphene and perylo[3,2,1,12-*pqrab*]perylene (*compounds 173* and *174*). The first structure has five fully benzenoid aromatic rings and two isolated π bonds, and the second has four fully aromatic rings and five isolated π bonds. Their UV maxima are at 417 and 482 nm, respectively, a difference of 65 nm for these two seemingly similar LPAHs. This large difference highlights the effect of having one fully-benzenoid ring versus three isolated π bonds.

An even more dramatic example of the addition of rings leading to a great decrease in the absorbance wavelengths is the series phenanthro[1,10,9,8]*opq-*

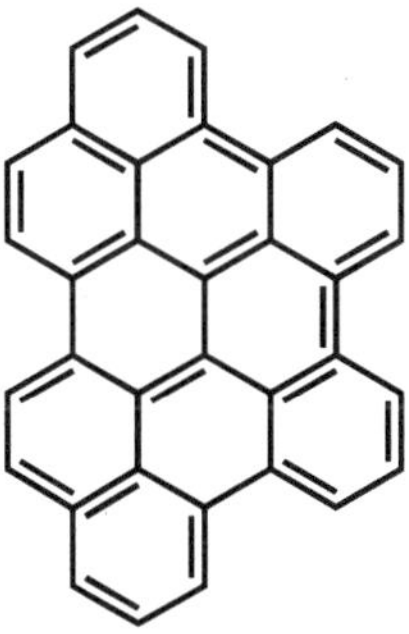

173

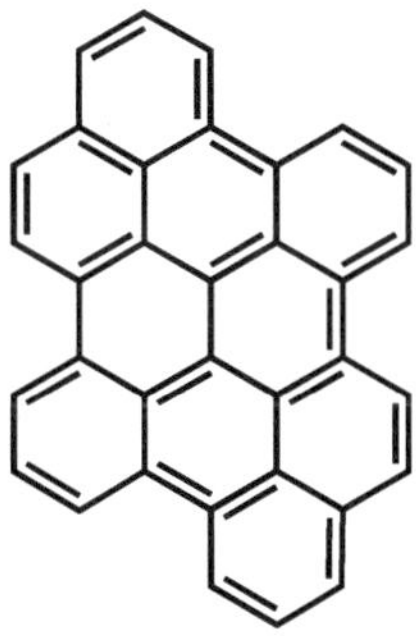

174

ra]perylene (bisanthene), dibenzo[*bc*,*ef*]coronene, and ovalene. The first has a band at 663 nm, the second at 536 nm, and the last at 456 nm. The Clar structures, however, reflect this change because these LPAHs have two, three, and four fully benzenoid sextets.

Gutman and Agranat (12) describe more fully the structure and π electron conjugation in the many benzologues of dibenzo[*cd.lm*]perylene. Their cyclic-conjugation model finds that the annellation of the rings is much like Clar's theory predicts, even though this later work is a more rigorous examination of delocalization and its effects on aromaticity.

The effect of the "fully benzenoid sextets" can be seen in the absorbance or fluorescence excitation spectra of certain very large PAHs. LPAHs of 42, 54 and 60 ring carbons were synthesized by Muellen and co-workers (21–25). Their highest wavelength maxima were found to be 444 nm, 455 nm, and 468 nm. These structures consist only of linked aromatic sextets, with no additional annellation due to isolated double bonds. These bands are much lower than those of most of the LPAHs that have fewer rings and much, much lower than LPAHs of the same carbon number. These smaller LPAHs, however, have isolated π bonds that extend their conjugation substantially, produce lower energy transitions to the excited state, and result in higher-wavelength absorbance bands.

An example of the predictive nature of annellation theory was the characterization of a new eight ring LPAH isolated from a catalytically cracked petroleum process deposit. Mass spectral analysis showed a molecular weight of 350, corresponding to $C_{28}H_{14}$. Two known isomers were isolated and identified from this material, but this third isomer did not correspond to any known LPAH. This specific structure is also hinted at by the production scheme found for the other LPAHs found. The seven ring precursor adds two carbons at its bay regions to produce the known isomers benzo[*a*]coronene and benzo[*pqr*]-naphtho[8,1,2-*bcd*]perylene.

The UV spectrum of this unidentified LPAH was very characteristic, having three prominent bands and an intensity pattern very different from most of the LPAHs. There are three prominent narrow bands at a somewhat low wavelength, with the middle one being much more intense than the others. From the fully benzenoid sextet structure of the possible isomers, phenanthro[5,4,3,2-*efghi*]perylene (*compound 48*) stood out as a strong candidate. This isomer has a structure similar to previously known 10-ring $C_{36}H_{18}$ LPAH dibenzo-[*fg,ij*]phenanthro[10,9,4,3,2-*pqrst*]pentaphene *(compound 175*). The latter structure is one with two ortho-fused rings on the former. Annellation theory predicted that their absorbance spectra should be very similar because the sextet structures are identical and the location of the two additional rings on the larger structure would not extend the π conjugation (and thus would not shift the spectral pattern to higher wavelength). Figure 102 shows the spectrum for the new LPAH and a comparison with one of the known compound. This structure also matched the proposed ring-addition scheme, which explained the occurrence of those three specific isomeric structures. This other LPAH results from closure at the third bay region of dibenzo[*b,pqr*]perylene.

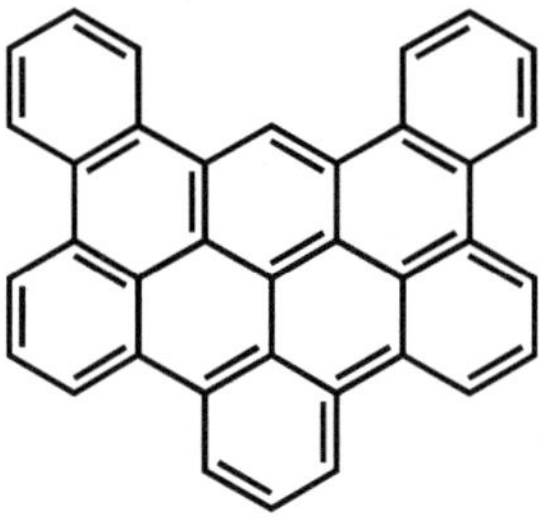

175

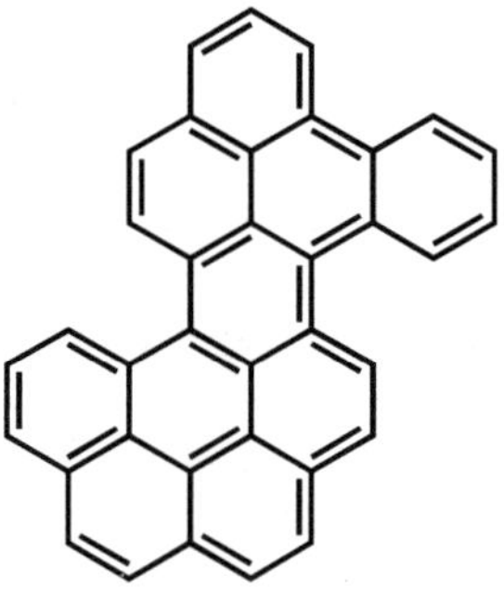

176

The main product of a Clar condensation of two ketones was the cross-product. This new LPAH, dibenzo[*j.lm*]phenanthro[5,4,3-*abcd*]perylene (*compound 176*), was compared to the two known and similar structures of one fewer and one more ring that resulted from each ketone reacting separately. Annellation theory predicts that the absorbance spectra should be very similar, but should display trends to higher wavelengths because of the additional rings. The observed spectrum is intermediate to those of the two other LPAHs. Figure 103 shows the three absorbance spectra, with the new one in the middle.

Tinnamens and Laarhoven (26) used annellation theory to select or eliminate certain isomers as possible products in a series of condensation reactions. Rings fused to fully benzenoid sextets or to isolated π bonds were the alternatives in some cases, and the measured UV absorbance bands indicated which was the case.

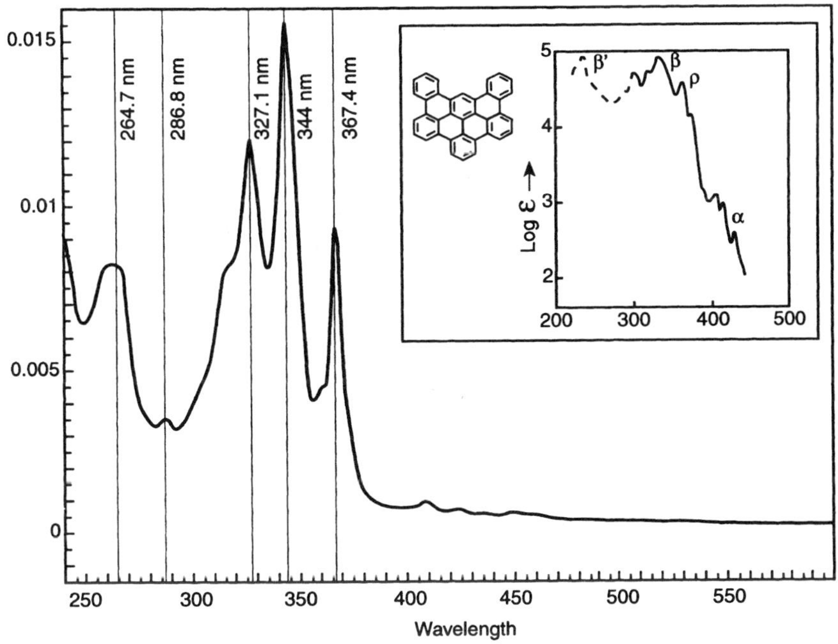

Figure 102. UV spectrum of phenanthro[5,4,3,2-*efghi*]perylene and (inset) spectrum of the dibenzo annellation analogue. From J. C. Fetzer and W. R. Biggs, *Polycyclic Arom. Cmpds.* **5** 197 (1994), reproduced with permission of Gordon and Breach Science Publishers.

Umemoto, Kawashima, Sakata, and Misumi synthesized pyrene, dibenzo[*cd,lm*]perylene, and benzo[*rst*]dinaphtho[8,1,2-*cde*:2′,1′,8′-*klm*]pentaphene (*compound 69*) from a series of metacyclophanes (27). This series of lengthening LPAHs gives a new annellation trend through addition of fused phenalene groups. The UV absorbance band difference for the first two previously known PAHs was found to be 101 nm (337 and 438 nm for the two β bands). The difference for the second pair was 99 nm. A value of 537 nm was found for the β band of the new LPAH, but the solvents used for the spectra unfortunately were not the same. Tetrahydrofuran was used for the first two and 1,2,4-trichlorobenzene for the third. The differences in solvents may shift the value by a few nanometers, but the red-shift for this series appears to be ~ 100 nm.

Zander (28) used perhydrocoronene (PHC) as the matrix to collect the UV absorbance and fluorescence spectra of the 16-ring dinaphtho[2,1,8,7-*defg*:2′,1′,8′,7′-*mnop*]pentapheno[3,4,5,6,7,8,9,10-*uvwxyza*$_1$*b*$_1$*c*$_1$*d*$_1$]heptaphene and 16-ring benzo[13,14] pentapheno[3,4,5,6,7,8,9,10-*uvwxyza*$_1$*b*$_1$*c*$_1$*d*$_1$]di-

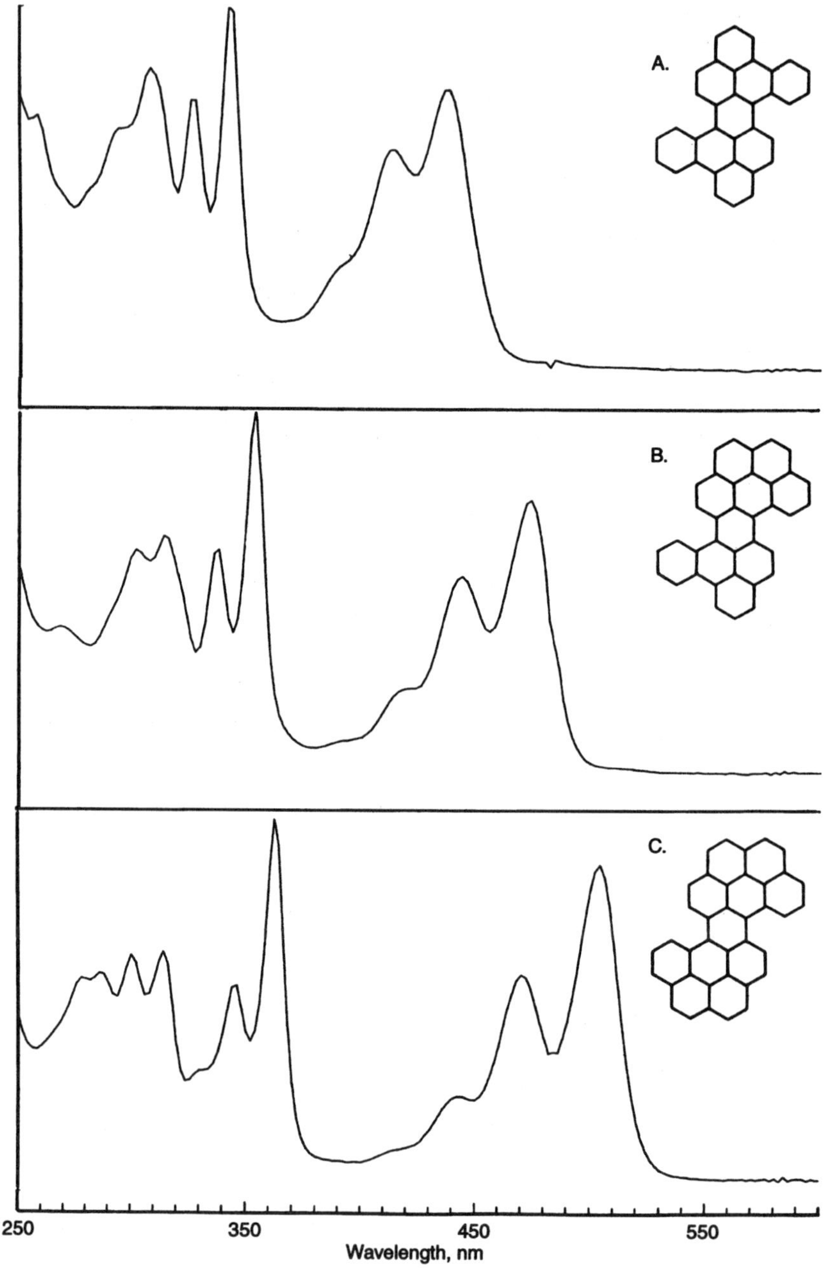

Figure 103. UV spectra of the series of LPAH series tetrabenzo[*a*,*cd*,*j*,*lm*]perylene, dibenzo[*j*,*lm*]-phenanthro[5,4,3-*abcd*]perylene, and dibenzo[*jk*,*uv*]dinaphtho[2,1,8,7-*defg*:2′,1′,8′,7′-*opqr*]pentace.

naphtho[2,1,8,7-*defg*:2′,1′,8′,7′-*mnop*]heptaphene (*compounds 31* and *32*). These new LPAHs were synthesized by maleic anhydride addition to benzo[1,2,3-*bc*:4,5,6-*b*′*c*′]coronene. With each additional ring the solubility, even in hot TCB, grows small enough to make spectroscopy difficult. The PHC matrix avoids this, and Zander obtained spectra for all three compounds. These compared well to the predicted band wavelengths for a series of similar structures with the same number of fully benzenoid sextets, but having one more than the 15-ring starting material. The highest-wavelength band for the two new LPAHs is at 455 and 458 nm, respectively. This is a ~50-nm blue-shift from the smaller starting material.

Broene and Dieterich (29) compared the predicted UV absorbance spectrum to the one found for the 40-carbon, 13-ring phenanthro[3,4,5,6-*bcdef*]ovalene (which they called by the trivial name of circumanthracene; *compound 177*). Figures 104 and 105 show the authentic spectrum and one predicted from the series pyrene, coronene, ovalene, and circumanthracene.

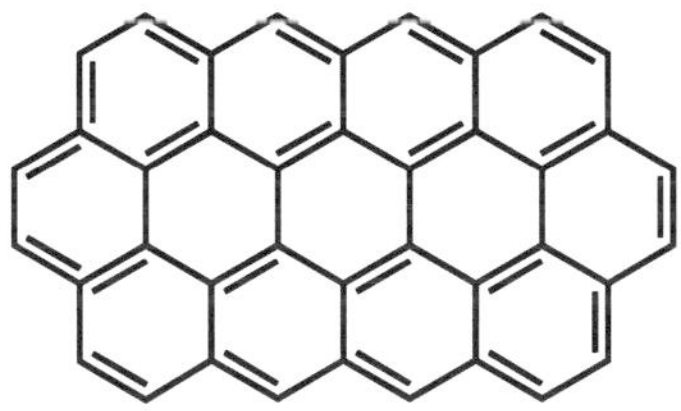

177

10.3 POLARIZED LIGHT SPECTROSCOPY

The electronic and physical mechanisms that yield the complex patterns in the UV absorbance and fluorescence spectra of the LPAHs can be used to determine their structure. These transitions arise from certain structural arrangements of the π electrons. Michl and Thulstrup (30, 31) have thoroughly discussed the principles and some applications of spectroscopy using polarized light in their two books.

The α, β, β′, and ρ bands are oriented along certain of the LPAH's axes. If the LPAH is planar, these can be simply predicted by a point-group analysis of the molecule's symmetry elements. Nonplanar LPAHs are a little more complex because the added dimensionality creates transitions that are not purely oriented in only the *x* and *y* axes. Polarized light can be used to differentiate these transitions and assign specific spectral bands to the three-dimensional orientation of structural features. This is known as linear dichroism (LC) spectroscopy.

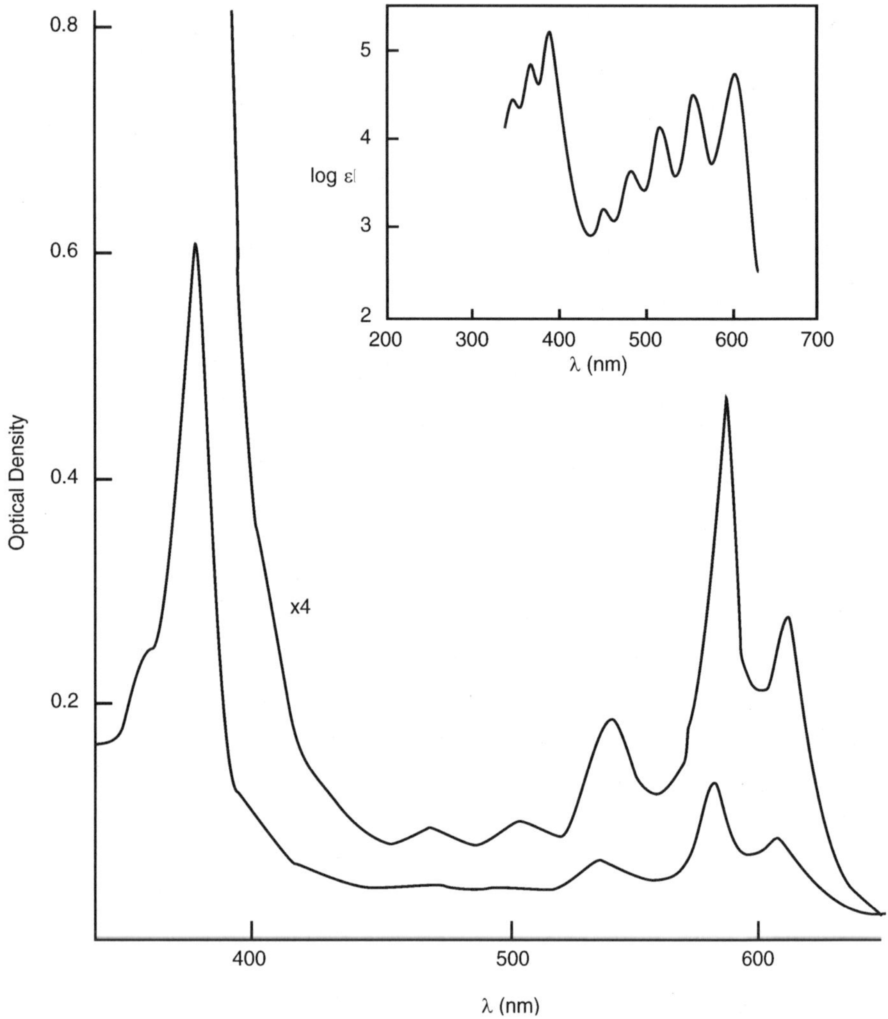

Figure 104. The absorbance spectrum of circumanthracene with the predicted one. From R. D. Broeme and F. Dieterich, *Tetrahedron Lett.* **32** 5230 (1991), reproduced with permission from Elsevier Science Publishers BV.

To perform polarized light spectrometry, both the orientation of incident light (and in the case of fluorescence, emitted light as well) and the sample molecules must be in specific orientations relative to each other. This is accomplished by trapping the molecule in a medium that has a specifically oriented structure. This

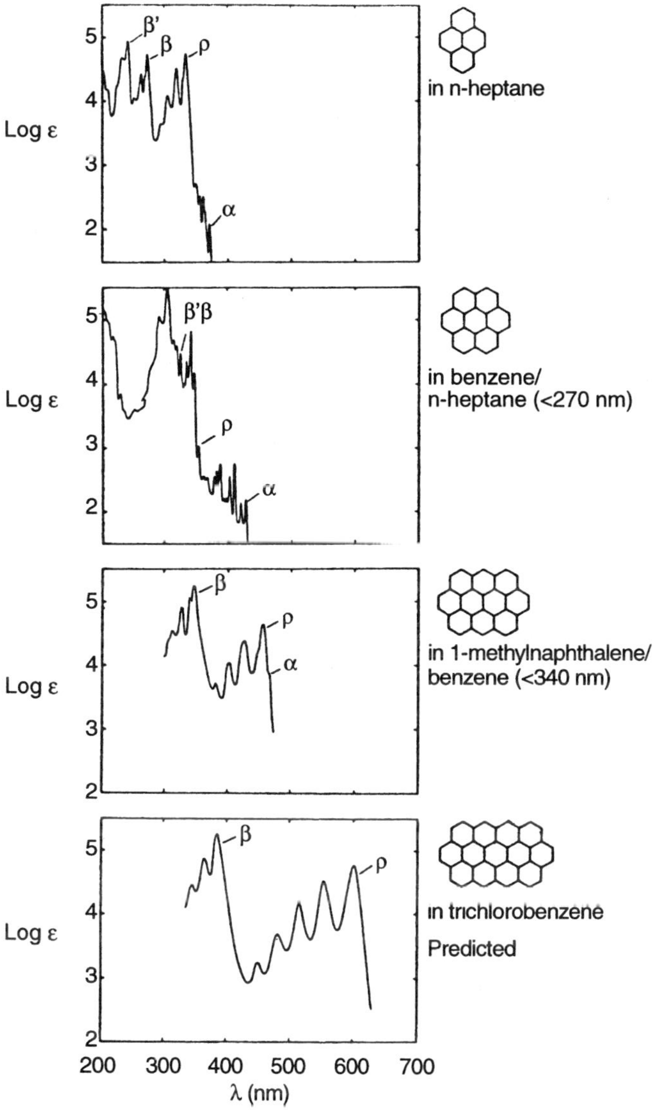

Figure 105. The original prediction of the absorbance spectrum of circumanthracene. From E. Clar, J. M. Robertson, M. Schloegl, and W. Schmidt, *J. Am. Chem. Soc.* **103** 1325 (1981), reproduced with permission from the American Chemical Society.

is commonly done using stretched films of polyethylene that have been impregnated with the LPAH molecules. A LPAH solution is used to impregnate the unstretched film. The stretching of the film causes the long hydrocarbon chains of the polymer to straighten and become parallel. The molecule to be studied then fits in between these highly oriented polyethylene chains.

The stretched film is then placed in the polarized light beam. It is moved so that the incident angle changes. The changes in the intensities of the spectral bands tell which bands are caused by transitions parallel and perpendicular to the light beam. This is especially simple for the two-dimensional structures of planar LPAHs. For nonplanar LPAHs, the data treatment is more complicated because the many portions of the LPAH ring structure have contributions in each of the three orientation axes.

The structure of several "peropyrene-type" LPAHs have been studied by polarized light spectroscopy (32–35). This approach showed that the structures were nonplanar because the absorbance bands were not oriented totally in the *x*- and *y*-axes. The planar "parent" structure dibenzo[*cd*,*lm*]perylene (peropyrene) was compared to three benzologues. Figure 106 shows the spectra for benzo[*j*,*lm*]phenanthro[5,4,3-*abcd*]perylene (*compound 176*).

Problems in collecting LC spectra of these nonplanar LPAHs were seen. The orientation in the stretched polyethylene may not be as regular as that for the planar PAHs. This leads to the LPAH molecules not fitting well between the polymer and gives spectra in which the orientations are harder to determine. The inherent low solubility of the LPAHs makes the impregnation more difficult, often leading to low-intensity spectra or ones in which trace-level impurities in the sample are more readily seen than the target LPAH.

10.4 NMR TECHNIQUES

The NMR spectra of LPAHs are usually very complex because all of the carbon or hydrogen atoms resonate in only a very narrow spectral region. Each nucleus will generally have a slightly different electronic environment than any of the others, and thus a slightly different shift. Symmetry can simplify the spectrum, so that the 26-carbon dibenzo[*cd*,*lm*]perylene has only eight peaks in the carbon-13 spectrum. Five peaks result from four carbons each, and each of the other three peaks results from two of the carbons. The signal for each nucleus will also be split by any adjacent carbon and hydrogen atoms, adding both to the complexity of the spectrum and to its potential utility in determining the carbon skeleton and peripheral hydrogens. This means that there will be individual contributions by each nucleus to the total spectrum.

This potential complexity for a single LPAH defines why NMR is not routinely used as an analytical tool in characterizing LPAH materials. In

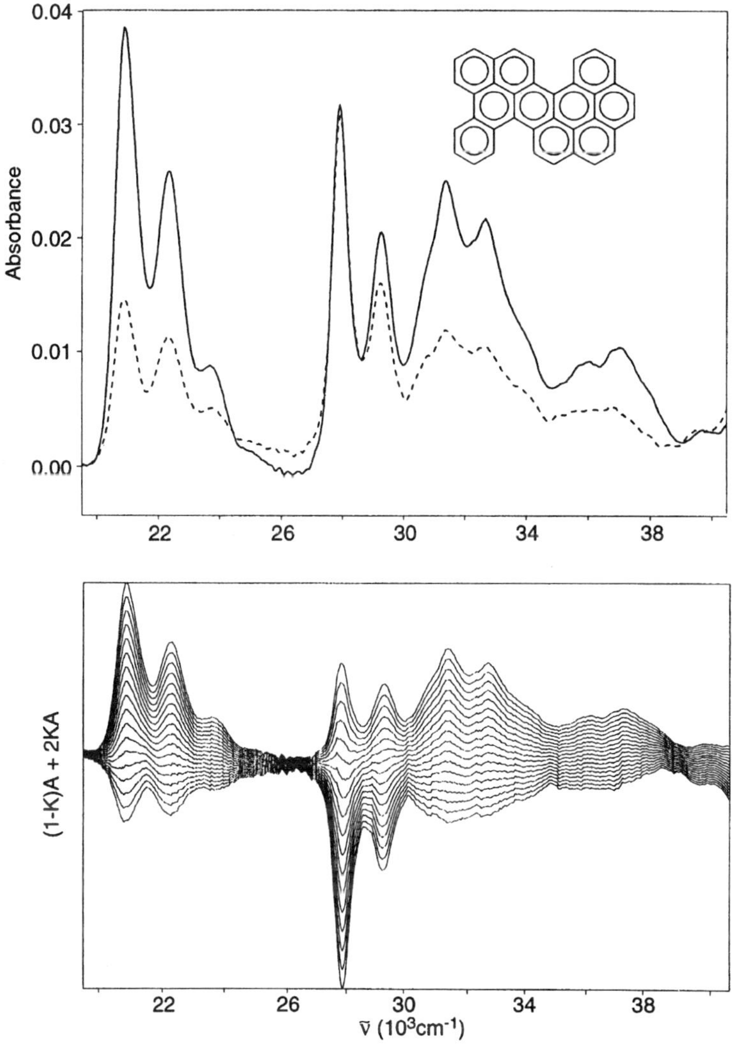

Figure 106. Polarized light spectrum of dibenzo[*j,lm*]phenanthro[5,4,3-*abcd*]perylene. The lower spectrum reflects the different intensities as the oriented sample is rotated in the polarized light beam. From J. Marczyk, J. Waluk, and J. Fetzer, *Acta Physica Polonica*, **88** 295–304 (1995).

traditional, low-field NMR signals overlap each other, as well as those of the PAHs and the other aromatic compounds, such as benzene, naphthalene, and their derivatives. In mixtures, little specific information about LPAHs can be gleaned from the complex spectra. Only unique signals, such as those from the interior carbon atoms in large peri-fused LPAHs like coronene or ovalene, can be easily seen. These, however, only can be measured as a total amount of "core" carbons, because they also overlap with each other.

The advent of very high-field (typically 300–600 MHz) instruments are capable of yielding spectra that spread out the complex signals of an individual LPAH. The resulting high resolution in these spectra have patterns that reflect the specific structure of the LPAH. Signals from each individual hydrogen or carbon atom are seen, including the splitting due to adjacent nuclei. Also, new multi-nuclear techniques yield 3-dimensional spectra, which show the coupling patterns of adjacent atoms. This allows a piecing together of the carbon skeleton of the LPAH, including those with hydrogen substituents. Morris has written an extensive review (36) of several of the multi-dimensional techniques amenable for structure determination. He describes 10 techniques that include several multi-pulse or spin coupling and decoupling approaches. Plummer (37) describes the use of some of these multi-dimensional techniques in the structural determination of several nonsymmetrical PAHs of up to 22 carbons. The examples in this work show the strength of NMR in building a carbon-hydrogen skeleton. Because the examples are not symmetrical, the LPAHs that are symmetrical could have many more carbon atoms and still give useful spectra.

Grant and co-workers (38–46) have examined a wide variety of crystalline pure PAHs, including naphthalene, pyrene, and perylene. They recently extended this series of studies to the LPAH coronene. The results of these experiments were used to determine the structural arrangement. They used high-field, multi-dimensional techniques to determine the electronic anisotropies of the individual nuclei in the crystal unit cell. The electronic densities will deviate from uniformity because of the bonding to the atom and the π electron density variations due to the delocalization. These results, in turn, are used in models to calculate the chemical shift (shielding) tensors of the structure. The extension of these techniques to a broader database of more LPAHs may yield information about the structural similarities to carbonaceous materials like soots. The technique has also been used at low temperature to study the structure of aromatic cations.

Zander (1) highlights the use of NMR techniques in aiding in the identification of PAH structures in fractions of coal tar pitch. The presence of solo, duo, trio, and quartet arrangements of the peripheral hydrogen atoms on rings is pointed out to be quantitative in NMR, where it is not in IR techniques.

In addition to the large amount of UV absorbance data in his book *The Aromatic Sextet*, Clar also has NMR data for many PAHs. The chemical shifts and hyperfine splitting found by Clar and his co-workers for many compounds is given there to augment Clar's points in discussing structural relationships among the PAHs (and particularly among the LPAHs).

The low solubility of LPAHs in many solvents can be a problem. The best solvents are deutero-substituted versions of the aromatic solvents like *o*-chlorobenzene. The carbon interferences from the ring carbons in these solvents prevent the collection of carbon-13 spectra, however. Fortunately, dichloromethane, chloroform, and 1,1,2,2-tetrachloroethane are all both reasonably good solvents and readily available in deuterated form. These solvents are not generally good for the larger high-condensed structures and require large sample sizes (1 cm or other larger wide-bore tubes).

Zander, Haase, and Dreeskamp (47) determined the structure of one of the products of perylene thermolysis, dibenzo[*jk*,*pq*]dinaphtho[8,1,2-*cde*:2′,1′,8′-*tua*]pentacene (*compound 80*), using "high-field" proton and carbon-13 NMR. This LPAH is asymmetrical, so none of the atoms are equivalent. This team used a 250-MHz instrument, so there was not complete spreading of all the nuclear lines into isolated peaks. Figure 107 shows the proton NMR and those of perylene and benzo[*ghi*]perylene used to assign some of the substructural signals.

Bax, Ferretti, Nashed, and Jerina (48) used homonuclear correlated spectroscopy (COSY) to differentiate the PAH isomers benzo[*c*]chrysene and benzo[*g*]chrysene. Although these are not LPAHs, their asymmetry yields complex NMR spectra. This work shows that this approach could readily be used for larger structures.

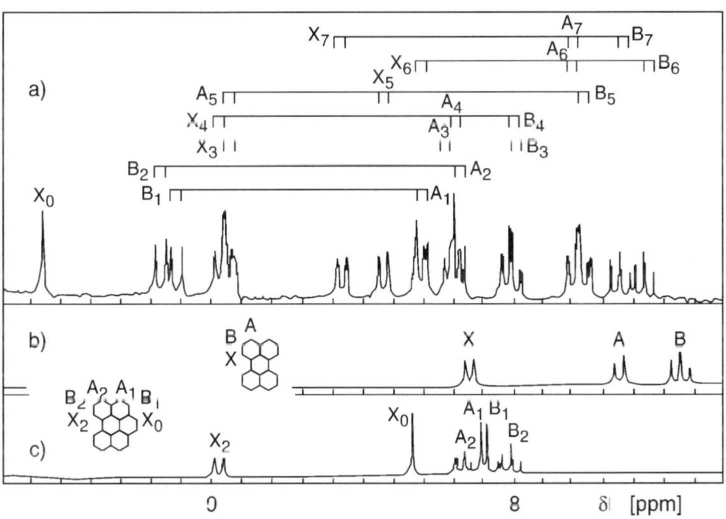

Figure 107. Proton NMR spectrum used to determine the structure of the unknown perylene dimer From M. Zander, J. Haase, and H. Dreeskamp, *Erdol Kohle Erdgas Petroch.* **35** 67 (1982).

Ueda, Iwashima, Aoki, and Takekawa (49) collected proton and carbon-13 spectra for six similarly structured LPAHs. They correlated the variations in structure to shifts in the peak locations. They used d_2-1,1,2,2-tetrachloroethane (TCE) as the solvent. Besides the planar dibenzo[*cd,lm*]perylene, two other of the compounds, benzo[*rst*]anthra[10,1,2-*cde*]pentaphene and benzo[*rst*]phenanthro[10,1,2-*cde*]pentaphene (*compounds 170* and *171*), are also planar and have low solubilities in many solvents. In TCE, this group reported the concentrations of their samples at 1 mg in 0.5 mL. They also observed that the solubility of the three nonplanar compounds was higher and spectra with high signal-to-noise ratio were easier to obtain. Figures 108 and 109 show the proton and carbon-13 NMR for some of these peropyrene-type LPAHs. The proton spectra show the typical patterns of multiple splittings due to other protons on adjacent carbons. The carbon-13 spectra are also typical, showing that these LPAHs have little symmetry and often one line for each carbon atom.

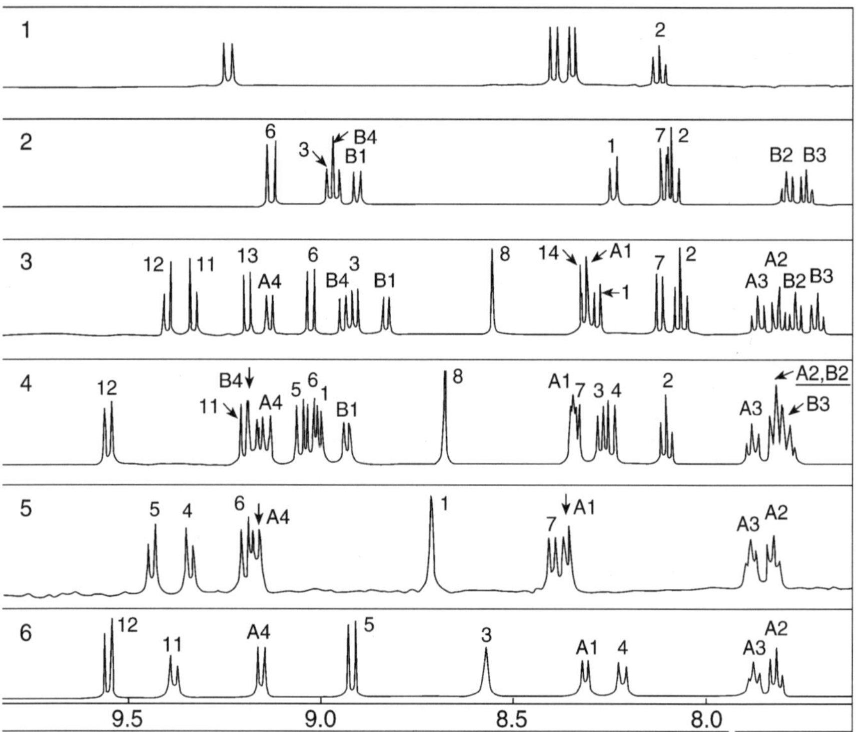

Figure 108. The spectra are, from top to bottom, of dibenzo[*cd,lm*]perylene, tetrabenzo[*a,cd,j,lm*]-perylene (*compound 121*), and its four isomers, *compounds 170*, *171*, *34*, and *33*. From T. Ueda, S. Iwashima, J. Aoki, and M. Takekawa, *Magn. Reson. Chem.* **33** 98 (1995), reproduced with permission from Wiley-Interscience.

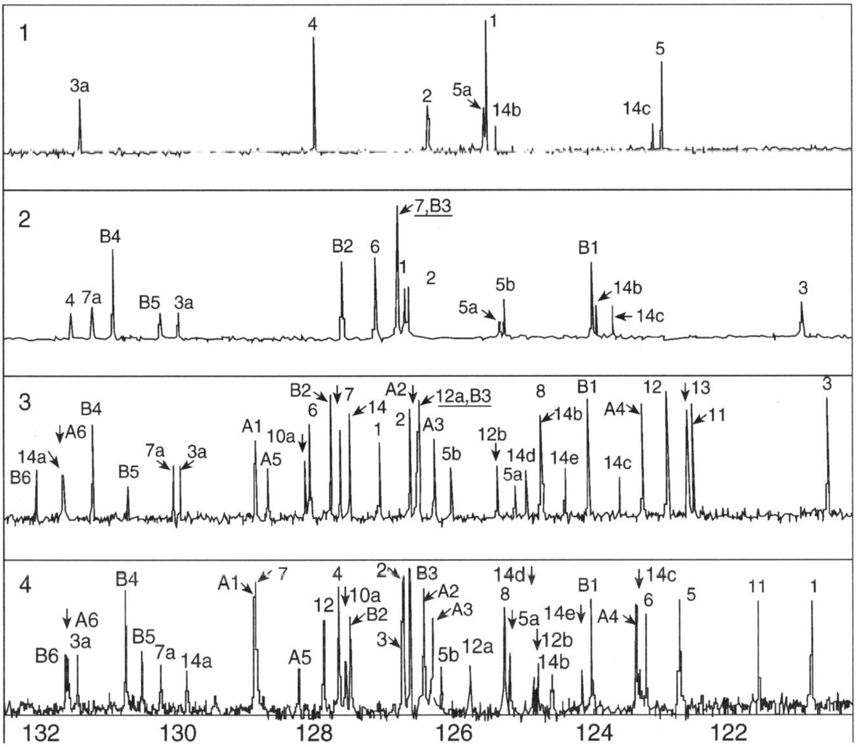

Figure 109. Carbon-13 NMR of dibenzo[*cd,lm*]perylene and the three isomers tetrabenzo-[*a,cd,j,lm*]perylene (*compound 121*) and *compounds 170* and *171*. From T. Ueda, S. Iwashima, J. Aoki, and M. Takekawa, *Magn. Reson. Chem.* **33** 98 (1995), reproduced with permission from Wiley-Interscience.

Proton NMR has shown that the nine-ring tetrabenzo[*a,cd,j,lm*]perylene changes conformation in different solvents (50). The steric hindrance of the structure causes it to be very nonplanar. The degree of this nonplanarity is apparently perturbed by solvation strength. The spectra in methanol and dichloromethane mixtures show differences in the fine splitting due to the spatially near hydrogen atoms of the cove regions.

Liu (51, 52) has published mathematical calculations that can be used to predict the proton and carbon-13 chemical shifts of LPAHs. The algorithms assume an additivity of contributions due to the s and π electrons and the effects of Van derWaal interactions. Cyranski and Krygowski (53) developed a program based on PAHs, including coronene. They assessed the geometric and energetic contributions to the electron environments around the nuclei.

10.5 PHOTOELECTRON SPECTROSCOPY

In photoelectron (PE) spectroscopy, the ionization potentials of the LPAHs are measured through the use of high-energy light passing through a volatilized sample. The onset of a LPAH ionization potential is often a very sharp band in the PE spectrum. This is mainly due to the fact that samples are examined under vacuum conditions, so the LPAHs are in a very diffuse gas phase. In this state, the broadening of bands due to solvation and collisions with other molecules does not occur. The ionization potentials for any LPAH structure can be calculated with a high degree of accuracy by many available modeling programs. Eilfeld and Schmidt (54) published an early collection of these values for a large number of LPAHs. Hites and Simonsick (55) published many values in their book that contained the calculated properties of many LPAHs. These calculated values should match those found to within 0.01 electron volts for planar LPAHs and 0.02 for the nonplanar LPAHs.

Clar and Schmidt published a large number of papers describing the PE spectra of many structural families of LPAHs (56–63). The spectral features due to each type of structure were assessed, collected, and then used to establish the utility of PE spectra for the reversed process of correlation of PE bands to PAH structural arrangements of the aromatic rings. Figure 110 shows their correlation between predicted and measured IP values.

The bands in the UV absorbance and ionization potentials measured by PE spectra of the alternant LPAHs can be directly correlated. For the nonalternant LPAHs, there is no set connection between absorbance wavelengths (or it may be better to think of these as being reflections of the excitation energies) and the ionization potentials. Thus PE spectra can easily determine whether an unknown structure is alternant or nonalternant.

Each PE spectrum of a LPAH is the summation of the spectral contributions due to each of the various substructures found in the whole molecule. This is known as substructurality. Bands resulting from the resonance structures for small, incised PAHs are readily seen in the spectra of larger PAHs. Thus ovalene shows spectral bands due to naphthalene, anthracene, phenanthrene, pyrene, perylene, coronene, and other substructures. This makes the spectra of the ortho-fused LPAHs much simpler than those of the peri-fused LPAHs.

The purity of samples is not a major concern because the heating of samples allows for the slow sublimation of impurities at different times than the LPAH being studied. Often the impurities are not seen at all because they are less volatile than the sample. The spectra collected at different temperatures should be similar for a pure sample.

The structures of several LPAHs that had been assumed from supposed reaction mechanisms determined to be in error by the PE spectra of the products. The compound shown in Clar's volumes to be the alternant six-ring LPAH

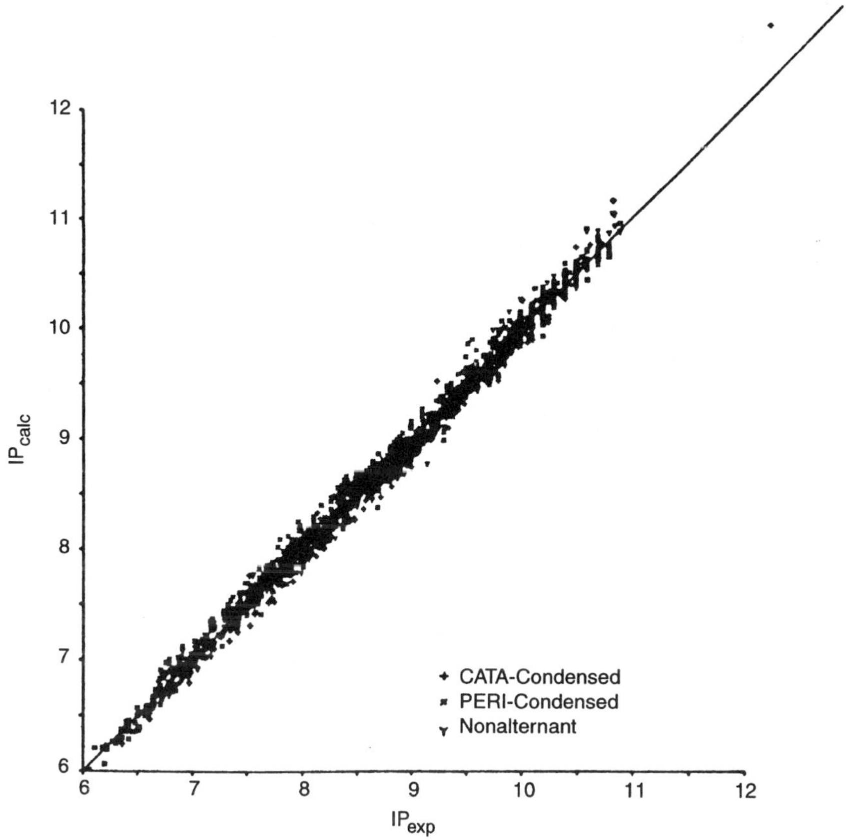

Figure 110. Comparison plot of the calculated and experimental first ionization potential used to predict LPAH photoelectron spectra. From E. Clar, J. M. Robertson, M. Schloegl, and W. Schmidt, *J. Am. Chem. Soc.* **103** 1321 (1981), reproduced with permission from the American Chemical Society.

dibenzo[*def,p*]chrysene was later found by PE spectroscopy to in fact be the nonalternant isomeric dibenzo[*b,k*]fluoranthene. The synthesis and isolation of the 13-ring "circumanthacene" had been reported by Clar, Kelly, Robertson, and Rossmann (63), but in later work by Clar, Robertson, Schloegl, and Schmidt (64) the subsequent PE spectra of this material showed it to be a nonalternant isomer. Figure 111 shows the series of PE spectra for pyrene, coronene, ovalene, and circumanthracene. The last did not resemble the sample from the earlier work and the authors concluded that the earlier reaction must have involved a carbon skeletal rearrangement during the Friedel-Crafts condensation. The authentic "circumanthracene" was not synthesized until Diederich and coworkers prepared it many years later in an involved multi-step synthesis.

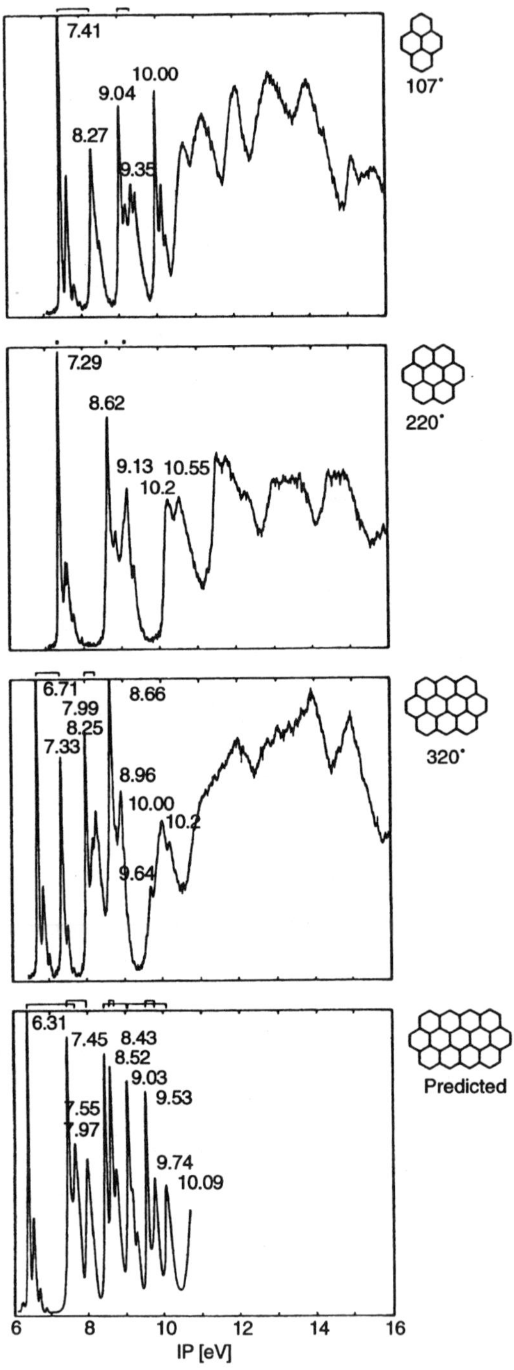

7.41
10.00
9.04
8.27
9.35
107°
7.29
8.62
9.13 10.55
10.2
220°
6.71
7.99
8.25
7.33
8.66
8.96
10.00
10.2
9.64
320°
6.31
7.45 8.43
8.52
9.03
9.53
7.55
7.97
9.74
10.09
Predicted
6 8 10 12 14 16
IP [eV]

Lempka, Obenland, and Schmidt (65) used the PE spectrum to determine the structure of Boente's hydrocarbon to be benzo[1,2,3-*bc*:4,5,6-*b*′*c*′]dicoronene and not the nonalternant bridged isomer. This material results from the aluminum(III) chloride condensation of coronene.

Sato, Inokuchi, Seki, Aoki, and Iwashima (66) studied the PE spectra of six 34-carbon LPAHs. They found that the nonplanar isomers had higher ionization potentials that the planar isomers. They ascribed this to the overcrowding due to cove and fjord regions. They also examined the PE spectra of solid samples of these six compounds and correlated the spectral energies to the conductivities of these LPAHs that show semiconductor behavior.

10.6 OTHER TECHNIQUES

In at least one example, that of the nine-ring LPAH tetrabenzo[*a,cd,f,lm*]perylene, Shpol'skii spectroscopy was used to help elucidate the structure and the molecule's photophysical properties (67). This nine-ring LPAH is nonplanar because of steric effects of spatially adjacent rings. Molecular modeling had shown that the reduction in strain could be accomplished by the central bridging ring either bending or twisting. These changes in conformation result in two conformers with almost identical energies, Both were seen when a sample dissolved in *n*-hexane was quickly frozen in liquid nitrogen to form a solid alkane matrix. The emission spectra of the two components were identical, indicating that the excited states were the same and the ground states were not. In contrast, when the sample was slowly cooled in the same alkane matrix solvent, only the more stable conformer was found.

REFERENCES

1. M. Zander, *Polycyclic Arom. Cmpds.* **7** 209–221 (1995).
2. M. F. C. Ladd and R. A. Palmer, *Structure Determination by X-ray Crystallography*, 3rd Ed., Plenum, New York, 1993.
3. J. P. Glusker and K. N. Trueblood, *Crystal Structure Analysis—A Primer*, 2nd Ed., Oxford University Press, Oxford, 1985.
4. G. H. Stout and L. H. Jensen, *X-Ray Structure Determination—A Practical Guide*, 2nd Ed., John Wiley and Sons, New York, 1989.

Figure 111. The measured photoelectron spectra of pyrene, coronene, and ovalene with the predicted one for circumanthracene. From E. Clar, J. M. Robertson, M. Schloegl, and W. Schmidt, *J. Am. Chem. Soc.* **103** 1324 (1981), reproduced with permission from the American Chemical Society.

5. J. D. Dunitz, *X-Ray Analysis and the Structure of Organic Molecules*, 2nd Ed., Cornell University Press, Ithaca, NY, 1995.
6. Y. Kohno, M. Konno, Y. Saito, H. Inokuchi, *Acta Crystallogr. B* **11** 2076 (1975).
7. I. Oonishi, S. Fujisawa, J. Aoki, and T. Danno, *Bull. Chem. Soc. Jpn.* **51** 2256–2260 (1978).
8. S. Fujisawa, I. Oonishi, J. Aoki, Y. Ohashi, and Y. Samada, *Bull. Chem. Soc. Jpn.* **55** 3424–3426 (1982).
9. I. Oonishi, S. Fujisawa, J. Aoki, and Y. Ohashi, *Bull. Chem. Soc. Jpn.* **59** 2233–2236 (1986).
10. S. Fujisawa, I. Oonishi, J. Aoki, and Y. Ohashi, *Acta Crystallogr. C* **42** 1390–1392 (1986).
11. I. Oonishi, S. Ohshima, S. Fujisawa, J. Aoki, Y. Ohashi, and T. M. Krygowski, *J. Mol. Struct.* **265** 283–292 (1992).
12. I. Gutman and I. Agranat, *Polycyclic Arom. Cmpds.* **2** 63–73 (1991).
13. R. Goddard, M. W. Haenel, W. C. Herndon, C. Krueger, and M. Zander, *J. Am. Chem. Soc.* **117** 30–41 (1995).
14. L. Barnett, D. M. Ho, K. K. Baldridge, and R. A. Pascal, *J. Am. Chem. Soc.* **121** 727–733 (1999).
15. E. Clar, *Polycyclic Hydrocarbons*, John Wiley and Sons, New York, 1964, Vols. 1 and 2.
16. E. Clar, *The Aromatic Sextet*, Wiley-Interscience, New York, 1972.
17. T. W. Armitt and R. Robinson, *J. Chem. Soc.* 1604 (1925).
18. J. C. Fetzer and W. R. Biggs, *J. Chromatogr.* **346** 81–92 (1985).
19. J. C. Fetzer and W. R. Biggs, *Org. Prep. Proc. Int.* **20** 223–230 (1988).
20. J. C. Fetzer, in *Polynuclear Aromatic Hydrocarbons*, L. Ebert, Ed., Advances in Chemistry 217, American Chemical Society, Washington, DC, 1988, p. 309–331.
21. M. Mueller, H. Mauermann-Duell, M. Wagner, V. Endelmann, and K. Muellen, *Chem. Eur. J.* **1** 1583-1586 (1995).
22. F. Morgenroth, C. Keubel, and K. Muellen, *J. Mater. Chem.* **1** 1207-1211 (1997).
23. V. S. Iyer, M. Wehmeier, J. D. Brand, M. A. Keegstra, and K. Muellen, *Angew. Chem. Int. Ed. Engl.* **36** 1604–1607 (1997).
24. M. Mueller, V. S. Iyer, C. Kuebel, V. Enkelmann, and K. Muellen, *Angew. Chem. Int. Ed. Engl.* **36** 1607–1610 (1997).
25. M. Mueller, C. Kuebel, F. Morgenrith, V. S. Iyer, and K. Muellen, *Carbon* **36** 827–831 (1998).
26. A. H. A. Tinnemans and W. H. Laarhoven, *J. Am. Chem. Soc.* **96** 4617–4622 (1974).
27. T. Umemoto, T. Kawashima, Y. Sakata, and S. Misumi, *Tetrahedron Lett.* **12** 1005–1006 (1975).
28. M. Zander, in *Polycyclic Aromatic Compounds: Synthesis, Properties, Analytical Measurements, Occurrence, and Biological Effects*, P. Garrigues and M. Lamotte, Eds., Gordon and Breach Science, Yverdon, Switzerland, 1993, p. 143–149.

29. R. D. Broene and F. Dieterich, *Tetrahedron Lett.* **39** 5227 (1991).
30. J. Michl and E. W. Thulstrup, *Spectroscopy with Polarized Light.* VCH, New York, 1986.
31. J. Michl and E. W. Thulstrup, *Elementary Polarization Spectroscopy*, VCH, New York, 1989.
32. J. Waluk, J. C. Fetzer, S. J. Hamrock, and J. Michl, *J. Phys. Chem.* **95** 8660–8663 (1991).
33. J. Marczyk, J. Waluk, and J. Fetzer, *Acta Physica Polonica* **88** 295–304 (1995).
34. J. Marczyk, J. Waluk, and J. C. Fetzer, *Polycyclic Aromat. Compd.* **9** 283–290 (1996).
35. J. Marczyk, J. C. Fetzer, J. Szubiakowski, and J. Waluk, *J. Luminesc.* **72–74** 517–519 (1997).
36. G. A. Morris, Magnet, *Magn. Reson. Chem.* **24** 371–403 (1986).
37. B. F. Plummer, *Polycyclic Arom. Cmpds.* **4** 25–39 (1994).
38. C. M. Carter, D. W. Alderman, J. C. Facelli, and D. M. Grant, *J. Am. Chem. Soc.* **109** 2839–2844 (1987).
39. M. H. Sherwood, J. C. Facelli, D. W. Alderman, and D. M. Grant, *J. Am. Chem. Soc.* **113** 750–753 (1991).
40. A. Soderquist, C. D. Hughes, W. J. Horton, J. C. Facelli, and D. M. Grant, *J. Am. Chem. Soc.* **114** 2826–2832 (1992).
41. A. M. Orendt, N. K. Sethi, J. C. Facelli, W. J. Horton, R. J. Pugmire, and D. M. Grant, *J. Am. Chem. Soc.* **114** 2833–2836 (1992).
42. R. J. Iuliucci, J. C. Facelli, D. W. Alderman, and D. M. Grant, *J. Am. Chem. Soc.* 117 2336–2343 (1995).
43. R. J. Iulliucci, C. G. Phung, J. C. Facelli, and D. M. Grant, *J. Am. Chem. Soc.* **118** 4880–4888 (1996).
44. R. J. Iulliucci, C. G. Phung, J. C. Facelli, and D. M. Grant, *J. Am. Chem. Soc.* **120** 9305–9311 (1998).
45. A. M. Orendt, J. C. Facelli, S. Bai, R. J. Pugmire, D. M. Grant, A. Rai, and L. T. Scott, *J. Phys. Chem.* **104** 149-155 (2000).
46. H. Strub, A. J. Beeler, D. M. Grant, J. Michl, P. W. Curtis, and K. W. Zinn, *J. Am. Chem. Soc.* **105** 3333–3334 (1983).
47. M. Zander, J. Haase, and H. Dreeskamp, *Erdol Kohle Erdgas Petroch.* **35** 65–69 (1982).
48. A. Bax, J. A. Ferretti, N. Nashed, and D. M. Jerina, *J. Org. Chem.* **50** 3029–3034 (1985).
49. T. Ueda, S. Iwashima, J. Aoki, and M. Takekawa, *Magn. Reson. Chem.* **33** 95–103 (1995).
50. K. Jinno, T. Ibuki, N. Tanaka, M. Okamoto, J. C. Fetzer, W. R. Biggs, P. W. Griffiths, and J. M. Olinger, *J. Chromatogr.* **461** 209–227 (1989).
51. C.-W. Liu, *Hua Hsueh Hsueh Pao* **39** 121–137 (1981), in Chinese.
52. C. Liu and H. Cao, *Huaxue Xuebao* **40** 673–687 (1982), in Chinese.

53. M. K. Cyranski and T. M. Krygowski, *Tetrahedron* **54** 14919–14924 (1998).
54. P. Eilfeld and W. Schmidt, *J. Electr. Spectrosc.* **24** 101–120 (1981).
55. R. A. Hites and W. J. Simonsick, *Calculated Molecular Properties of Polycyclic Aromatic Hydrocarbons*, Elsevier, Amsterdam, 1987.
56. W. Schmidt, *J. Chem. Phys.* **66** 828–845 (1977).
57. E. Clar and W. Schmidt, *Tetrahedron* **31** 2093–2097 (1975).
58. E. Clar and W. Schmidt, *Tetrahedron* **31** 2263–2271 (1975).
59. E. Clar and W. Schmidt, *Tetrahedron* **33** 2093 (1977).
60. E. Clar and W. Schmidt, *Tetrahedron* **34** 3219–3224 (1978).
61. E. Clar and W. Schmidt, *Tetrahedron* **35** 1027–1032 (1979).
62. E. Clar and W. Schmidt, *Tetrahedron* **35** 2673 (1979).
63. E. Clar, W. Kelly, J. M. Robertson, and M. G. Rossmann, *J. Chem. Soc.* 3878 (1956).
64. E. Clar, J. M. Robertson, M. Schloegl, and W. Schmidt, *J. Am. Chem. Soc.* **103** 1320–1328 (1981).
65. H. J. Lempka, S. Obenland, and W. Schmidt, *J. Chem. Phys.* **96** 349–360 (1985).
66. N. Sato, H. Inokuchi, K. Seki, J. Aoki, and S. Iwashima, *J. Chem. Soc. Faraday Trans. 2* **78** 1929–1936 (1982).
67. F. J. Morgan, R. Lamotte, P. Garrigues, and J. C. Fetzer, *Polycyclic Aromat. Compds.* **2** 141–153 (1991).

CHAPTER

11

SAMPLE PREPARATION

Most materials that contain LPAHs are complex mixtures. With a wider range in carbon number, a greater range in the degrees of condensation for each carbon number, and a much larger number of isomers for each empirical formula, sample preparation or fractionation is almost an absolute necessity. Very seldom can these samples be directly analyzed without some pretreatment. To analyze such complex mixtures, either fractionation into a more manageable set of samples or the use of series of complimentary methods is needed. The first option is usually the choice because the analytical chemist only has a limited number of the potentially useful instruments at hand.

Sample fractionation is done commonly by at least one of three main approaches, if not a combination of them. These approaches are that the sample is extracted through the use of a series of solvents, the sample is distilled or sublimed to generate fractions based on volatility and boiling points, or the sample is fractionated by the use of chromatography.

Solvent extraction, particularly with Soxhlet extractors, is commonly used because it is cheap, efficient, and requires little expertise or attention. In Soxhlet extraction, a sample is loaded into a porous thimble. The thimble is open at the top and has a cylindrical body with a closed, rounded bottom. They are usually made from high-purity cellulose or sintered glass. The thimble is then placed in the extraction chamber of the Soxhlet apparatus. The solvent charge sits below this in a round-bottom flask within a heating mantle. The solvent is heated to a gentle boil. The vapors rise through a sidearm whose upper end connects to the apparatus above the extraction chamber. A condenser is attached above so that all the vapors collect and percolate through the sample. The extract then siphons off down to the round-bottom flask.

For LPAH-laden materials, a good weak solvent is an alkane, like *n*-hexane or 2,2,4-trimethylpentane (iso-octane). This weak solvent will remove most of the smaller PAHs. For oily samples, this first step will also remove the saturated hydrocarbons. The LPAHs are extracted with dichloromethane if there is no strong adsorption with the sample matrix.

Large (C>—24) Polycyclic Aromatic Hydrocarbons: Chemistry and Analysis, by John C. Fetzer.
ISBN 0-471-36354-5

For soils and other samples that contain water, a good weak solvent would be methanol. This removes both the water and smaller PAHs, as well as many polar molecules of low molecular weight. The LPAHs are not very soluble in methanol, particularly for carbon numbers greater than 24. Once the water is removed, then stronger solvents can be readily used.

If there is strong adsorption, then a better desorbing solvent than dichloromethane is needed. Aromatic solvents, like toluene or xylene, are good, but the chlorinated aromatic solvents are even better. Chlorobenzene is preferred because it has a relatively low-boiling point of 132°C. *o*-Dichlorobenzene and 1,2,4-trichlorobenzene (TCB) have an even stronger affinity to break up the adsorption between LPAHs and sample matrices, but have very high-boiling points of 180°C and 213°C.

Fetzer and Rechsteiner (1), for example, extracted a carbon black exhaustively with dichloromethane (as evidenced by a lack of any fluorescence from 300 to 600 nm). This fraction was 251.1 mg. When the solvent flask was charged with chlorobenzene, a red color was immediately observed as this fresh solvent contacted the carbon black. A further 89.7 mg were extracted. After solvent removal, this second fraction was completely redissolved in dichloromethane for HPLC analysis. This analysis showed the same species as in the first fraction, PAHs from pyrene through those of 36 carbons. The distribution was shifted to higher carbon number, reflecting the stronger adsorption. TCB yielded 22.5 mg more, which was also completely soluble in dichloromethane and composed of the same PAHs. Interestingly, thermal treatment of the residual carbon black yielded small amounts of the same PAHs, showing that there was some very strongly adsorbed material.

The LPAHs have very high boiling or sublimation points. Coronene, one of the smallest of the LPAH, has a reported sublimation temperature of well over 500°C. These high temperatures necessitate the use of very low-pressure vacuums to make the working temperature range lower. The sublimation apparatus often has both vacuum pressure and temperature controls that can be varied in small increments. LPAHs mixtures can be fractionated, but more importantly class separations can be readily achieved. Polar components often have much higher sublimation points because of the hydrogen bonding or acid-base interactions. This is particularly advantageous in synthetic reaction mixtures where many starting materials and byproducts are polar. A concentrated LPAH fraction can be further separated by adsorption chromatography into carbon-number range fractions or by high-resolution chromatography into individual components (or isomers).

Melhuish described a very simple micro-sublimation device (Ref. 2, Figure 112). A gradient tube heater was constructed by wrapping a heating filament around a copper tube. The gradient temperature, running from the highest temperature at the inlet, was obtained by gradually reducing the density in the

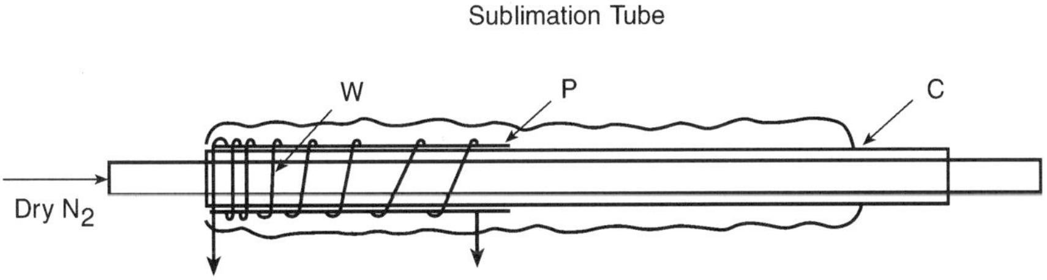

Figure 112. A simple vacuum sublimation apparatus suitable for purifying small samples of LPAHs. W are two plugs of glass or quartz wool used to keep the sample powder in place, C is a copper tube to sheath the quartz tube, and P is a sheath to insulate the wire from the copper tube. From W. H. Melhuish, *Nature* **184** 1933 (1959).

number of turns of the filament. A long quartz tube was inserted into the copper tube as the actual sublimation chamber and sample collection device. The sample was held within this inner tube by quartz wool plugs. A very slow stream of inert gas passed through the quartz tube, and after insertion the outlet could also be attached to a vacuum. The LPAH fractions collect either in the cooler portion of the heated tube or at the outlet where the quartz tube is cooled rapidly by ambient air. The sublimation zones can be spread further by slowly moving the front end of the tube toward the inlet. The fractions appear as colored bands either in normal or under UV light. Fetzer and co-workers used this type of device to isolate several LPAHs (3–5).

Schmidt used an apparatus with very precise, small temperature increments to separate commercial coronene from the 1% LPAH impurities (6). After the bulk of the sample was removed as pure coronene, individual LPAHs were isolated. These included benzo[*pqr*]naphtho[8,1,2-*bcd*]perylene, naphtho[8,1,2-*abc*]coronene, and ovalene. Several even higher-carbon-number fractions were obtained as mixtures of ultra-trace unidentifiable samples. Some of the lack of identification was due to a lack of standard compounds that match as well as the small amounts obtained.

The most common of the chromatographic separations used for fractionaltion of LPAHs is adsorption chromatography (1, 7, 8). In adsorption chromatography there are two somewhat independent physical phenomenon, desorption and solubilization. This is similar to the effects encountered in extraction. An elutropic series of solvent (ones with increasing desorption strengths) will result in a set of fractions. For those samples containing saturated hydrocarbons, the use of an alkane solvent will elute these while retaining the PAHs.

Of the common adsorbents, aluminas work better than silica at fractionating LPAHs because of their stronger retaining power. For fractionating the LPAHs themselves, there is little difference between acidic, basic, or neutral aluminas. If the sample has heteroatom-containing molecules, particularly oxygen- and nitrogen-containing molecules, then acidic or basic aluminas will retain these stronger and give cleaner LPAHs fractions. For example, toluene will elute the smaller PAHs from basic alumina, and dichloromethane will elute the LPAHs without much coelution of similar heteroatom-containing analogues. The LPAHs of 40 or more carbons either will elute with the heteroatom molecules or not at all because of solubility limitations.

To ensure complete recovery, however, a final step is needed that involves a solvent mixture. A strong desorbing solvent, such as methanol, ethanol, or acetonitrile, is blended with a good LPAH solvent, such as toluene or chlorobenzene.

Other combinations of chromatographic separations have been used to fractionate LPAH samples. Matsushita, Esmi, Suzuki, and Handa used thin-layer chromatography (TLC) on alumina. Each spot was then scraped off and extracted (9). These were then separated on cellulose TLC plates. McKay and Latham (10) used size-exclusion chromatography. The elution order, however, shows that adsorption was probably the mechanism. In size exclusion, the largest molecules elute first and the others elute sequentially in decreasing size. The order in this separation was from smaller to larger. Wise, Benner, Liu, Byrd, and Colmsjo (11) used a sequence of normal-phase HPLC to fractionate a coal tar, with reversed-phase HPLC of each of the fractions from that for the analysis of individual LPAHs. The first mode of HPLC separated the sample into "ring-number" fractions.

The matching of sample solvent and HPLC mobile-phase solvents can be tricky. For normal-phase separations, the best solvents are aromatic and chloro-aromatic solvents. These are all relatively weak mobile phases, so there are few difficulties. For reversed-phase HPLC, which is necessary when information on individual isomers is needed, the opposite is true. The best solvents are the strongest mobile-phase components. If the sample is soluble in dichloromethane, a mobile-phase of only a small proportion of dichloromethane in the base solvents methanol or acetonitrile may be required to get good peak separation. A severe mismatch between sample solvents and mobile phase solvents will result in nonequilibrium conditions as the plug of sample solvent moves down the column. This results in fronting and misshapen peaks. In some instances the peak for one component may appear as a split peak with two maxima and a valley in between. One solution for a volatile solvent, like dichloromethane, is that a small amount of the base solvent may be added to the dissolved sample. Through gentle evaporation, some of the dichloromethane may be removed without sample precipitation.

Because the LPAHs are so nonvolatile, the sample handling precautions that must be used to prevent volatility losses for the small PAHs are not needed. The high biological activity of some of these, however, does require that care similar to that used for the smaller PAHs be taken to reduce exposure.

REFERENCES

1. J. C. Fetzer and C. E. Rechsteiner, in *Polycyclic Aromatic Hydrocarbons: Eleventh International Symposium*, M. Cooke, K. Loening, and J. Merritt, Eds., Battelle, Columbus, OH, 1991, p. 259–270.
2. W. H. Melhuish, *Nature* **184** 1933–1934 (1959).
3. J. C. Fetzer and W. R. Biggs, *J. Chromatogr.* **322** 275–286 (1985).
4. R. F. Sullivan, M. M. Boduszynski, and J. C. Fetzer, *Energy Fuels* **3** 603–612 (1989).
5. J. C. Fetzer and W. R. Biggs, *Polycyclic Arom. Cmpds.* **5** 197 (1994).
6. W. Schmidt, in *Polycyclic Aromatic Hydrocarbons and Astrophysics*, A. Leger, L. d'Hendecourt, and B. Boccara, Eds. Reidel, Dordrecht, The Netherlands, 1987, p. 149–164.
7. J. C. Fetzer, B. R. T. Simoneit, P. Garrigues, and H. Budzinski, *Polycyclic Arom. Cmpds.* **9** 109–120 (1996).
8. B. R. T. Simoneit and J. C. Fetzer, in *Organic Geochemistry: Developments and Applications to Energy, Climate, Environment and Human History*, J. O. Grimalt and C. Dorronsoro, Eds., A. I. G. O. A., Donostia, San Sebastian, Spain, 1995, p. 414–417.
9. H. Matsushita, Y. Esmi, A. Suzuki, and T. Handa, *Bunseki Kagaku* **21** 1471 (1972).
10. J. F. McKay and D. R. Latham, *Anal. Chem.* **45** 1050–1055 (1973).
11. S. A. Wise, B. A. Benner, H. Liu, G. D. Byrd, and A. Colmsjo, *Anal. Chem.* **60** 630–637 (1988).

CHAPTER 12
COMPARISON OF METHODOLOGIES

After the preceding chapters on individual analytical tools, an overview is needed so that the researcher can wisely choose techniques that are best suited for his or her needs. In some cases one technique will be better than another for determining composition in a particular type of sample matrix, or perhaps it would be better to use a combination of complementary techniques separately or in sequence (such as a separation before spectroscopy). Another possibility is that perhaps the researcher is trying to determine the structure of an unknown, newly isolated LPAH and is unsure which approach will be most useful.

These two main areas, compositional analysis and structural determination, have different requirements. This leads to different choices in methodologies. First, the choosing of approaches to characterize a complex sample will be described in detail. Then the latter scenario will be reviewed from the aspect of choosing approaches to determine structure, which was not covered in the separate chapter.

12.1 CHOOSING AN APPROACH FOR COMPOSITIONAL ANALYSIS

When planning an analysis scheme for determining LPAH composition, the two most important questions that need to be first answered are, How complex is the sample matrix? and How much detail of the composition is desired from the information?

Even before any analysis is performed on a material, certain bits of information about the sample's history can give clues to the first question, for example, at what conditions and temperatures it was formed under. The second question can also be simplified if the aims of the analysis are defined. Is the analysis targeting specific compounds, such as How much ovalene might be present? or Is there any evidence of dibenzo[*h*, *rst*]pentaphene? Specific targets such as these, require less-involved schemes than a search for the full compositional information required of the question, What LPAHs are present in this sample and in what amounts?

Large (C>–24) Polycyclic Aromatic Hydrocarbons: Chemistry and Analysis, by John C. Fetzer.
ISBN 0-471-36354-5

Some of the answers to these questions can be gained through some efforts in the initial sample preparation steps. Zander (1) describes the general approach for coal tars, and the philosophy is similar for most materials. Fractionation of the sample by sublimation or adsorption chromatography can yield several separate samples each of which may be much easier to characterize than the bulk material. If the bulk of the material is smaller PAHs, then sublimation can remove these and leave a concentrate of the LPAHs. Adsorption chromatography can be used for further fractionation of the LPAHs or removal of any interfering heteroatom compounds (2). An elutropic series of solvents with small increments in strengths, such as *n*-hexane, toluene, chlorobenzene, methyl-*t*-butyl ether, ethyl acetate, and dichloromethane, can give fractions that differ in carbon-number ranges. For this work, it is best to use solvents of high-purity, such as HPLC grade, because large-scale fractionations can readily gain solvent impurities as artifacts that cannot be readily recognized as being alien to complex samples (petroleum, coal tar pitch, etc.).

For complex mixtures, some higher-resolution chromatography is necessary after the initial fractionation. GC has the high resolution needed, especially if a stationary phase that can separate isomers is used. The liquid-crystal phases are very good for separating PAH isomers. The LPAHs are not volatile enough to make GC analysis generally amenable for most of them. The limit of GC is only up through those LPAHs of $\sim$30 ring carbons. If the focus is only on the presence and amount of LPAHs at or below this size, then GC is a very good choice because of its high resolution. A second weakness in the use of GC is the lack of detectors that can differentiate the specific LPAH species. MS is the most informative common detector, but only the molecular weight is obtained without elaborate chemical ionization techniques. A FID detector gives no information other than the retention time and thus is virtually useless in characterization work.

The two main modes of liquid chromatography, normal and reversed phases, differ greatly in their separation mechanisms. They thus differ greatly in how LPAHs are separated. Normal-phase mode relies on the polarizability of the molecules. LPAHs differ by their degrees of condensation and carbon number, which is reflected in the number of π bonds each has. Because of the valency of carbon, the number of π bonds in an LPAH results in an even number of carbon atoms. The polarizability of the LPAHs, therefore, increases in discrete steps based on this two-carbon increment. The normal-phase separation is dominated by this, so that LPAHs generally elute in separated clusters defined by the increasing number of π bonds. For the planar LPAHs, there is only a moderate effect due to the differences in the structure of isomers. For nonplanar LPAHs, there is a larger spread among isomers because the interacting area between the LPAH and the chromatographic packing differs much more.

Normal-phase liquid chromatography can separate LPAHs by the number of π bonds (3, 4). Much of the time this is equivalent to separating them by the number of rings (which is how it is referred to in much of the published literature). As the carbon number increases, the addition of a π bond has less of an effect on this type of separation (the change from 14 to 15 π bonds in the LPAHs is less than the change from 9 to 10 in the PAHs). The gap between carbon-number clusters narrows, and more overlap occurs between LPAHs of different carbon numbers.

The use of normal-phase HPLC, however, can be a very useful step in detailed characterization if it is used with a further fractionation step. Subsequent compound-specific information is then easier to obtain in a reversed-phase separation that could not be gained by the latter separation alone. Thus, to gain the optimal separation of individual LPAH species, normal-phase HPLC is a very good sample preparation step before reversed-phase HPLC if the sample contains a wide range of isomers of different carbon numbers and degrees of condensation.

The reversed-phase separation separates mainly by the contact area between an LPAH and the bonded-phase moieties. For the polymeric phases that separate the isomeric LPAHs so well, the contact area is a three-dimensional one. This results many times in either a complete separation of all the isomers present in a sample or in one where there is only moderate coelution (in many cases this can be overcome by the use of a DAD).

Because the mechanism is not governed by LPAH carbon number but rather by the sizes in the three-dimensional area, severe overlap of very dissimilar PAHs and LPAHs is common. Thus the very nonplanar 11-ring LPAH hexabenzo[*a*,*cd*,*f*,*j*,*lm*,*o*]perylene elutes before the 4-ring planar PAH pyrene. This capability to separate the LPAH isomers makes this mode of HPLC the method of choice for most separations. If prior normal phase fractionation has been used, then the spread of isomers is large and any overlap with other carbon-number LPAHs is not an issue because they are in other fractions.

As was pointed out in the chapter on chromatography, these separations are powerful. The use of HPLC, however, is not adequate in itself to identify and quantitate the LPAHs. Isomer specific detection, either on-line or off-line, is necessary. The best current on-line detectors are the DAD absorbance detectors (5–7). For off-line detection, fluorescence and mass spectroscopies are useful, especially if both are used in concert.

The spectroscopic techniques by themselves usually are not very informative for LPAH mixtures. The infrared spectra of LPAHs are very similar to each other, and so only bulk information on the structural features of the mix can be obtained. The same is true for NMR spectra. In some very limited instances, UV absorbance can be useful in compositional analysis. For example, if a component has a specific high-wavelength absorbance band, then it might be seen beyond the lower-wavelength absorbances of the other components. Generally the overlap of individual LPAHs in the UV absorbance spectra precludes their use for quantitation.

There is a similar possibility for fluorescence spectroscopy. This instance is much more likely for this type of spectroscopy because the LPAH with the highest excitation bands can often be seen.

As described in the section on fluorescence, synchronous-scanning (SSF) or three-dimensional fluorescence techniques can often see several individual LPAHs in mixtures. Because each fluorescing LPAH in dilute solutions essentially acts independently of the others, these techniques can see spectral information for an individual LPAH without interference from the others. Barring inner-system filtering effects, quantitation and identification can both be done. The width of the fluorescence bands in a SSF spectrum, typically 20–30 nm, limits the number of analytes that can be seen. If two LPAHs have peaks that are closer than this, overlap occurs. Estimation of the amounts of each can be made using standard deconvolution techniques, however, if the overlap is not too severe. The peaks in an SSF spectrum are typically gaussian in shape and even small shoulders on large adjacent peaks can be measured.

The data in the mass spectra of mixtures of LPAHs must be carefully used. Not only do isomers usually give both the same molecular ion and fragment ions, they, like any other molecules with the same carbon number, can greatly interfere. The isotope peaks of, and the ready gain or loss of hydrogen atoms by, LPAH ions usually gives clusters of ions for each carbon number. This means that even comparison of the relative amounts of each molecular mass cannot usually be used to compare proportions of LPAHs with degrees of condensation. The use of MS data is complicated even further by the large differences in fragmentation rates or the prevalence of multiply charged ions for slightly different environments in the mass spectrometer.

12.2 CHOOSING AN APPROACH FOR DETERMINING STRUCTURE

Ideally, crystallography is the best method for structure determination. It, however, has two major requirements. First, there must be enough pure material to be used in forming crystals. This means sample sizes in the milligram range and purities of greater than 95%. The second requirement is a scheme to form large-enough crystals. The practical range in size is often quoted as being 1 mm in length and breadth of the crystals.

To make crystals of this size without deformations or the formation of polymorphous material requires slow growth. Rapid formation leads to amorphous material or crystalline samples with mixed crystal types. Solvent inclusion can also occur during rapid crystal formation, especially if the high-boiling strong solvents 1-methylnaphthalene or 1,2,4-trichlorobenzene are used. These two solvents should be avoided when trying to form crystals because although they are the best ones for initial solubilization of the LPAH sample, they are very difficult to remove totally. This chance of residual solvent and the presence of impurities also lead to mixed crystals or amorphous material.

NMR methods also require large sample sizes, as do the very specialized uses of polarized and Shpol'skii spectroscopies. These latter two techniques are so specific and require such specialized equipment that they are not of general utility in characterizing LPAH structures. NMR gives very detailed structural information and with the specific three-dimensional tickling methods can determine the hydrocarbon skeleton (with the attached hydrogens). High-resolution instruments capable of doing this are becoming more common, but their cost still makes them less accessible than the much less useful low-field instruments routinely found in most laboratories.

The use of annellation theory is not sample size or purity limited. Collection of an UV absorbance spectrum in a HPLC-DAD analysis can be sufficient. The structural assignment relies on other information, such as a molecular weight or in the case of a synthesis, knowledge of the reactants and reaction route. Many of the aromatic cores that can be used for this approach are known, but this is not necessarily the case. For large structures in which no structure of a known LPAH of only one or two fewer rings can be found, extrapolation from a much smaller PAH may lead to very erroneous assignments.

Annellation theory works best when only one or two benzo fusions are made onto the core that is then used to extrapolate the spectrum of the desired LPAH. Ortho-fused naphtho groups can also be estimated, particularly if the fusion creates a linear arrangement of rings. Bent ortho-fusion and peri-fusion in particular are not as accurate. Unless there are analogous fusions from ring systems that are very similar to those being tried, annellation theory must be used with caution. For example, the peri-fused naphtho group in going from pyrene to benzo[*ghi*]perylene (a 50-nm red shift) is not like the peri-fused naphtho group in going from perylene to naphtho[8,1,2-*bcd*]perylene (a 35 nm red shift). The numbers of fully benzenoid rings and isolated π bonds must be correctly accounted for and compared in the starting and final structures.

Zander compared IR and NMR techniques in the determination of solo, duo, trio, and quartet hydrogen atoms attached to the rings in a LPAH structure (1). NMR is quantitative but requires the sample to be fairly soluble in a deutero NMR solvent. The IR techniques are not quantitative, but samples can readily be made into potassium bromide pellets. This will at least indicate the presence or absence of certain types of rings.

REFERENCES

1. M. Zander, *Polycyclic Arom. Cmpds.* **7** 209–221 (1995).
2. J. C. Fetzer and C. E. Rechsteiner, in *Polycyclic Aromatic Hydrocarbons: Eleventh International Symposium*, M. Cooke, K. Loening, and J. Merritt, Eds., Battelle, Columbus, OH, 1991, p. 259–270.
3. R. F. Sullivan, M. M. Boduszynski, and J. C. Fetzer, *Energy Fuels* **3** 603–612 (1989).
4. S. A. Wise, B. A. Benner, H. Liu, G. D. Byrd, and A. Colmsjo, *Anal. Chem.* **60** 630–637 (1988).
5. J. C. Fetzer and J. R. Kershaw, *Fuel* **74** 1533–1536 (1995).
6. J. C. Fetzer, B. R. T. Simoneit, P. Garrigues, and H. Budzinski, *Polycyclic Arom. Compds.* **9** 109–120 (1996).
7. B. R. T. Simoneit and J. C. Fetzer, *Organic Geochemistry: Developments and Applications to Energy, Climate, Environment and Human History*, J. O. Grimalt and C. Dorronsoro, Eds., A. I. G. O. A., Donostia, San Sebastian, Spain, 1995, p. 414–417.

APPENDIX

The UV absorbance spectra of these compounds are generally in Erich Clar's *Polycyclic Hydrocarbons*, a two-volume set, published by John Wiley and Sons in 1964. Exceptions are naphtho[8,1,2-*abc*]coronene [J. C. Fetzer and W. R. Biggs, *J. Chromatogr.* **322** 277 (1985)] and phenanthro[5,4,3,2-*efghi*]perylene [J. C. Fetzer and W. R. Biggs, *Polycyclic Arom. Cmpds.* **5** 197 (1994)].

FLUORESCENCE SPECTRA OF SELECTED LPAHs

These spectra are provided for identification of some of the more common LPAHs, as well as a variety of others. For many of the LPAHs that are not included here, including a large majority of the peri-fused LPAHs, the UV absorbance spectrum and fluorescence excitation spectrum are identical for the highest set of wavelength bands. The corresponding fluorescence emission spectrum is a mirror image of the excitation spectrum, with the Stokes shift typically being 5–10 nm. Some LPAHs with spectra with this pattern, such as dibenzo[*cd,lm*]perylene, are included as examples.

The spectra of several common LPAHs of 24 carbons have recently been published in a three-volume set (*Spectral Atlas of Polycyclic Aromatic Compounds*, W. Karcher, R. J. Fordham, J. J. Dubois, P. D. J. M. Glaude, and J. A. M. Ligthart, Eds., Reidel, Dordrecht, The Netherlands, 1983, Vol. 1; *Spectral Atlas of Polycyclic Aromatic Compounds*, W. Karcher, Ed., Kluwer, Dordrecht, The Netherlands, 1988, Vol. 2; *Spectral Atlas of Polycyclic Aromatic Compounds*, W. Karcher, J. Devillers, and P. Garrigues, Eds., Kluwer, Dordrecht, The Netherlands, 1991, Vol. 3). This collection also contains UV absorbance spectra.

All spectra were collected in 1,2,4-trichlorobenzene (EM Science) to eliminate the possible band height variations in using solvents of different polarities or the band shifts due to solvation energies. The LPAHs were weighed on a Mettler model microbalance. Although the concentrations of these solutions were known, these could only be taken as only approximate because the possibility of nonfluorescing impurities or incomplete solubilization was not ensured. The heights of each spectrum were normalized to approximately full scale to highlight the smaller features.

Large (C>=24) Polycyclic Aromatic Hydrocarbons: Chemistry and Analysis, by John C. Fetzer.
ISBN 0-471-36354-5

A Hitachi model F-4500 spectrofluorometer was used with 2.5-nm slits, a scan rate of 60 nm/min, and the medium-gain setting on the photomultiplier. Selection of the excitation and emission wavelengths was not rigorous. These were only chosen to ensure similar responses in the corresponding spectra.

The standard compounds were collected from a variety of sources; the large majority came from synthesis and isolation in the author's laboratory, or from Dr. W. Schmidt (Griefenburg am Ammersee, Germany), Dr. M. Zander (Castrop-Rauxel, Germany), or Dr. K. Muellen (Mainz, Germany).

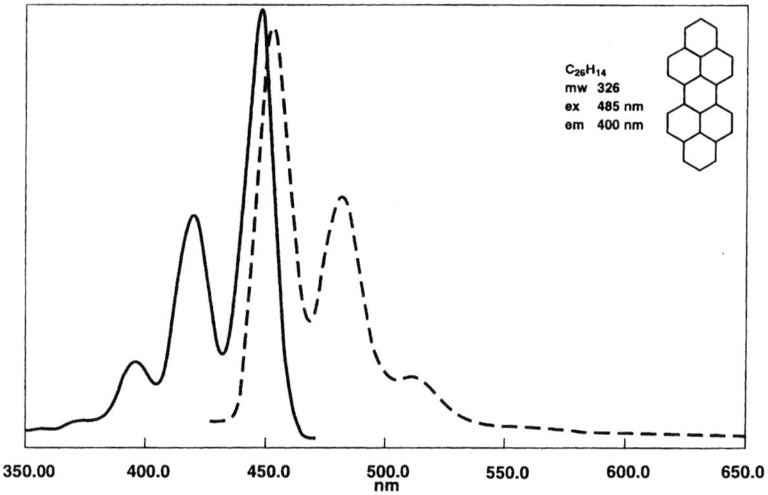

Dibenzo[*cd*,*lm*]perylene

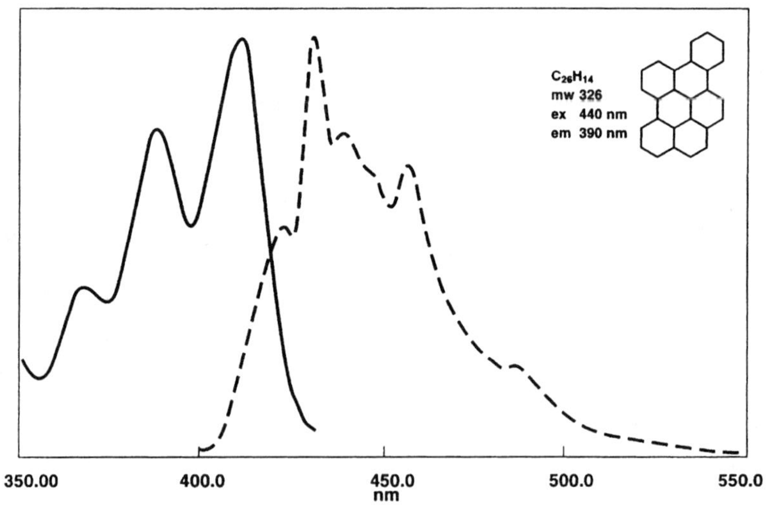

Dibenzo[*e*,*ghi*]perylene

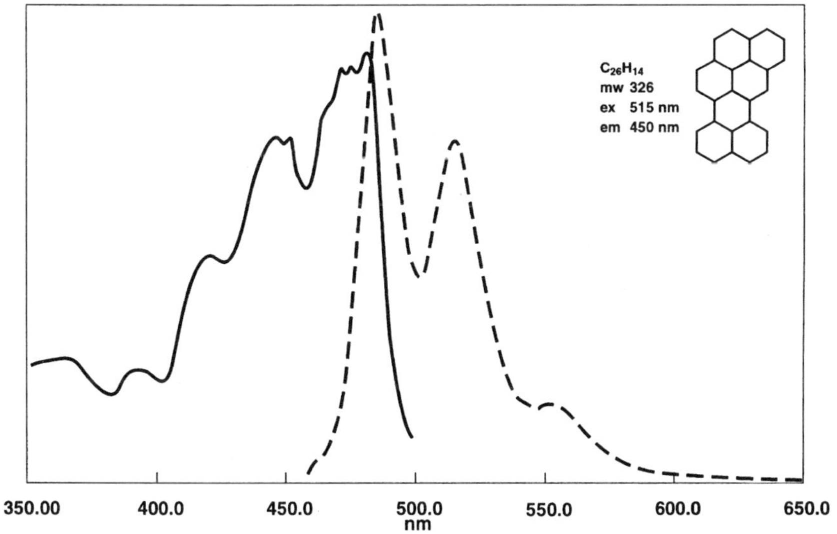

Naphtho[8,1,2-*bcd*]perylene

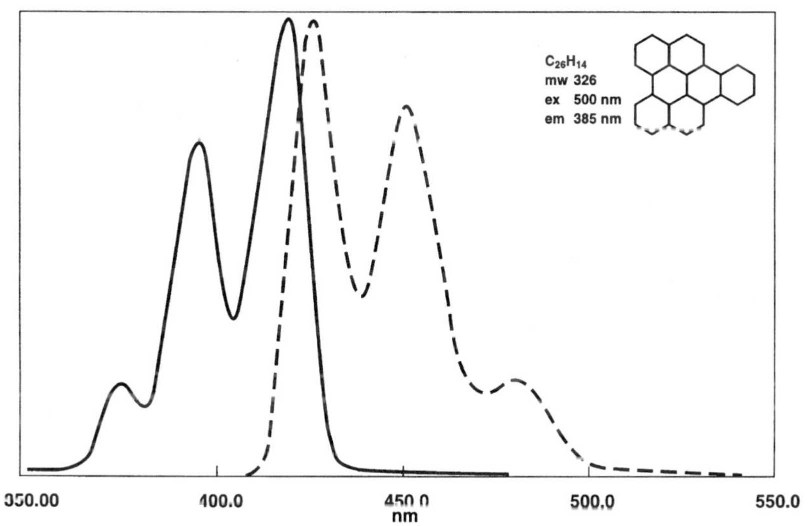

Naphtho[1,2,3,4-*ghi*]perylene

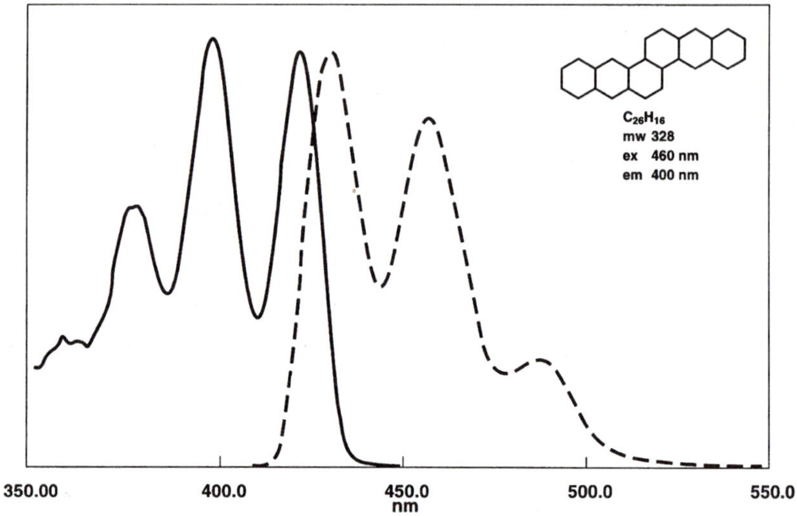

Dibenzo[*b*,*k*]chrysene

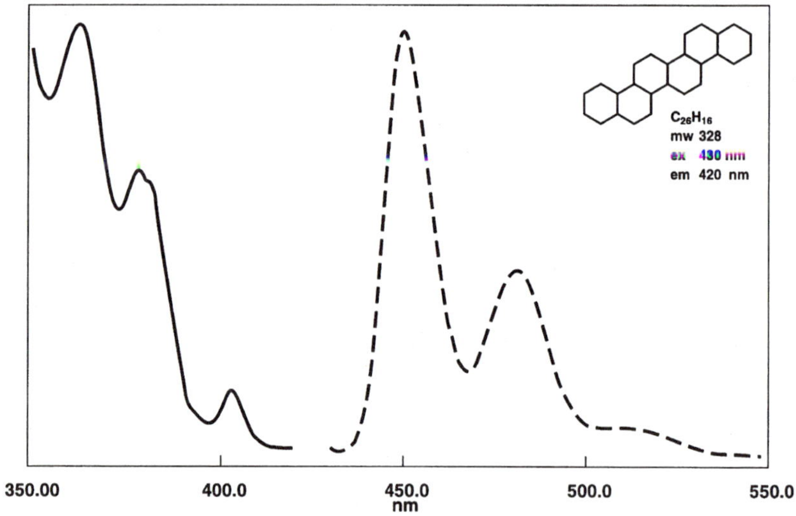

Benzo[*c*]picene

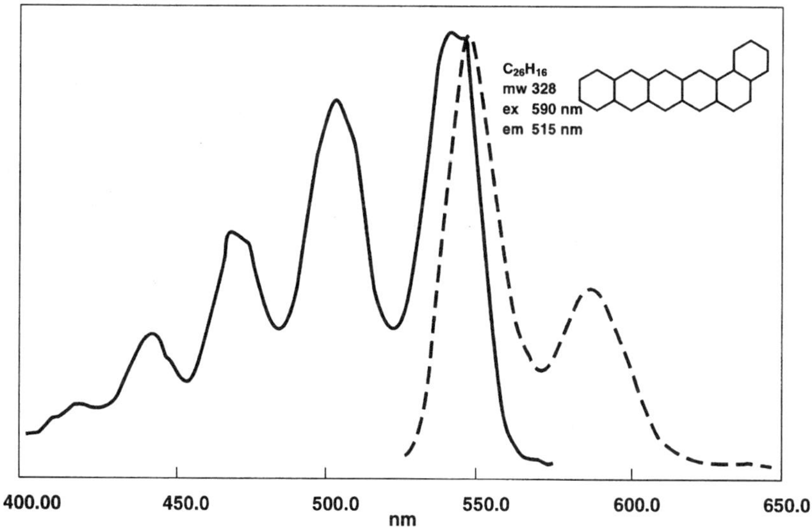

Benzo[*a*]pentacene

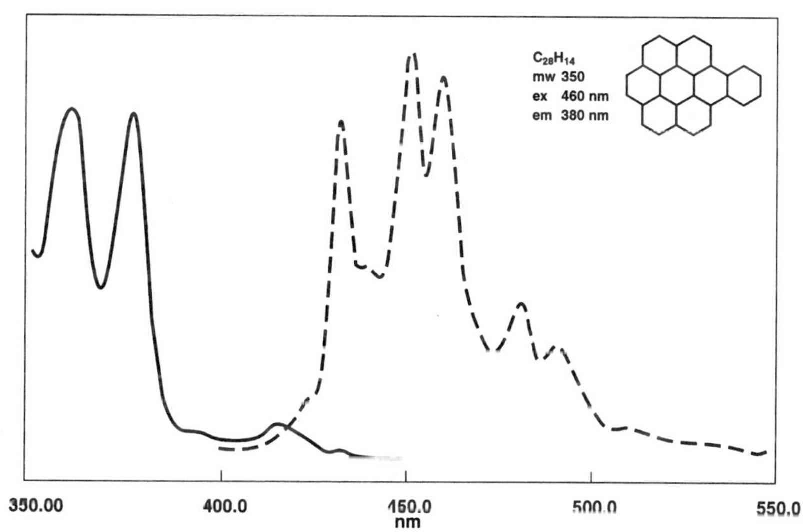

Benzo[*a*]coronene

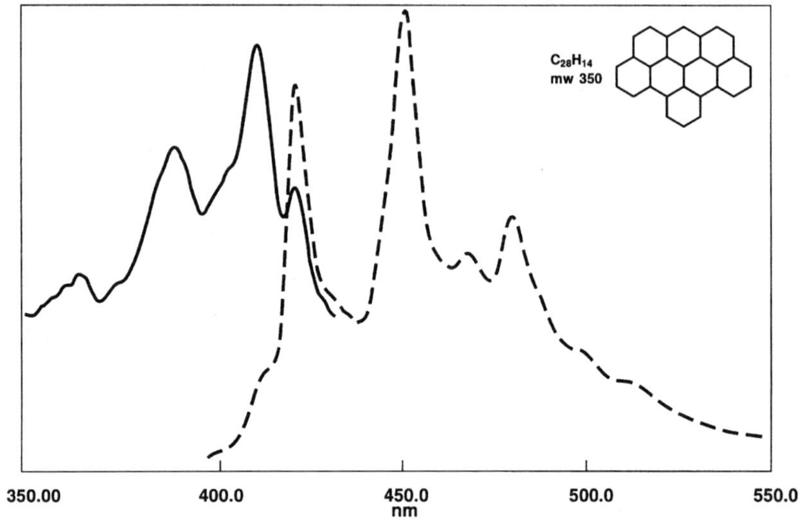

Phenanthro[5,4,3,2-*efghi*]perylene

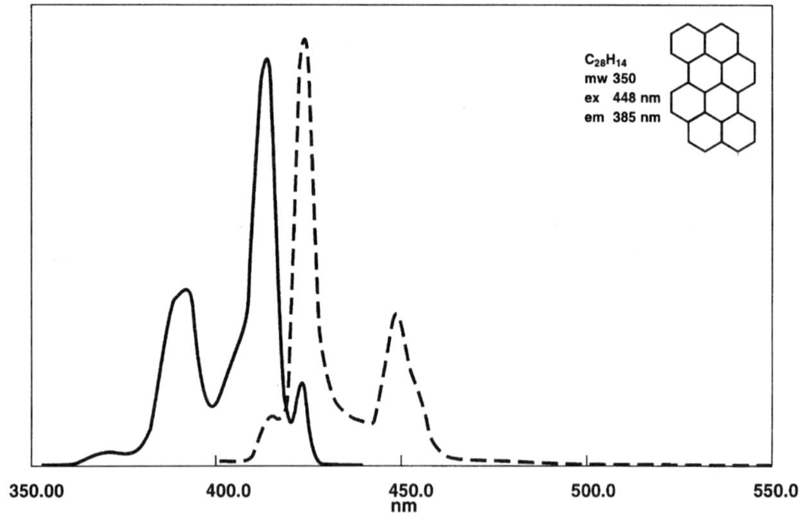

Benzo[*pqr*]naphtho[8,1,2-*bcd*]perylene

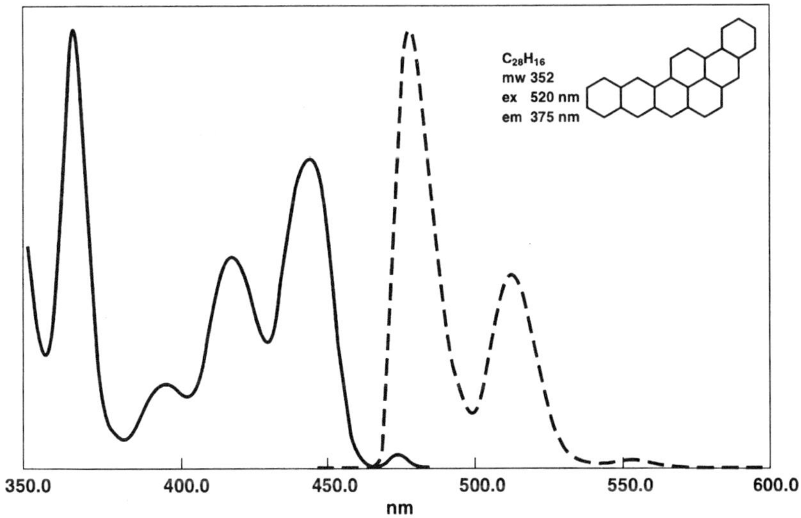

Benzo[*vwx*]hexaphene

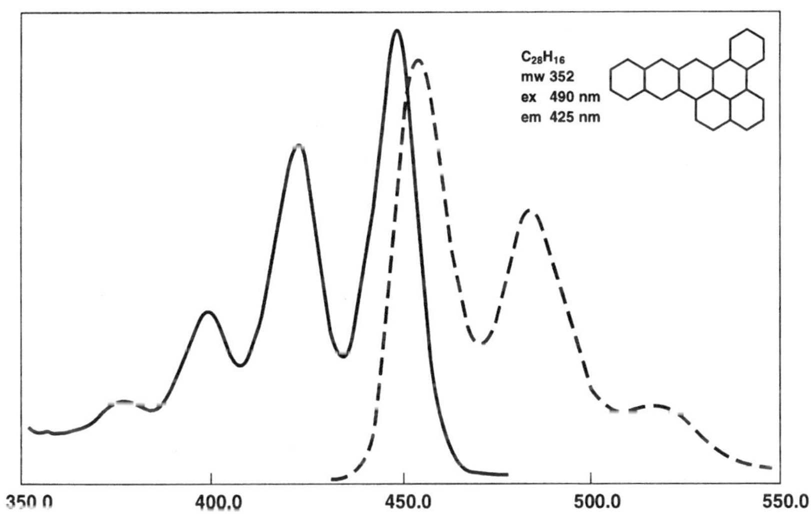

Benzo[*c*]naphtho[2,1,8-*qra*]tetracene

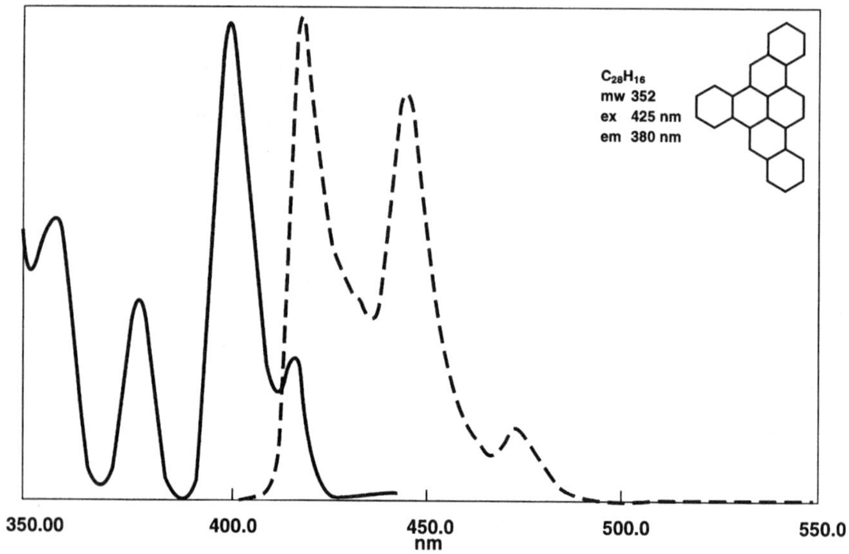

Dibenzo[*h*,*rst*]pentaphene

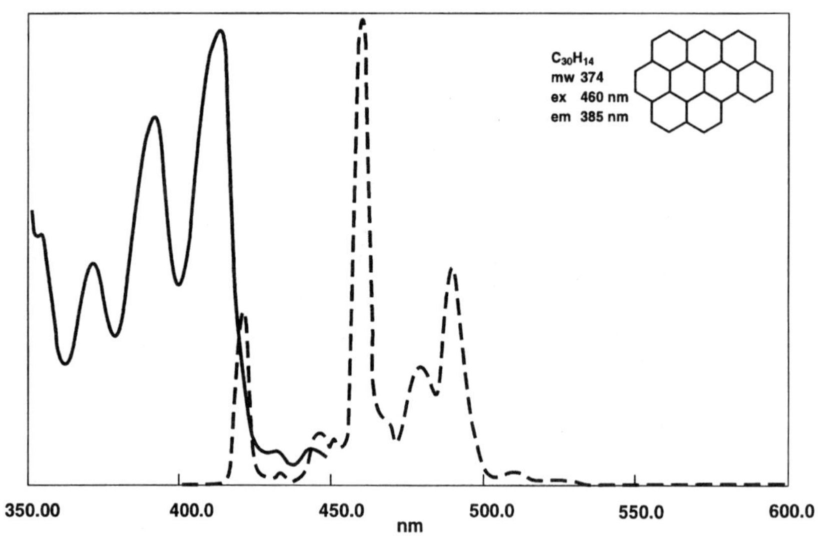

Naphtho[8,1,2-*abc*]coronene

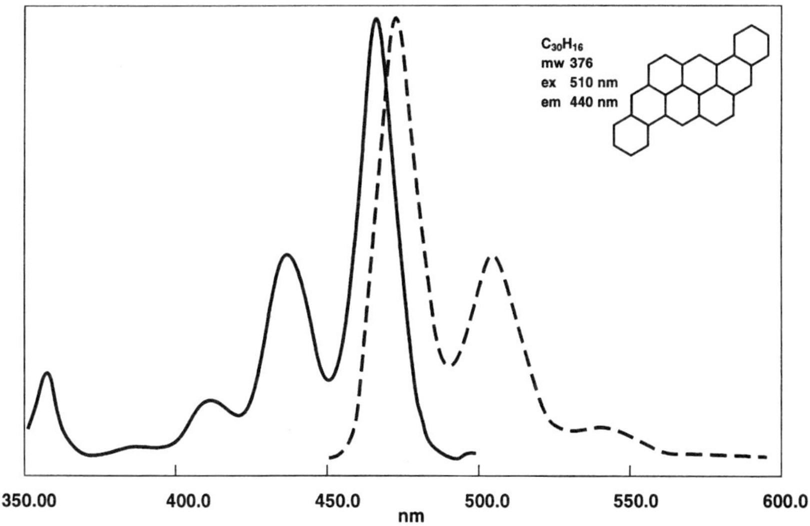

Pyranthrene

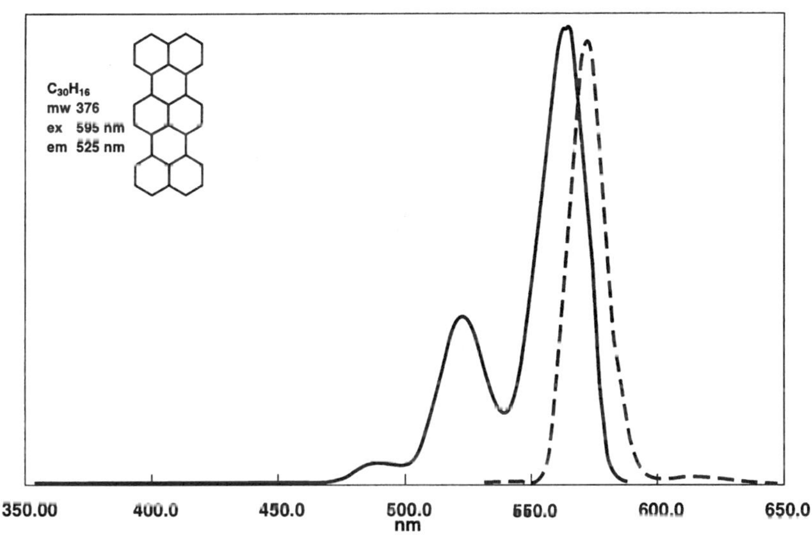

Tribenzo[*de*,*kl*,*rst*]pentaphene

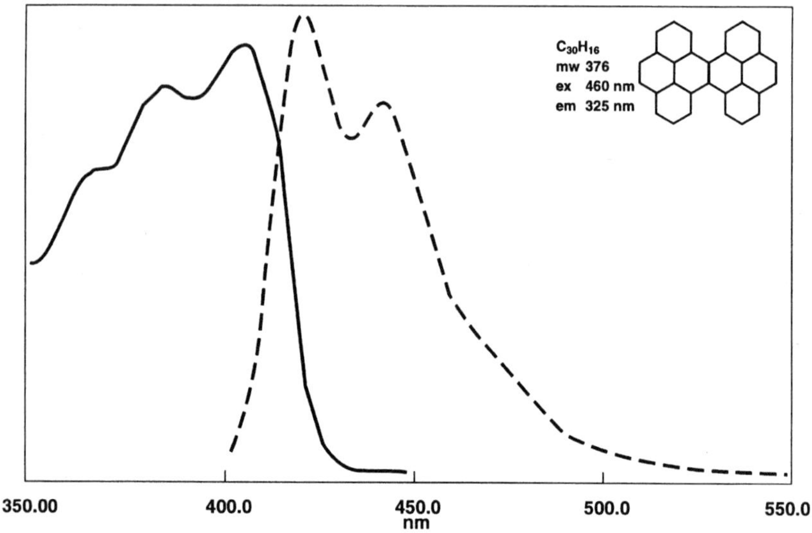

Tetrabenzo[*de*,*hi*,*mn*,*qr*]tetracene

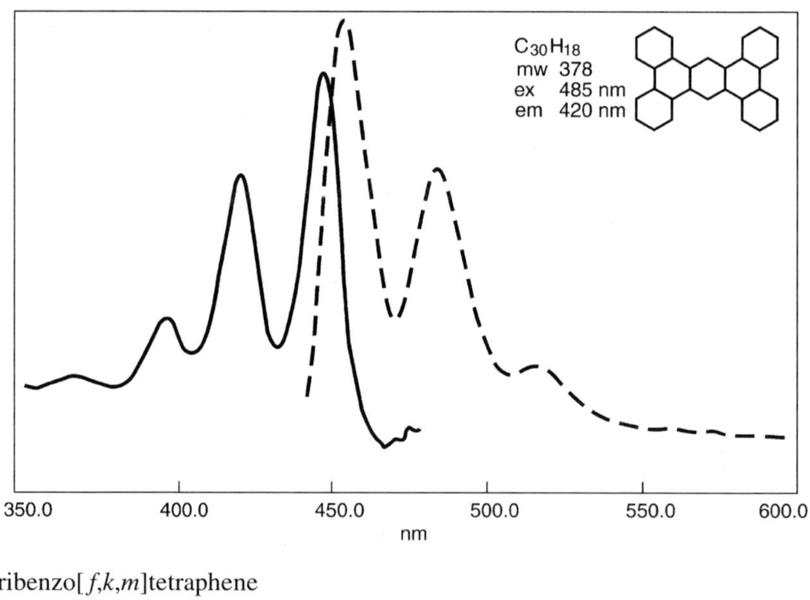

Tribenzo[*f*,*k*,*m*]tetraphene

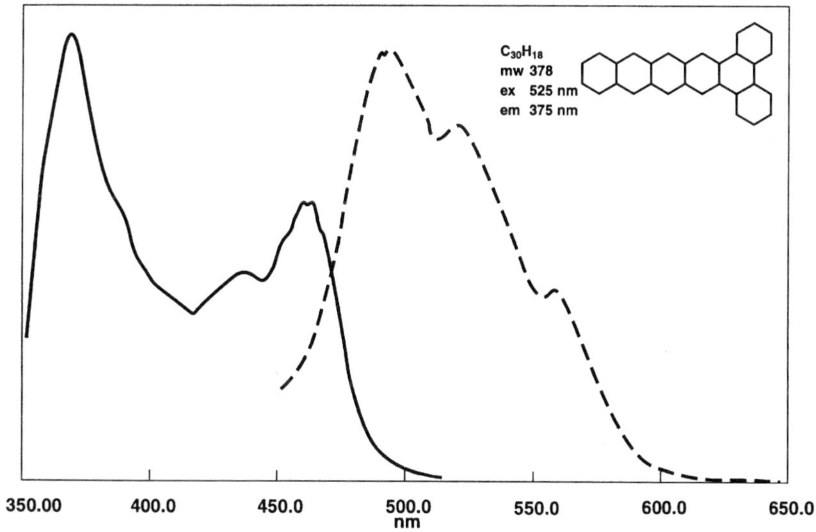

Dibenzo[*a*,*c*]pentacene

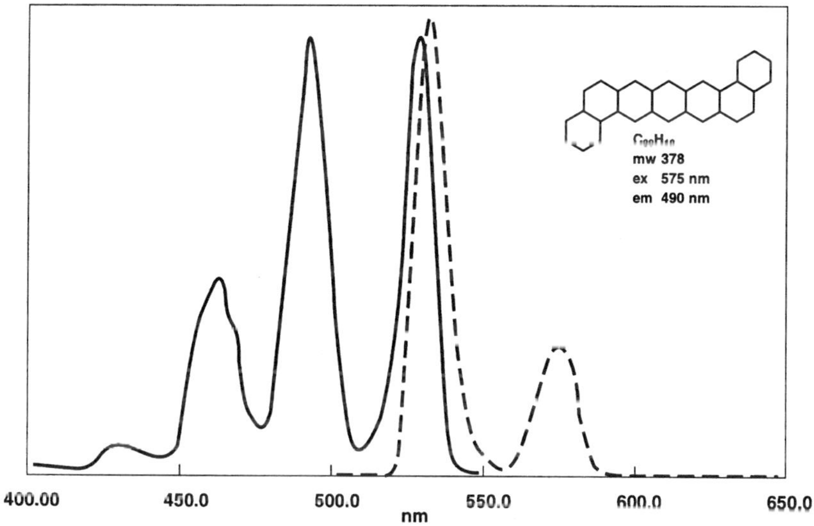

Dibenzo[*a*,*l*]pentacene

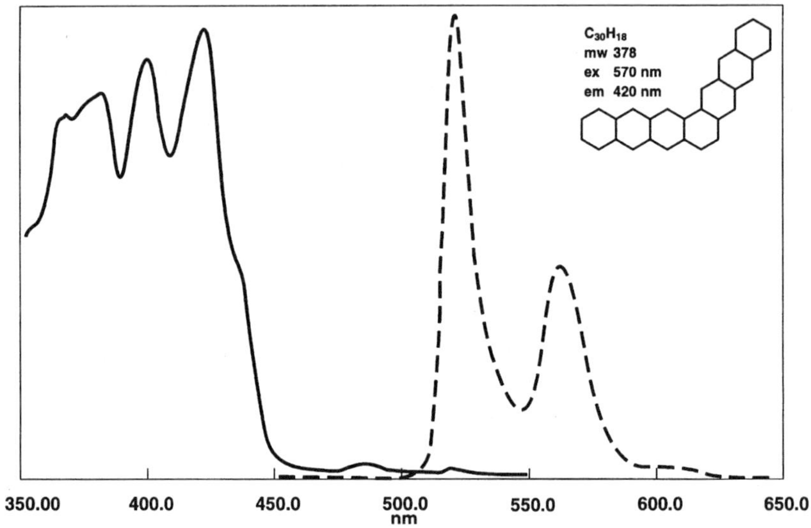

Heptaphene

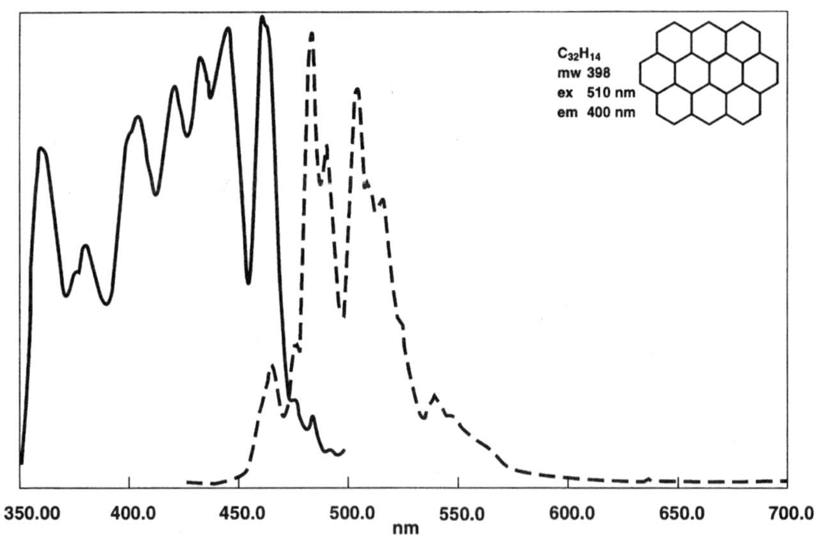

Ovalene

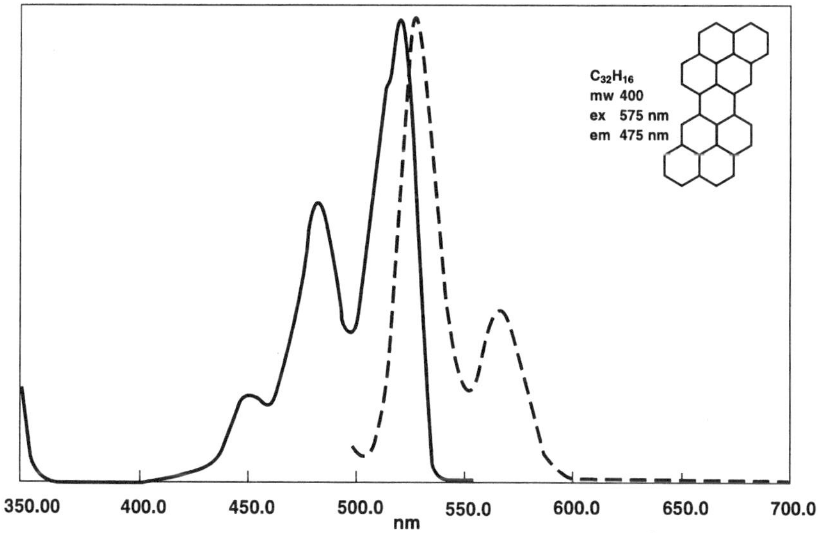

Dinaphtho[2,1,8,7-*defg*:2′,1′,8′,7′-*opqr*}pentacene

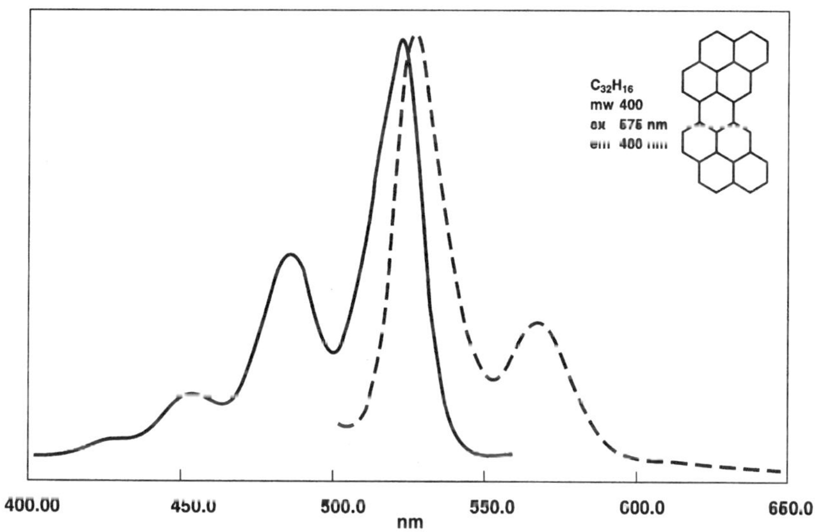

Dinaphtho[2,1,8,7-*defg*:2′,1′,8′,7′-*ijkl*]pentaphene

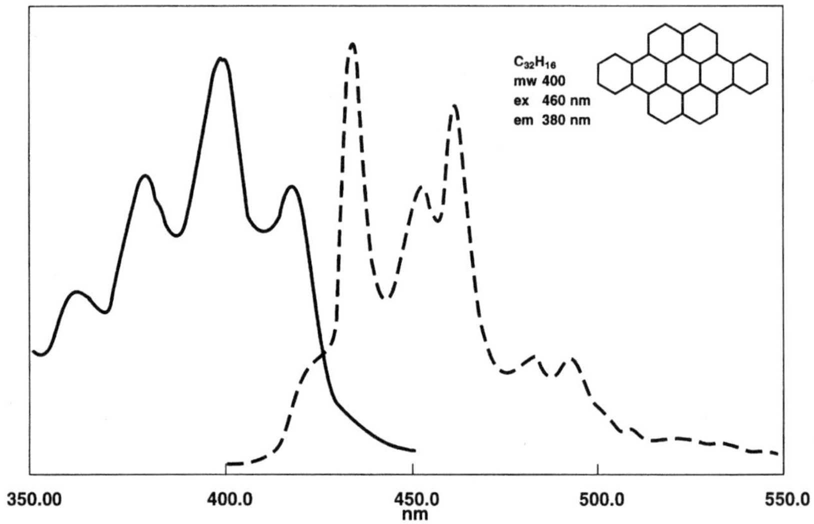

Dibenzo[*a*, *j*]coronene

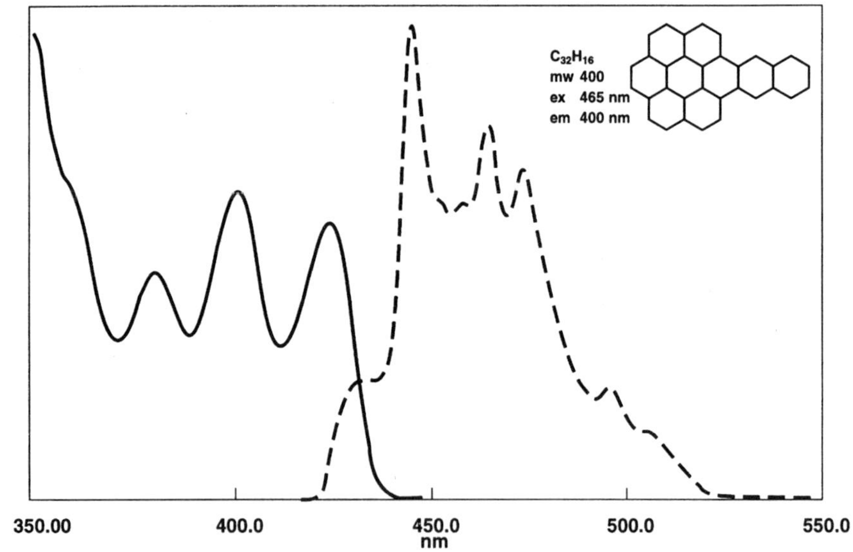

Naphtho[1,2-*a*]coronene

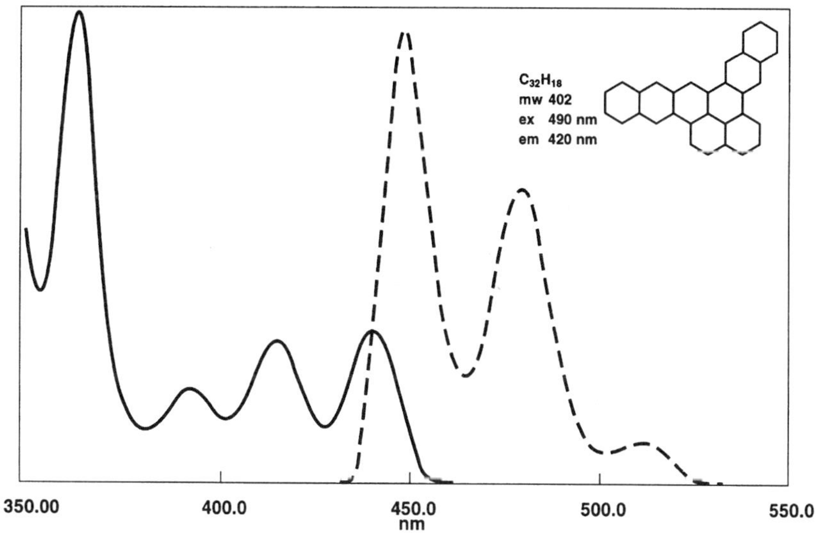

Naphtho[8,1,2-*hij*]hexaphene

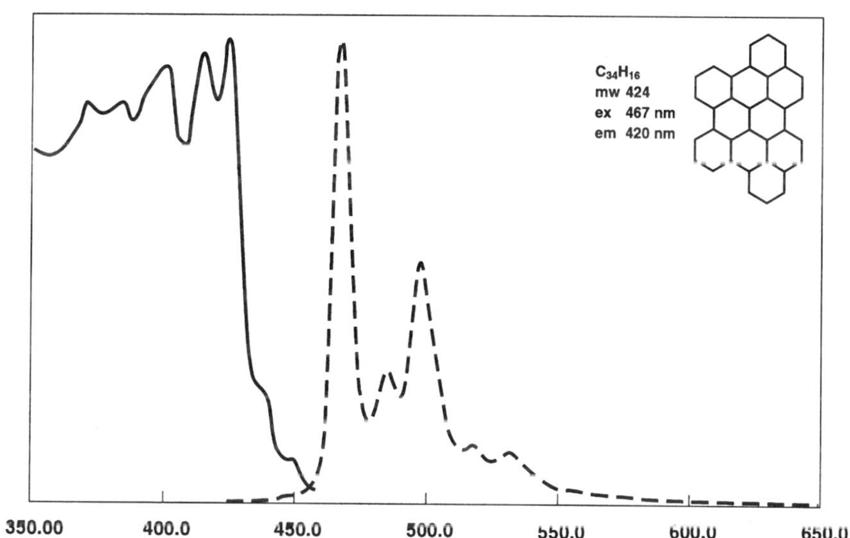

Dibenzo[*fg,ij*]phenanthro[2,1,10,9,8,7-*pqrstuv*]pentaphene

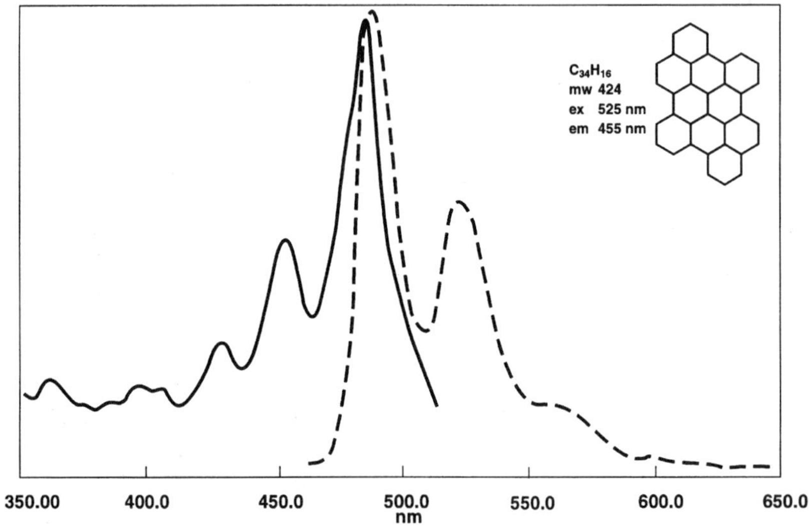

Peryleno[3,2,1,12-*pqrab*]perylene

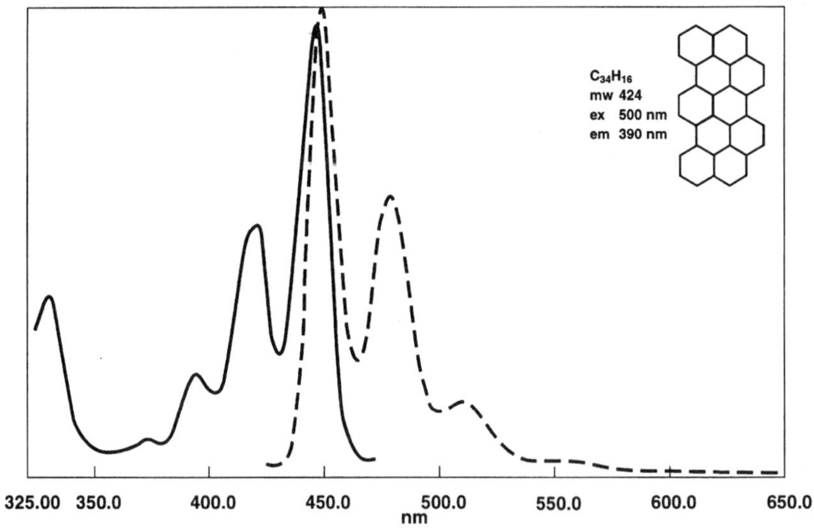

Benzo[*rst*]dinaphtho[2,1,8,7-*defg*:2′,1′,8′,7′-*ijkl*]pentaphene

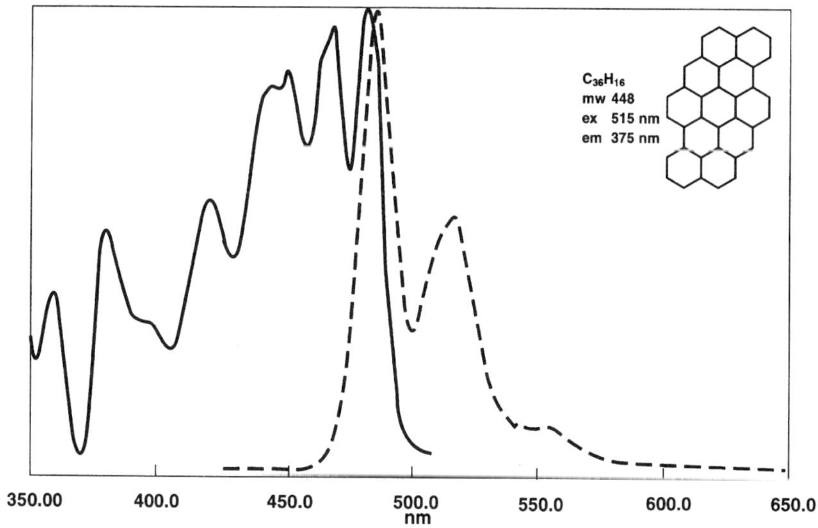

Dinaphtho[8,1,2-*abc*:8′,1′,2′-*jkl*]coronene

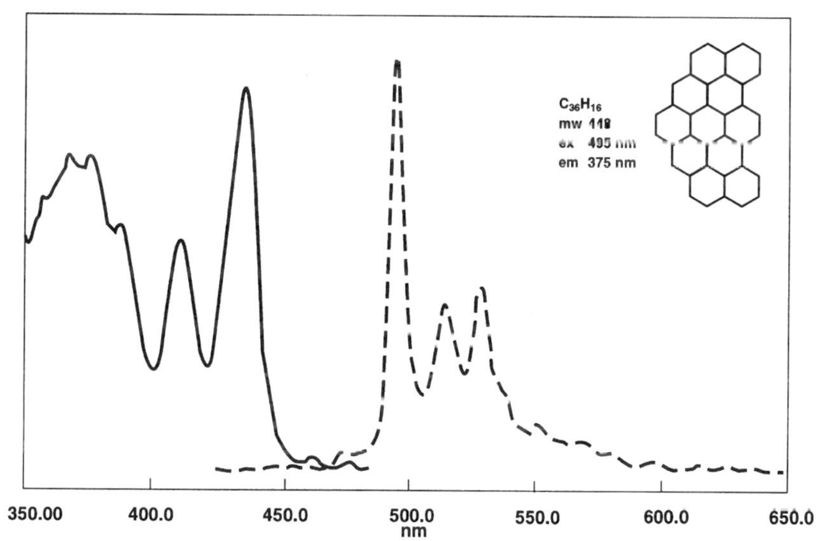

Dinaphtho[8,1,2-*abc*:2′,1′,8′-*klm*]coronene

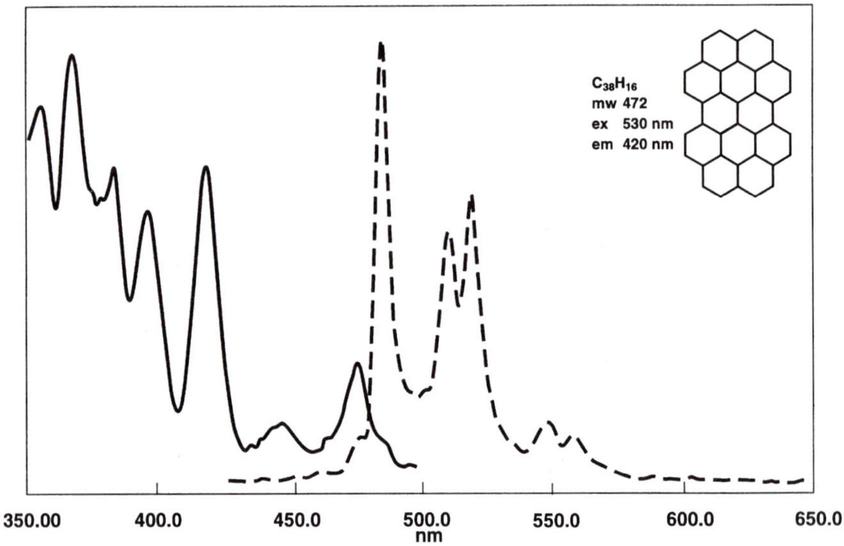

Naphth[2′,1′,8′,7′:4,10]anthra[1,9,8-*abcd*]coronene

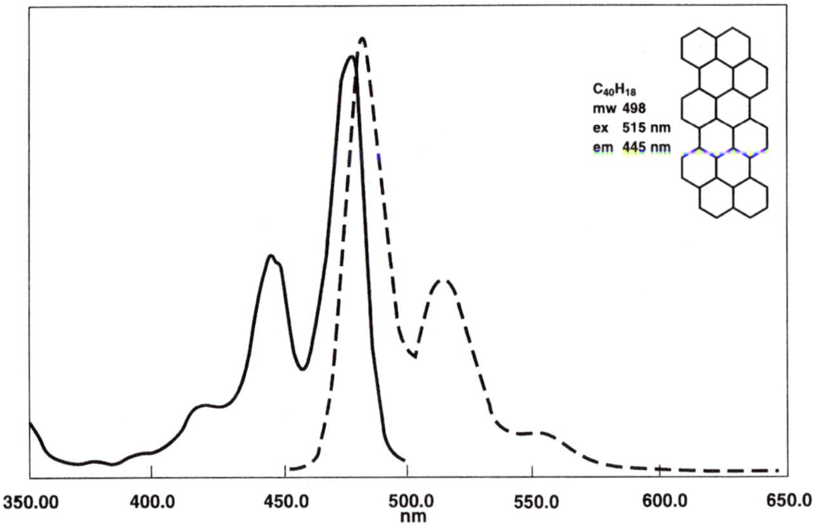

Tetrabenzo[*def*,*lm*,*qrs*,*yz*]pyranthrene

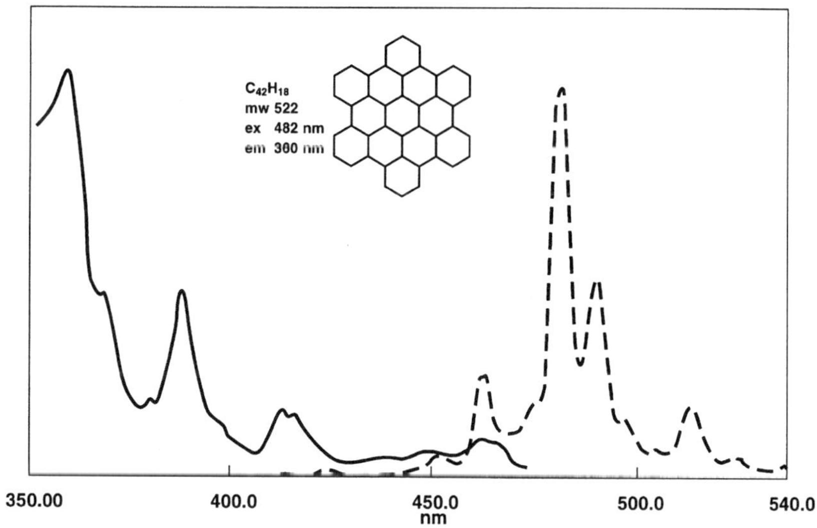

Hexabenzo[*bc*,*ef*,*hi*,*kl*,*no*,*qr*]coronene

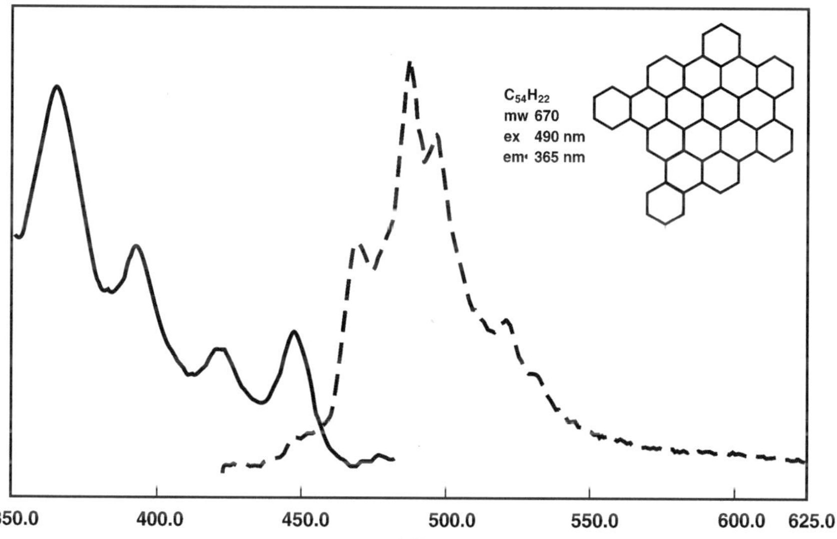

Tetrabenzo[*bc*,*ef*,*o*,*uv*]phenanthro[2,1,10,9-*hijk*]ovalene

INDEX